Agricultural Engineering

Agricultural Engineering

Edited by **Harvey Parker**

SYRAWOOD
PUBLISHING HOUSE

New York

Published by Syrawood Publishing House,
750 Third Avenue, 9th Floor,
New York, NY 10017, USA
www.syrawoodpublishinghouse.com

Agricultural Engineering
Edited by Harvey Parker

© 2016 Syrawood Publishing House

International Standard Book Number: 978-1-68286-139-4 (Hardback)

Printed in the United States of America.

Contents

Preface

This book has been a concerted effort by a group of academicians, researchers and scientists, who have contributed their research works for the realization of the book. This book has materialized in the wake of emerging advancements and innovations in this field. Therefore, the need of the hour was to compile all the required researches and disseminate the knowledge to a broad spectrum of people comprising of students, researchers and specialists of the field.

Agricultural engineering is an innovative branch of engineering as it brings together concepts from chemistry, engineering, physics and biology, and seeks to apply them in the field of agriculture. This book aims to equip students and experts with the advanced topics and upcoming concepts in this area. Included in this book are extensive researches on topics such as agricultural genomics, fertilizers, sustainable farming, etc. This book aims to serve as a resource guide for students and experts alike and contribute to the growth of the discipline.

At the end of the preface, I would like to thank the authors for their brilliant chapters and the publisher for guiding us all-through the making of the book till its final stage. Also, I would like to thank my family for providing the support and encouragement throughout my academic career and research projects.

Editor

Assessment of the genetic diversity in a germplasm collection of cowpea (*Vigna unguiculata* (L.) Walp.) using morphological traits

T. Stoilova[1]* and G. Pereira[2]

[1]Institute of Plant Genetic Resources, Sadovo, Bulgaria.
[2]Instituto Nacional Recursos Biológicos (INRB/INIA), P. O. Box 6, 7350-951 Elvas, Portugal.

Cowpea is considered a minor crop with generally low grain yield; nevertheless production area is spread all around the world, mainly in marginal areas with poor soil and limited rainfall. It is a multipurpose crop grown for green pods as vegetable, dry seeds as pulse and green fodder. The aim of this study was to evaluate 48 accessions of cowpea by 24 morphological descriptors in order to identify accessions with specific behaviour that could be exploited by plant breeders. The accessions with best performance for the development of new varieties and more interesting for inclusion in cowpea breeding programme are 87-052, 95-017, A4E007 and 98210005. The descriptors pod length, number of seeds per pod, seed thickness and 100 seed weight were found the most stable traits over the three years.

Key words: Cowpea, *Vigna unguiculata,* landraces, breeding lines, genetic resources, morphological diversity.

INTRODUCTION

Cowpea (*Vigna unguiculata* (L.) Walp.), a member of the family *Fabaceae*, is a crop grown throughout the tropics and the substropics covering Africa, Asia, South America, parts of Southern Europe and the United States (Singh et al., 1997). It has been estimated that the total production of cowpeas for dry seeds is 5.5 million tones and the total area grown was 10.5 million (www.faostat.fao.org. 2010). Cowpea is a heat loving crop which tolerates drought and lower soil fertility (Coetzee, 1995; Mortimore et al., 1997).

Due to its high protein content (20 to 25%), cowpea plays a major role in human nutrition (Singh et al., 1997). The grain is valued for its flavor and short cooking time and the plant is especially favored by farmers because of its ability to maintain soil fertility through its ability to fix nitrogen (Blade et al., 1997).

The cultivated cowpea consists of two main *subsp.* *unguiculata* and *subsp. sesquipedalis*. Landraces or traditional old varieties have played important role for in the introduction of improved adaptive characteristics (Hawtin et al., 1996). A large number of landraces are still grown in the gardens and small farms mainly for dry seeds. The farmers maintain old populations for their family subsistence and for the nearest local market.

Physiological, morphological, or phenological criteria could be implemented to select the improved adaptation to dry environments (Blum, 1988). Traditionally, diversity is estimated by measuring variation in phenotypic or qualitative traits (starts flowering, time to maturity, plant type, flower color, seed type, seed color, seed size, hilum color) and quantitative agronomic traits However, this approach is often limited and expression of quantitative traits is subject to strong environmental influence (Kameswara, 2004).

Increasing major components of grain yield such as pods/plant, pod length, seed/pod and seed size will allows improving cowpea yield potential. The variability of these morphological traits has been reported from

---

*Corresponding author. E-mail: tz_st@abv.bg

Table 1. Cowpea accessions subjected to morphological characterization.

Accession	Status of sample	Accession	Status of sample	Accession	Status of sample
77	Breeding line(Hungary)	95-025	Breeding Line (USA)	A4-093	Breeding line (IITA)
98210001	Landrace (Portugal)	95-030	Breeding Line (USA)	A4-094	Breeding line (IITA)
98210003	Landrace (Portugal)	95-042	Breeding Line (USA)	A4-096	Breeding line (IITA)
98210004	Landrace (Portugal)	95-045	Breeding Line (USA)	A4E-007	Landrace (Bulgaria)
98210005	Landrace (Portugal)	95-057	Breeding Line (USA)	A4E-008	Landrace (Bulgaria)
2005-01	Landrace (Bulgaria)	95-073	Breeding Line (USA)	A7E-0735	Landrace (Bulgaria)
87-003	Breeding Line (IITA)	95-081	Breeding Line (USA)	A8E-0523	Landrace (Bulgaria)
87-007	Breeding Line (IITA)	95-095	Breeding Line (USA)	A8E-0542	Landrace (Bulgaria)
87-026	Breeding Line (IITA)	97-001	Breeding Line (Japan)	A8E-0551	Landrace (Bulgaria)
87-052	Breeding Line (IITA)	A4-080	Breeding Line (IITA)	A8E-0554	Landrace (Bulgaria)
87-058	Breeding Line (IITA)	A4-081	Breeding Line (IITA)	A8E-0562	Landrace (Bulgaria)
87-060	Breeding Line (IITA)	A4-083	Breeding Line (IITA)	A8E-0563	Landrace (Bulgaria)
91-010	Breeding Line (IITA)	A4-084	Breeding Line (IITA)	A8E-0492	Landrace (Bulgaria)
92-002	Breeding Line (VMW)	A4-086	Breeding Line (IITA)	BOE07	Landrace (Bulgaria)
95-017	Breeding Line (USA)	A4-087	Breeding Line (IITA)	BOE08	Landrace (Bulgaria)
95-023	Breeding Line (USA)	A4-088	Breeding Line (IITA)	St	Landrace (Bulgaria)

different authors, as Patil and Baviskar (1987), Sardana et al. (2001), Mishra et al. (2002), Carnide et al. (2007).

Knowledge of phenotypic variation and relationships among genotypes will assist breeders to develop appropriate breeding strategies and to create the most adaptive and productive cultivars. The study on landraces variation in morphological, phonological and agronomic traits would be useful in the development in new varieties with better adaptation to biotic and abiotic stress factors, as well as for high yield potential.

In Bulgaria, cowpea is regarded as a minor crop, as in other European countries (Negri et al., 2000) and no statistics are available for our country, where this crop is often mistaken for *Phaseolus vulgaris*, the common bean. In many countries (Portugal, Spain, Italy, Bulgaria) cowpea is cultivated for both, the seeds and green pods (Negri, 2009).

The cowpea collection maintained at the Institute of Plant Genetic Resources in Sadovo, consists 336 introduced accessions and landraces. The bigger number of accessions gives opportunity to select the most appropriate of them for the respective breeding objectives after a complex evaluation of samples, best adaptation capacity to certain growing conditions, as accessions with manifested tolerance to abiotic stress factors (Hamidou et al., 2007; Agbicodo et al., 2009).

The landraces included in our investigation were represented by 14 accessions collected mainly from Southeastern part of the country. Most of foreign accessions were introduced from IITA (17), USA (10), Portugal (4), Vietnam (1), Japan (1) and Hungry (1).

The main objective of this study was to determine the variation among cowpea landraces collected from different agro-ecological zones of Bulgaria based on morphological and phenotypic characterization as well as

with introduced accessions with different geographical origin.

MATERIALS AND METHODS

A total of 48 accessions of cowpea were analyzed in this study. These accessions included 18 landraces from Bulgaria and Portugal and 30 advanced breeding lines with different origins (Table 1). The experimental work was carried out at the Institute for Plant Genetic Resources (IPGR, Sadovo, Bulgaria), during 3 years. The field trials were in a randomised complete block design with three replications. Each accession was represented by 15 plants.

The genotypes were described based on the descriptors for cowpea of IBPGR (1983). In each accession, 5 were randomly chosen for biometric measurements. Nineteen quantitative characters (Table 2) and five qualitative characters (flower and seed color, seed shape, hilum color and testa texture) were recorded during the vegetative and reproductive stages. Days to flowering were determined as the number of days from sowing to 50% of the plants has begun to flower and days to maturity as the number of days from sowing to 90% of plants have mature pods. Plant height and height to first pod was evaluated at the end of flowering.

Data were analysed by numerical taxonomy techniques, using NTSYS-pc package, version 2.01 (Rohlf, 1997). An unweighted pair-group method of the arithmetic average clustering procedure (UPGMA) was employed to construct dendrograms. Principal component analysis was also performed to establish the importance of different traits in explaining the total variation.

RESULTS

The majority of the cowpea accessions have white (47.9%) or lilac flowers (47.9%). Only 4.2% of the accessions present flowers with others colors (Table 3). Seeds have predominant cream colour and kidney shape. The majority of seeds have colored hilum. The

Table 2. Quantitative descriptors used in the characterization of the 48 accessions.

Morphological descriptor	Abbreviation	Morphological descriptor	Abbreviation
Days to flowering	DFL	Number of pods per plant	Npod/pl
Flowering duration	DurFL	Weight of pods per plant (g)	Wpod/pl
Days to maturity	DM	Number of seeds per plant	NSeed/pl
Plant height (cm)	Height	Seed length (cm)	Lseed
Plant weight (g)	WPI	Seed width (cm)	Wseed
Height to first pod (cm)	H1stpod	Seed thickness (cm)	Tseed
Pod length (cm)	Lpod	Seed weight per plant (g)	Wseed/pl
Pod width (cm)	Wpod	100 seed weight (g)	W100S
Pod thickness (cm)	Tpod	Yield (g/m^2)	Yield
Number of seeds per pod	NS/pod		

Table 3. Qualitative traits observed on 48 accessions of *V. unguiculata*

Accession	Flower colour	Seed colour	Seed shape	Hilum colour	Testa texture
77	**White**	**White**	**Rhomboid**	**Black**	**Smooth**
98210001	Lilac	Cream	Kidney	Brown	Rough
98210003	Light lilac	White	Kidney	Black	Smooth
98210004	White	Cream	Kidney	Dark brown	Rough
98210005	White	Cream	Kidney	Green	Rough
2005-01	White	Cream	Kidney	Black	Rough
87-003	Light lilac	Cream	Kidney	Dark brown	Rough
87-007	Dark lilac	Redish	Kidney	-	Smooth
87-026	White	Cream	Rhomboid	Dark brown	Rough
87-052	Light lilac	Cream	Globose	Black	Smooth
87-058	White	Cream	Rhomboid	Beige	Rough
87-060	Light lilac	Cream	Globose	Dark brown	Smooth
91-010	Dark lilac	Cream	Globose	Green	Smooth
92-002	Lilac	Black	Ovoid	-	Rough
95-017	White	Cream	Kidney	Black	Rough
95-023	White	Cream	Kidney	Beige	Rough
95-025	Dark lilac	Cream	Kidney	Dark brown	Smooth
95-030	White	White	Kidney	Black	Smooth
95-042	Lilac	Cream	Rhomboid	Beige	Rough
95-045	White	Cream	Kidney	Dark brown	Rough
95-057	Dark lilac	Cream	Kidney	Dark brown	Rough
95-073	Lilac	Cream	Rhomboid	Green	Smooth
95-081	Light lilac	Cream	Rhomboid	Dark brown	Smooth
95-095	Dark lilac	Cream	Rhomboid	-	Smooth
97-001	yellow	Red	Rhomboid	White	Smooth
A4-080	Light yellow	Cream	Ovoid	-	Smooth
A4-081	white	White	Kidney	Dark brown	Rough
A4-083	white	Brown	Rhomboid	-	Rough
A4-084	white	Cream	Kidney	-	Rough
A4-086	Light lilac	Brown	Kidney	-	Smooth
A4-087	white	Brown	Kidney	-	Rough
A4-088	white	Cream	Rhomboid	Dark brown	Rough
A4-093	white	Cream	Kidney	Dark brown	Rough
A4-094	Light lilac	Cream	Kidney	-	Smooth
A4-096	Dark lilac	Cream	Rhomboid	Green	Smooth
A4E-07	white	White	Kidney	Black	Rough

Table 3. Contd.

A4E-0008	white	Cream	Kidney	Dark brown	Rough
A7E-0735	lilac	Cream	Kidney	Green	Smooth
A8E-0492	Light lilac	White	Kidney	Black	Rough
A8E0523	lilac	Cream	Rhomboid	Dark brown	Rough
A8E0542	Light lilac	Cream	Rhomboid	Green	Smooth
A8E0551	Light lilac	Cream	Rhomboid	Green	Smooth
A8E0554	Light lilac	Cream	Rhomboid	Green	Smooth
A8E0562	White	Cream	Kidney	-	Smooth
A8E0563	White	White	Kidney	Black	Rough
BOE-07	White	Cream	Kidney	Green	Rough
BOE-08	White	Cream	Kidney	Dark brown	Rough
St	White	White	Kidney	Black	Rough

Table 4. Mean, ranges and coefficients of variation for the descriptors observed on 48 accessions of cowpea.

Variable	Mean	Min.	Max.	CV (%)
Days to flowering	50.81	40.00	60.70	9.58
Flowering duration (days)	18.03	14.00	23.50	11.90
Days to maturity	86.12	72.70	94.00	4.98
Plant height (cm)	74.46	37.70	122.20	31.70
Plant weight (g)	43.98	21.20	125.80	39.27
Height to first pod (cm)	19.08	9.00	28.40	23.19
Pod lenght (cm)	14.49	9.80	17.70	10.05
Pod width (cm)	0.79	0.60	0.90	9.46
Pod thickness (cm)	0.61	0.50	0.70	8.87
Number of seeds per pod	10.54	7.70	13.80	13.43
Number of pods per plant	14.53	7.10	36.20	32.30
Weight of pods per plant (g)	18.52	10.50	34.50	29.36
Number of seeds per plant	93.68	40.20	200.50	37.28
Seed lenght (cm)	0.98	0.70	4.40	52.37
Seed width (cm)	0.67	0.50	0.80	8.81
Seed thickness (cm)	0.54	0.40	0.60	11.93
Seed weigth per plant (g)	14.93	7.70	59.90	52.79
100 seed weight (g)	19.05	9.10	27.20	23.61
Yield (g/m^2)	131.15	40.50	218.10	34.70

color of hilum can range from white, beige, green, brown and black. Regarding the testa texture, the seeds can be classified in two groups: smooth (44%) or rough (56%).

Minimum and maximum values and the coefficient of variation for each quantitative trait are presented in Table 4. The data indicate that there is considerable morphological variation among the accessions. The most variable characters were weight of seeds per plant, seed length, plant weight and number of seeds per plant. The characters with less variation included number of days to maturity and to flowering, pod and seed width.

The dendrogram based on morphological data is shown in Figure 1. The accessions can be separated in seven clusters, one of which contains only one accession.

The status of the sample does not determine the cluster pattern of the accessions, that is, it is not possible to discriminate landraces from the advanced breeding lines in distinct clusters. Except for group A, group D and G, all of others groups are constituted by landraces and advanced breeding lines. In group D, all accessions are advanced breeding lines and almost of them are from IITA. The group E is composed mainly by landraces from Bulgaria. In this study, the accessions with lower similarity level with regard to others are the advanced breeding line 97-001 from Japan and line 77 from Hungry. The line 97-001 is the only accession that has yellow flowers and red seeds with white hilum.

In the principal component analysis for the 19 quantitative

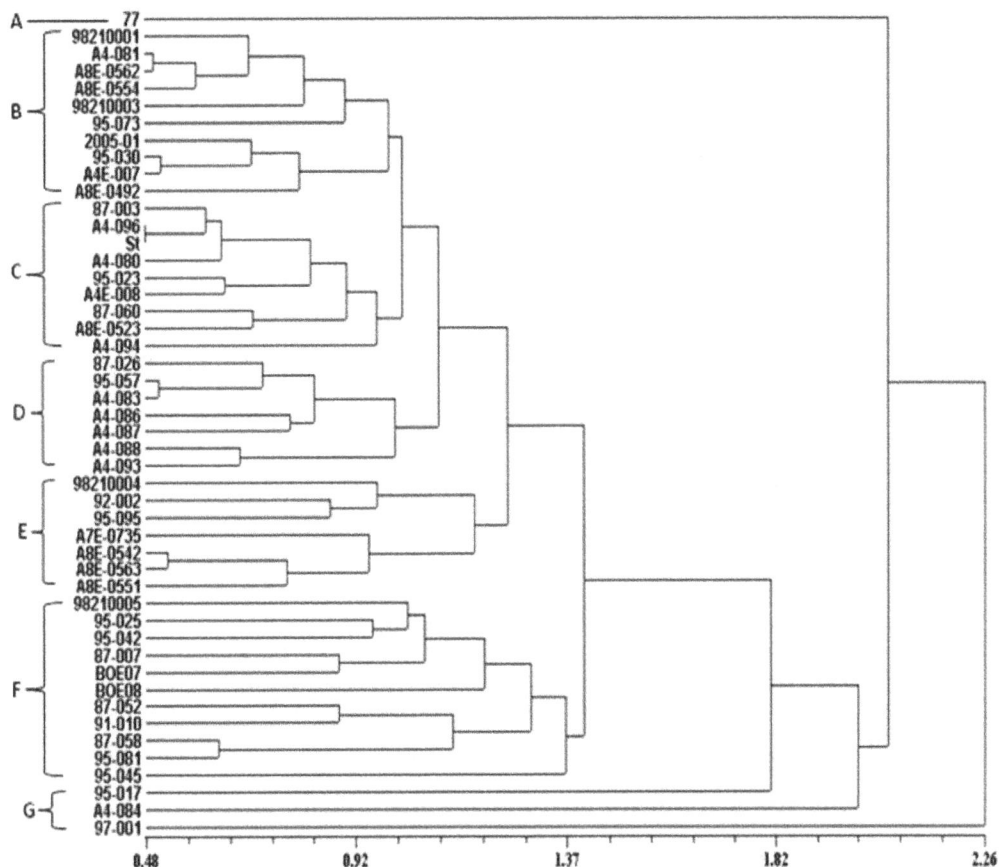

Figure 1. Dendrogram of 48 cowpea accessions obtained by analysis of quantitative morphological data.

quantitative traits, the first three principal components explained 56.46% of the total variance. The characteristics responsible for accessions separation along the first principal component were the number of seeds and pods per plant, the weight of the plant and 100 seed weight. The accessions were grouped according to their productive capacity. Vural and Karasu (2007) reported similar results. Accessions placed more to the right in the Figure 2 (87-052, 87-058, 95-081, 98210005, 91-010 and 77 - Group A) are those with higher yield potential and these accessions are in group F of the dendrogram. It can also be observed that the accessions with higher number of pods and seeds have small seed size, meaning that these characteristics are negatively correlated (Figure 3).

The second principal component grouped the accessions according to the pod width, pod thickness and the number of days to flowering. The accessions that are on the top of Figure 2 (group B) are those that have bigger pods and the plants required more days to reach the flowering period. In contrast, the accessions included in group C are characterized by plants that start flowering earlier and have small pods. The number of days required to reach the flowering is negatively correlated with

the duration of the period of flowering and production of seeds. The early flowering accessions were those that have larger flowering period and higher yield.

As it was already verified in the dendrogram, the advanced breeding line 97-001 from Japan is isolated from the others accessions. This accession differs from the others mainly by having plants that produce a very high number of pods and seeds with small size. The 100 seed weight is 9.5 g while the average of the others accessions is 19.1 g.

This study allowed us to point out which the characteristics remained stable over the three years. Of the 19 characteristics analyzed, pod length, number of seeds per pod, seed thickness and 100 seed weight (Figure 4) are the most stable. These traits were not influenced by environmental conditions and give us the same high level of information over the 3 years.

DISCUSSION

Genetic diversity is a prerequisite for the genetic improvement of a crop. But rational use of the genetic diversity present in germplasm collections requires a

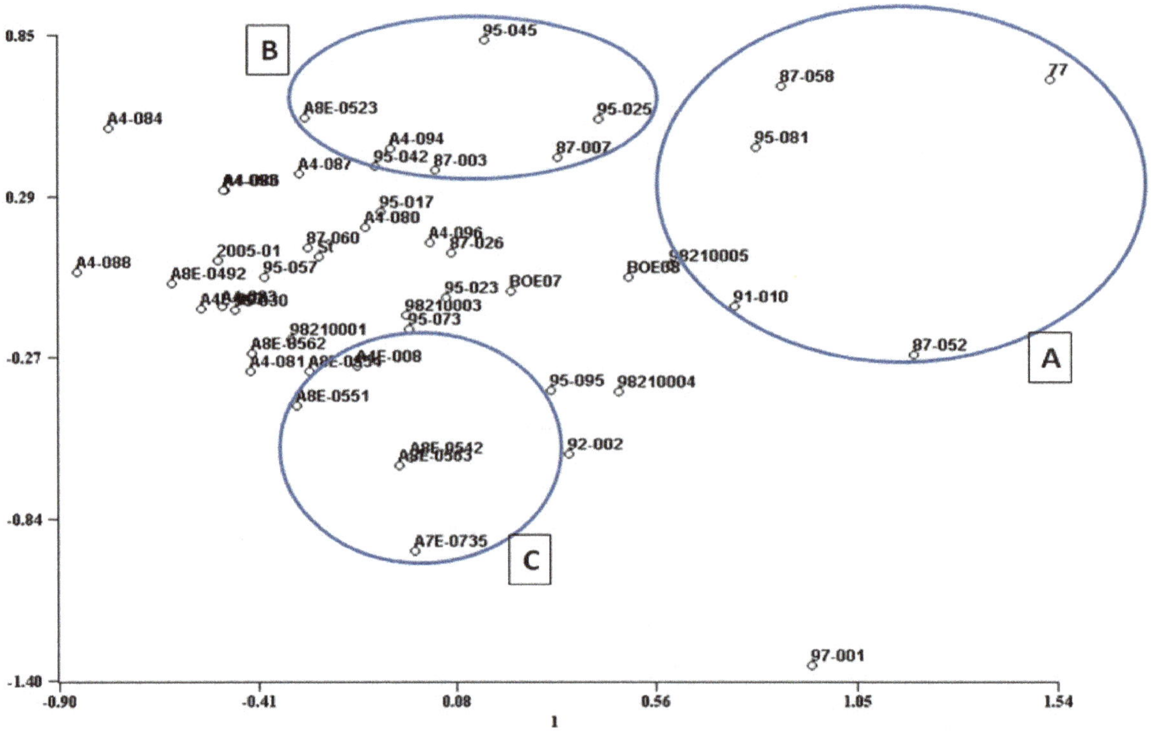

Figure 2. Projection of the 48 cowpea accessions into the plane defined by the two first principal components.

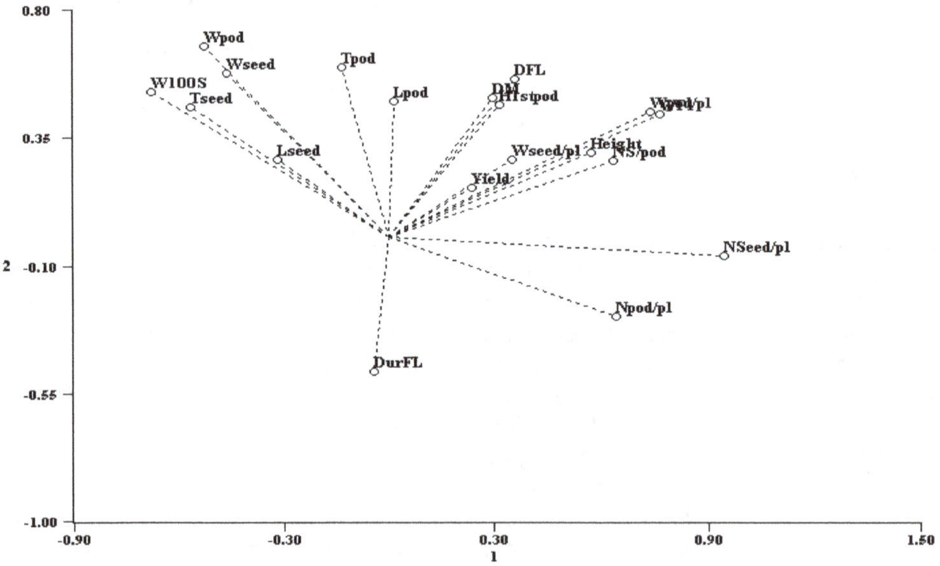

Figure 3. Projection of the 19 characteristics in axe 1 and 2.

good knowledge about their characteristics. Characterization of accessions is traditionally based upon morphological and agronomic traits, which is of high interest for plant breeders.

Our study confirms the existence of a large morphological variability in the collection of cowpea in IPGR (Sadovo, Bulgaria). The accessions evaluated in this present research represent an interesting starting point for a future plant breeding programme aimed at the development of new varieties. For the consumers and farmers, the most important characters of cowpea seed as in all food legumes are seed colour, seed size and seed coat. The most preferred colour is white and cream seeds with black hilum and smooth or smooth to rough

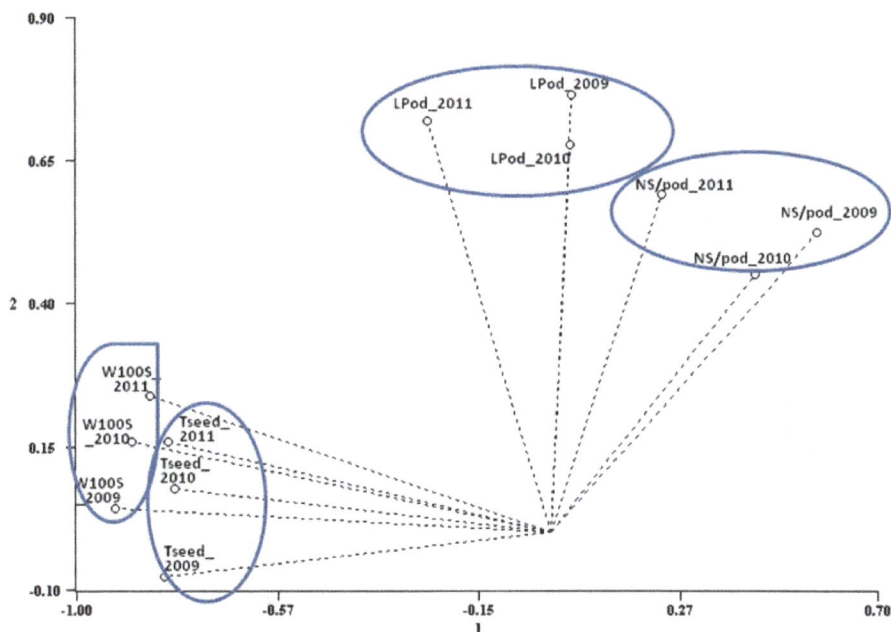

Figure 4. Projection of 4 characteristics observed during 3 years in axe 1 and 2.

coat. The more interesting accessions with cream colour of seeds, black hilum and big seed size with rough coat are the Bulgarian landraces A4E007 and 2005-01 and the breeding lines from IITA A4-084 and A-088 with cream colour of seeds, dark brown hilum and rough coat. The accessions with bigger seeds were A4-084, A4-088, A4-083, 2005-01 (more than 22 g/100 seeds weight). As has been verified by Omoigui et al. (2006), in this study the seed size was not influenced by the duration of reproductive phase, it is governed by other genetic factors. The phenological stages (days to flowering and days to maturity) showed a lower coefficient of variation. Similar results were reported from Hedge and Mishra (2009). The earliest accessions started flowering 40 days after germination comparing with latest ones needed 60.7 days for beginning of this phase. The shortest period to start flowering is advantage as high temperatures and low air humidity may be avoided. The results published by Hamidou et al. (2007) confirmed higher drought susceptibility in the 5 varieties of cowpea at flowering stage than during vegetative stage. In this study the accessions with earlier flowering are in group C of the Figure 2. Bulgarian landraces were characterized by an earlier maturity (72 to 75 days) in relation to the advanced breeding lines with different geographical origin (90 to 94 days).

Most of the studied quantitative traits were influenced by the environment conditions. But the our data have shown that pod length, number of seeds per pod, seed thickness and 100 seed weight were less affected by external conditions. These results suggest that these four descriptors should be useful in the coming evaluation studies.

Conclusion

The results obtained here provided useful knowledge about the diversity and breeding value of the Bulgarian landraces and all introduced accessions and give useful information for the selection of the most promising accessions to be used in future breeding programmes. It was not observed a clear distinction between landraces and accessions with foreign origin.

The most important components of yield are the number of pods and seeds per plant. High values for these traits were found in one Portuguese landrace (98210005) and in three advanced breeding lines (87-052, 87-058). The accessions with higher weight of seeds/plant, weight of pods/plant and weight of 100 seeds were the Bulgarian landraces A4E007 and 2005-01.

ACKNOWLEDGEMENT

This work was carried out under the project: "*Enrichment diversity of vigna and phaseolus germplasm collections-evaluation, maintenance and better utilization in correspondence with global climate change*". Ref. No GSP09GRD2_2.4_01, awarded by a Global Crop Diversity Trust.

REFERENCES

Agbicodo ME, Fatokun AC, Muranaka S, Visser FGR, Linden van der GC (2009). Breeding drought tolerant cowpea: constraints, accomplishments, and future prospects. Euphytica 167:353-370.
Blade SF, Shetty RVS, Terao T, Singh BB (1997). Recent

developments in cowpea cropping systems research. In: Advances in cowpea research, In Singh BB, Mohan Raj DR, Dashiell KE, Jackai LEN (eds). Copublication of IITA – JIRCAS. IITA, Ibadan, Nigeria, pp. 114-128.

Blum A (1988). Plant breeding for stress environments. CRC Press, Boca, Florida, USA, pp. 220-223.

Carnide V, Pocas I, Martins S, Pinto-Carnide O (2007). Morphological and genetic variability in Portuguese populations of cowpea (*Vigna unguiculata* L.). 6th European Conference grain legumes, 12-16 November, Lisbon. Book of Abstracts. p. 128

Coetzee JJ (1995). Cowpea: A traditional crop in Africa. Africa crop info 95 Leaflet. Vegetable and Ornamental Plant Institute and the Grain crops Institute, Agric. Resea Council, Pretoria.

Hawtin G, Iwanaga M, Hodgkin T (1996). Genetic Resources in breeding for adaptation. Euphytica 92:255-266.

Hamidou F, Zombre G, Diouf O, Diop N, Guinko S, Braconnier S (2007). Physiological, biochemical and agromorphological responses of five cowpea genotypes (*Vigna unguiculata* (L.) Walp.) to water deficit under glasshouse conditions. Biotechnol. Agron. Soc. Environ. 11(3):225-234

Hedge SV, Mishra KS (2009). Landraces of cowpea, *Vigna unguiculata* (L.) Walp. As potential sources of genes for unique characters in breeding. Genetic Resour. Crop Evolut. 56:615-627.

IBPGR (1983). Descriptors for cowpea. IBPGR, Rome.

Kameswara RN (2004). Biotechnology for Plant Resources conservation and use. Principles of seed handling in Genebanks Training course, Kampla, Uganda.

Mishra SK, Singh BB, Chand D, Meene KN (2002). Diversity for economic traits in cowpea. In Henry A, Kumar D, Singh NB (eds). Recent advances in arid legumes research for food, nutrition security and promotion of trade, CCH Haryana Agricultural University, Hissar, May 15-16, 2002. Indian Arid Legumes Society, CAZRI, Scientific Publishers, Jodhpur, India, pp. 54-58.

Mortimore JMA, Singh BB, Harris F, Blade FS (1997). Cowpea in traditional cropping systems. In: Singh BB, Mohan Raj DR, Dashiell KE, Jackai LEN (eds). Advances in Cowpea Research. Copublishing of IITA - JIRCAS, IITA, Ibadan, Nigeria, pp. 99-112

Negri V, Tosti N, Falcinelli M, Veronesi F (2000). Characterization of thirteen cowpea landraces from Umbria (Italy). Strategy for their conservation and promotion. Genetic Resour. Crop Evolut. 47:141-146.

Negri V (2009). Fagiolina" (*Vigna unguiculata subsp. unguiculata* (L.) Walp.) from Trasimeno lake (Umbria Region, Italy). In: Vetelãinen M, Negri V, Maxted N (eds). European landraces on-farm conservation management and use. Bioversity International. pp. 177-182.

Omoigui OL, Ishiyaku FM, Kamara YA, Alabi OS, Mohammed GS (2006). Genetic variability and heritability studies of some reproductive traits in cowpea (*Vigna unguiculata* (L.) Walp.). Afr. J. Biotechnol. 5(13):1191-1195.

Patil RB, Baviskar AP (1987). Variability studies in cowpea. J. Maharashtra Agric. Univ. 12:63-66.

Rohlf J (1997). NTSYS-pc: Numerical Taxonomy and Multivariate Analysis System. New York. Exeter Publishing.

Sardana S, Mahajan RK, Kumar D, Singh M, Sharma GD (2001). Catalogue on cowpea (*Vigna unguiculata* L. Walp.) germplasm. National Bureau of Plant Genetic Resources, New Delhi, India, p. 80.

Singh BB, Chambliss OL, Sharma B (1997). Recent advances in cowpea breeding. In Singh BB, Mohan Raj DR, Dashiell KE, Jackai LEN (eds). Advances in Cowpea Research. Copublishing of IITA - JIRCAS, IITA, Ibadan, Nigeria, pp. 30-49.

Vural H, Karasu A (2007). Agronomical characteristics of some cowpea ecotypes (Vigna unguiculata L.) grown in Turkey; vegetation time, seed and pod characteristics. Not. Bot. Hort. Agrobot. Cluj. 35(1):43-47.

Development of nursery raising technique for "system of rice intensification" machine transplanting

P. Dhananchezhiyan[1], C. Divaker Durairaj[1] and S. Parveen[1]

[1]Agricultural Engineering College and Research Institute, Tamil Nadu Agricultural University, Coimbatore - 641 003, Tamil Nadu, India.
[2]Department of Food Processing and Engineering, Karunya University, Coimbatore - 641 003, Tamil Nadu, India.

This study was aimed to develop the spaced mat nursery to suit the available transplanter for System of Rice Intensification (SRI) method of cultivation. To achieve 100% seed germination, enough root networks to provide enough rigidity for the mat and to offer conducive growth environment, the soil medium was optimized. Nine treatment media were prepared namely vermisoil (field soil+vermicompost-1:1, 2:1 and 3:1), soil+farm yard manure (FYM) soil (field soil+farm yard manure-1:1, 2:1 and 3:1), field soil+coirpith (1:1 and 2:1) and field soil alone. After 14 days of sowing, seedling height and root length were measured in all trays. Among the nine treatment media studied the maximum nursery height and root length of 17.06 and 10.75 cm was observed in FYM soil and vermisoil, respectively prepared in 1:1 ratio. For the same treatment media when ratio was changed to 2:1 it recorded 16.26 and 10.14 cm respectively. For the stiffness studies, field soil was mixed with decomposed sieved coirpith and fibrous coirpith each in the ratio of 1:1 and 2:1 and tested with and without base layer and measured the stiffness force. The mat stiffness was found to be maximum for a media mixture of field soil and coirpith at 1:1 and 2:1 ratios with a corrugated sheet base layer. From the results, the soil medium for growth and stiffness was optimized as field soil, FYM and fibrous coirpith in the ratio of 2:1:1.

Key words: System of rice intensification (SRI), mat nursery, rice transplanter, soil medium, mat stiffness.

INTRODUCTION

System of rice intensification (SRI) is being adopted in many states in India and the response from farmers has been overwhelming seeing the benefits of the method. Square planting method helps in operating weeders in check-rowed geometry to obtain maximum weeding efficiency and better soil aeration (Hameed and Jaber, 2007). In SRI, rice seedlings are transplanted by labourers. Square markings are made on the puddled fields either by ropes or by iron roller type marker. Each node of the square marking is transplanted with single rice seedling at precise spacing, usually 25 × 25 cm, about 16 plants per square meter. This method of planting requires careful planting on the grid which is difficult for the workers, who do not normally follow proper spacing in planting and maintain seedling population per hill. So this is the need of the hour for the transplanter to plant 14 days single seedling in a square planting. Instead of developing a new transplanter for this purpose the existing transplanter can be used with modification of existing mat nursery for planting in SRI method. These transplanters pick more than five seedlings per pick (Dewangan et al., 2005; Sahay et al., 2002). Study on various transplanters showed that Chinese make Yanji 8 row self propelled rice transplanter is most commonly used in India and the following parameters namely affordable for the majority of the farmers, simple mechanism, riding type, wider float, less weight and less cost given a way to this transplanter to suit SRI method. The main component in the development of SRI transplanter is suitable nursery raising method. The SRI transplanter will be successful only when the nursery raising method is modified to suit the SRI transplanter. Spaced mat nursery found to be suitable for SRI transplanter.

In SRI, single seedling is required to be planted per hill at spacing of 240 × 240 mm which would amount to 1,74,000 single seedlings per hectare. Therefore for planting one hectare, 345 trays of seedlings grown in the said configuration would be needed. Only 3.5 to 5 kg seed is required to plant one hectare as against 60 to 80 kg in conventional transplanting, assuming a 100% viable seeds. Hence, even in the absence of yield advantage, SRI is superior to conventional transplanted rice in terms of seed, labour and time required during transplanting (Sharma and Masand, 2008). When the seeds are sown at a sparse seed rate, with one seed placed at the each of the 504 nodes of the grid on the tray, the following points were of concern in implementing such a procedure; (i) The seed germination when not 100% would create lot of nodal voids after germination, (ii) The single seedling when grown for 14 days may not have enough root networks to provide enough rigidity for the mat to be handled by man and machine and (iii) The growth medium that is being extensively used for rice mat nursery production, may or may not provide conductive growth environment to the seeds to grow. Keeping these observations in mind the research was undertaken with the following objectives namely preliminary studies on transplanter for development of spaced mat nursery, optimization of soil medium for seedling health and mat stiffness.

MATERIALS AND METHODS

Laboratory experiment on Yanji 8 row self propelled rice transplanter

Determination of number of picks per row of mat

Since the width of the seedling box of the transplanter is 220 mm, a carton sheet of 220 × 440 mm size was used as a template. This template was kept inside the seedling box in the transplanter with seedling gate is kept in the starting position. When the engine crank was rotated manually the seedling box moved from one side to the other side by 220 mm and the transplanting needle made 14 marks on template. The distance between each marking was found to be 15.7 mm.

Determination of needle reaches into mat

The distance travelled by the separating needle to cut and pick from the nursery mat in seedling box through the seedling gate was found to vary between 7 and 20 mm by using the adjusting knob.

Development of spaced mat nursery

Since the requirement for the SRI transplanter is to pick and place single seedlings, it was planned to grow the seedlings themselves in a grid like sparse pattern on a typical transplanters' tray. Hence, it was contemplated that the tray surface area from which the transplanting finger picks the seedling need be divided into an imaginary grid in each of the nodes a single seedling is to be grown.

Based on the study with templates, the number of picks per row of seedling tray was determined as 14 and the reach length of picking into the tray as 7 to 20 mm. The typical seed tray width of 'Yanji' transplanter being 220 mm, the width from which one pick is drawn is hence 220/14=15.7 mm. The length of reach for the picker needle though adjustable between 7 and 20 mm, it was prefixed as 12 mm, since it is to be kept greater than the typical paddy seed length of 7 to 10 mm. Hence if the seedlings are grown in a grid pattern of 15.7 × 12 mm with one seedling occupying each node, the picking finger of that row would pick a single seedling to transplant. A typical transplanter tray of 440 × 220 mm size would then hold $\frac{432 \times 220}{15.7 \times 12} = 504$ grid nodes with one seedling in each node (Figure 1). Special nursery trays were developed to grow such sparsely sown mats. Stainless steel trays of 440 × 220 × 25 mm size were fabricated with the shutter to provide for easy removal of mat from tray. The side walls were provided with 5 mm beading bent outwards to avoid sharp edges as well to draw the mat out from tray.

Trials on development of sparsely sown seedling mats

Preliminary exploratory study on the sparse mat development

Field soil was sieved through 2 mm size sieve (BSS8) to remove stones, stubbles and lumps and was mixed with DAP powder at the rate of 15 g m^{-2} area. It was filled in the developed tray with an effective mat area of 432 × 220 mm to the brim with the shutter placed. Then water was sprinkled over the surface of the soil and leveled. The soil surface was brought to a wax like condition so that any marked line if made will not get erased. Then grid markings were formed using a foot rule (Figure 2).

Certified good quality seeds were selected and single seeds were sown in each node of the grid manually and covered using sieved field soil. Water was sprinkled using rose can. On the 5th day after sowing, germination was relatively less and on the 9th day the seedlings were observed to yellow and wither. To rectify this, 0.5% urea + 0.5% zinc sulphate were sprinkled, but no significant changes were noticed and the seedlings withered and did not survive (Figure 3). This preliminary trial was repeated twice and the results were similar. This proved that the methods of raising conventional mat nursery and raised bed nursery were not appropriate for spaced mat nursery.

Selection of suitable soil medium nursery

Selection of quality seed to ensure seedling count in sparse nursery experiment

Since SRI insists on single seedling per hill and young seedlings of 12 to 14 days old, obtaining healthy and robust seedlings from quality seeds is obligatory and thus every seed counts. Quality seeds ensure vigorous seedling growth, absolute establishment in the field, uniform plant population, accelerated growth rate, resistance against pest and diseases and uniform maturity at harvest. Most importantly a quality seed was selected to have above 90% germination rate.

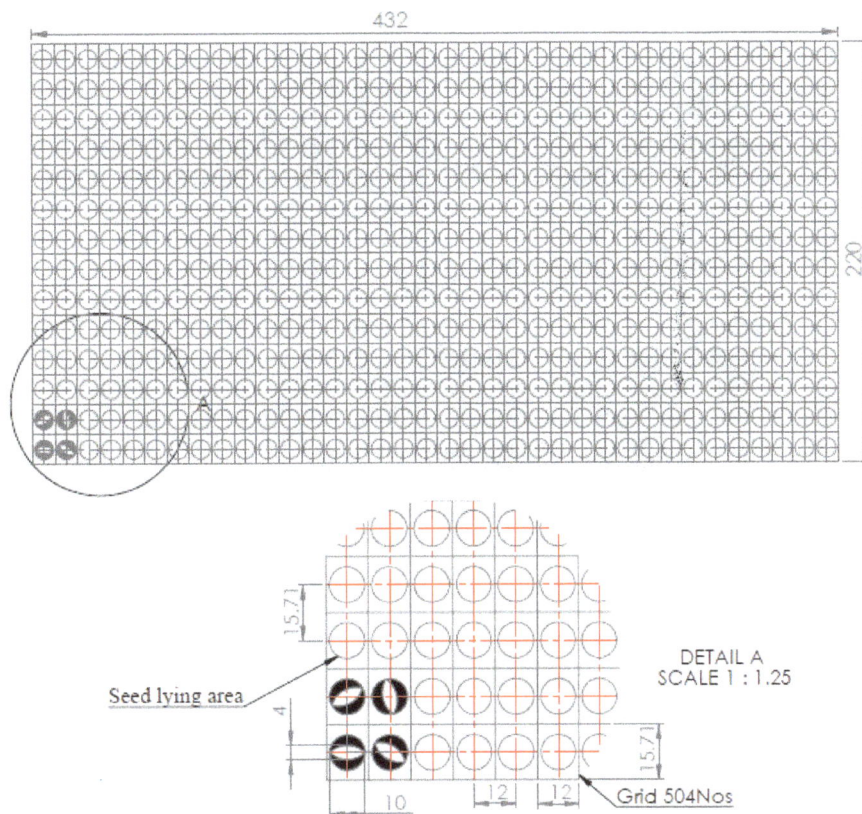

Figure 1. Imaginary grid on typical transplanter tray.

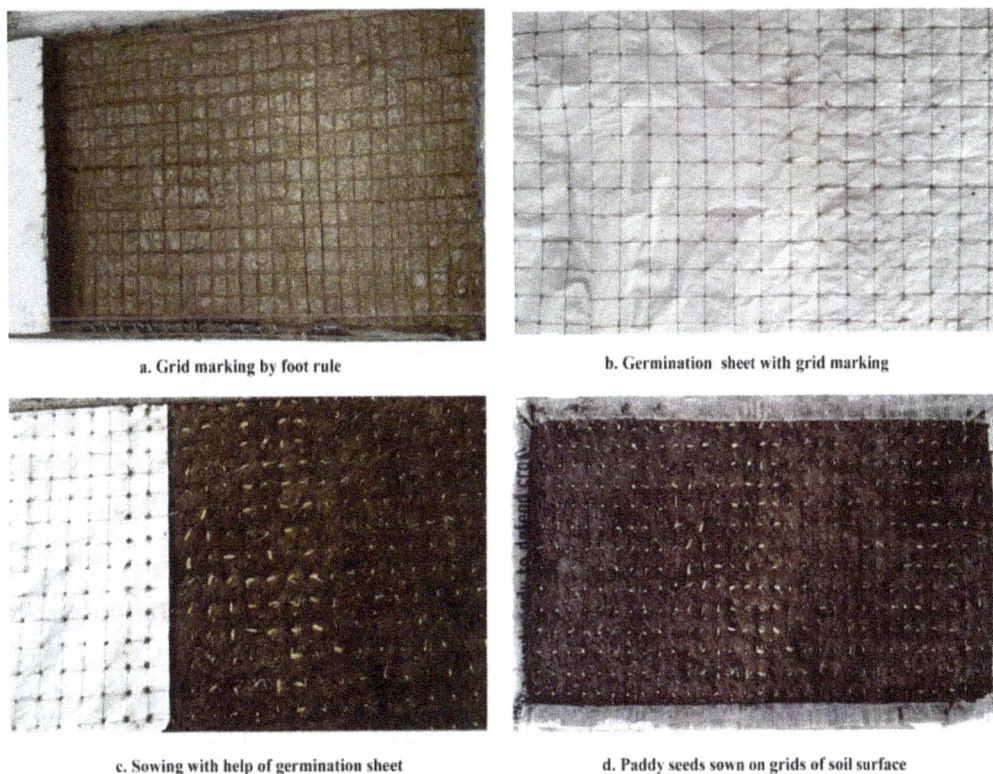

a. Grid marking by foot rule

b. Germination sheet with grid marking

c. Sowing with help of germination sheet

d. Paddy seeds sown on grids of soil surface

Figure 2. Manual sowing on mat

Figure 3. Poor growth of sparse nursery on conventional mat medium.

Table 1. Optimization of soil medium for better crop growth.

S/N	Medium	Mixing ratio
1	Field soil	-
2	Field soil + vermicompost	1:1
3	Field soil + vermicompost	2:1
4	Field soil + vermicompost	3:1
5	Field soil + FYM	1:1
6	Field soil + FYM	2:1
7	Field soil + FYM	3:1
8	Field soil + Coirpith	1:1
9	Field soil + Coirpith	2:1
10	Field soil	Control

Replication: 3.

Seed treatment: The certified paddy seeds with 90% germination rate were soaked in salt water (20%). Floating seeds over the water surface were removed and seeds that settled down at the bottom of the container were collected and washed thrice with fresh water. Then the seeds were treated in a mixture of *Pseudomonas, Azophos, Thirum 75 SD and Carbendazim* at 30, 400, 5 and 5 g, respectively for every 2 kg of seeds (Nghiep and Gaur, 2005). The seeds were soaked and treated thus for 12 h. Soaked seeds were transferred into a gunny bag and left for 24 h (Junsripibul, 1988). When white root or radicals emerge from the seeds, the seeds were used for sowing on the nursery bed.

Experiments on optimization of better soil medium

Nine treatment mediums were selected each with different combination of growth media as mentioned in Table 1, The media was added with DAP powder, *pseudomonas, VAM and Azophos* at the rate of 50, 6, 50 and 40 g m^{-2}, respectively in all the treatments (Ahamed and Ravi, 2006). The prepared soil media were separately filled in trays and leveled after sprinkling with water. Because of the tediousness involved in marking and dividing the soil surface into grids, grid pattern was drawn on 432 × 220 mm size germination sheet and kept on the wet soil in the tray, and the treated seed was sown in the grid one per node and covered with soil.

Conventional dense mat nursery was also prepared as a control treatment for the experiment. A seed rate of 160 g and 12 g per tray was used for conventional mat nursery and spaced nursery, respectively. Water was sprinkled twice (morning and evening) a day. After 14 days, seedling height, root length and germination percentage were measured in all trays. Statistical modeling (GNU 'R' statistical package) was done on this data to optimize the soil medium. Though the growth of the plant was good, the mats were not stiff enough to handle them while taken out from the tray. Insufficient root network was observed as the cause of such weak mat stiffness. Therefore the factors responsible for mat stiffness were studied next.

Method for assessment of mat stiffness

Before evolving the methods for improving the stiffness of the sparsely sown mats, it was necessary to develop a methodology for quantifying the stiffness of the mat. One easy and viable index indicative of the mat stiffness would be the rip or tear strength of the mat as measured by shearing the mat with a prong or claw. Rip force for separating a single seedling determines the stiffness of

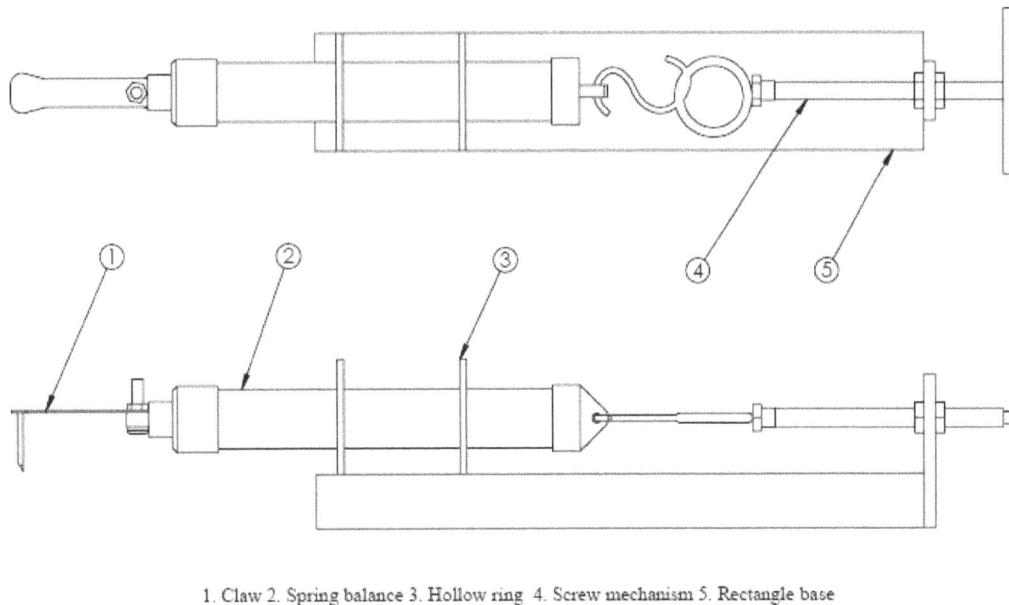

1. Claw 2. Spring balance 3. Hollow ring 4. Screw mechanism 5. Rectangle base

Figure 4. Test rig for rip force measurement.

nursery mat. A similar mechanism of transplanter picker was developed. A claw fitted with spring balance was used to find out the force required for separating seedlings from nursery mat made up of different soil mixture, different surfaces and different mat thicknesses. Rip strength was expressed in g cm^{-2}.

For finding ripping strength, a device was developed (Figure 4). The device contains a rectangle base (5), spring balance (2) (make: SALTER), a screw mechanism (4) and hollow rings (3) to support the spring balance. The spring end of the balance was fitted with a claw (1) and the other end was connected to the screw through a swivel. Minimum count of the spring balance was 5 g and maximum count was 1 kg. The manually driven screw mechanism is of 110 mm length and 6 mm diameter with the thread on the holding post. The base of device has two 2 mm thick hollow rings of 20 mm inner diameter to support the spring balance. Test rig and the nursery tray were kept on the same plane. The claw of the device was first made to pierce the mat from top at the required distances from edge of the mat and was kept aligned to facilitate pulling of single seedling along with medium. When the screw is manually tightened, the claw separates the seedling along with medium. The spring balance reading is read as the rip force in g. When divided by the ripping area (2 × thickness × length of rip) the rip strength in g cm^{-2} could be found.

Optimization of growth medium for better stiffness on spaced mat nursery

An experiment was carried out to test the stiffness of the mat composition devoid of any root network. To give the necessary stiffness to the medium, field soil was mixed with 4 year old decomposed coir pith in different ratios and filled in tray. Stiffness was measured at different settling time when it was taken out from the tray. This study was carried out in the soil medium without growing the seedling at all. The stiffness force was measured for the 8 types of soil medium with and without base layer of corrugated sheet at 4 different settling times.

Corrugated sheet was used as base layer as spread at the bottom of the tray and soil was filled after wetting. Two types of

coirpith material namely sieved coirpith in 3 mm sieve and fibrous coirpith were added with soil. Fibrous coirpith was mixed with soil in the ratio of 1:1 and 2:1 by volume. The length of the fibre ranged between 10 and 35 mm. In this way the prepared soil media were filled in tray, made wet and for every 12, 24, 36 and 48 h the stiffness force was measured and analyzed for optimization of the medium devoid of root network. The GNU 'R' statistical package (The R foundation www.r-project.org) was used for the statistical modeling mentioned above.

The optimized soil medium for growth and that for stiffness were judiciously combined for this study. Composition of medium for nursery bed was varied in two different levels in the experiment. The two different compositions of media are namely, vermin soil and FYM soil. Vermi soil contains field soil, vermicompost and fibrous coirpith and FYM soil contains field soil, farm yard manure and fibrous coirpith.

Assessment of root length and plant height

After 14 days of growth, the plants in the nursery were measured for root length. A part of the mat was gently washed in water and each single plant was meticulously separated. The total root lengths and height of the plant were measured by foot rule. The influence of the soil medium of mat on the health of the nursery was also to be monitored and optimized.

RESULTS AND DISCUSSION

Optimization of soil medium for better crop growth on sparsely sown mat

Influence on nursery height

The experiment on optimization of better soil medium for growing sparse nursery was explained. Nine different

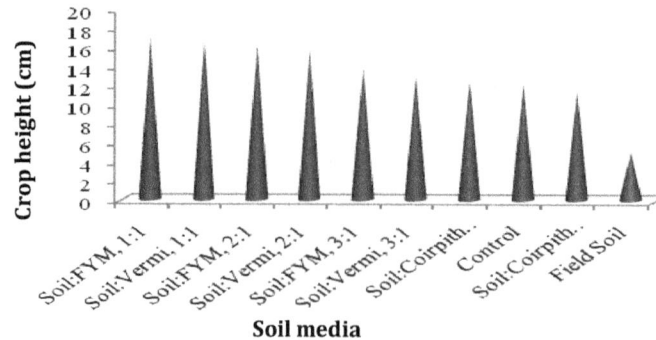

Figure 5. Effect of different soil media on nursery height.

combinations of growth medium were used to grow the selected/treated paddy seeds and on the 14th day, biometric observations on nursery height, root length and percentage germination were recorded. Single factor ANOVA was used to analyses the recorded data (**=$P<0.01$).

The treatment effect was found to be significant. The nursery height ranged from 15.69 to 17.06 cm. Ahmed et al. (2008) confirmed these results. The presence of vermicompost, FYM which were basically organic soil nutrients and coirpith an inert filler material mixed at different percentages had caused this difference in nursery height. The nursery height may be considered as an indicator of the health of the nursery that was grown. While comparing the treatment means by LSD (Least Significant Difference), it was found that a plain field soil reported the least effect on nursery height (5.07 cm), whereas the field soil mixture with organic composites such as vermicompost or FYM provided the best nursery growth (Figure 5). When the content of field soil was increased to a 3 parts in one ratio, the nursery growth drastically dropped to a lower level of about 13 cm. This fact provided further evidence that organic matter at an appropriate ratio is to be present in the medium for proper development of sparse nursery. The field soil/coirpith mixture was the next higher statistically on par group of nursery height levels (11 to 12 cm). Surprisingly, the control, namely the densely grown nursery on plain field soil was on par with this group in crop growth indicating that the medium selection is crucial for development of sparsely sown nursery. In the experiments on the suitability of soil medium for sparse nursery, the nursery heights observed in field soil with vermicompost soil medium were 16.49, 15.69 and 13.03 cm with respect to the ratio of 1:1, 2:1 and 3:1 and that observed in field soil with farm yard manure soil medium were 17.06, 16.26 and 13.8 cm with respect to the ratio of 1:1, 2:1 and 3:1.

Effect on root growth

The data on root length were analyzed using analysis of variance. Here also, the root length was significantly influenced by the treatment media (**=$P<0.01$). The reason for the above must be the same explained already for crop height. Since the root matrix development was anticipated to be the key component in forming the stiffness of the nursery, this factor was observed and analyzed separately. The mean comparison through LSD indicated that here again just similar to the ANOVA on nursery height, the presence of organic composites in the medium provided the best root growth of 10 cm which was higher than the values reported by Ahmed et al. (2008) and similar to the values reported by Vijayakumar et al. (2004). The observations clearly indicated a similar trend as in the case of nursery height. The only difference observed here was that the dense nursery in field soil (control) provided an on par root growth with that of the media with coirpith. Here also, the lowest root length of 4.69 cm was recorded on field soil with sparse sowing (Figure 6).

The sparse nursery root lengths observed in field soil with vermicompost based soil medium were 10.75, 10.14 and 9.02 cm with respect to the ratios of 1:1, 2:1 and 3:1 and that observed in field soil with farm yard manure based soil medium were 10.69, 10.47 and 8.49 cm for the ratios 1:1, 2:1 and 3:1 respectively. FYM soil and vermin soil at 2:1 ratio medium was chosen as the optimized medium for the production of sparse nursery with the lesser vermicompost content making it cheaper.

Optimization of soil medium for better stiffness of the sparsely sown mat

An experiment was carried out to assess the stiffness of medium devoid of any root network, as influenced by the ratio of mixtures, nature of medium, settling time after watering and provision of a base stiffening layer. A factorial analysis was attempted and a linear model was built on the mat stiffness. On analyzing the raw data of stiffness for normalcy, it was found that the distribution was not normal. But normalcy of data is mandatory for building a linear model. Various transformations such as square root and power transformations were tried on the

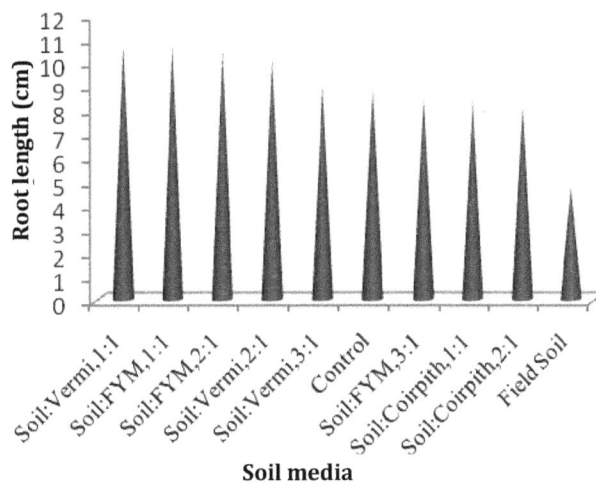

Figure 6. Effect of different soil media on root growth.

Table 2. Optimization of soil medium for better stiffness on spaced mat nursery.

S/N	Settling time (h)	With base layer		Without base layer	
		Soil : Sieved coirpith	Soil : Fibrous coirpith	Soil : Sieved coirpith	Soil : Fibrous coirpith
1	12	1:1	1:1	1:1	1:1
		2:1	2:1	2:1	2:1
2	24	1:1	1:1	1:1	1:1
		2:1	2:1	2:1	2:1
3	36	1:1	1:1	1:1	1:1
		2:1	2:1	2:1	2:1
4	48	1:1	1:1	1:1	1:1
		2:1	2:1	2:1	2:1

data with no avail (Table 2). Therefore a box-cox analysis was attempted on the model to find an appropriate transformation coefficient (λ). Figure 7 illustrate the log likely hood for different λ values for the full model being considered. The transformation coefficient was found to be -0.63, implying that the transformation necessitated on the stiffness (ripping strength, g cm^{-2}) is stiffness to the power of -0.63. Using the above transformation for the stiffness response, a full model involving the mixing ratio, type of soil medium, settling time after watering and the presence of stiffening base layer was built.

The linear model was stepped through by systematic removal of factors one by one based on AIC (Akaike Information Criteria) value and the minimal adequate model was searched through using the "step" function of the 'R' statistical tool. The best simple model was arrived as that with only layer, medium and mixing ratio in additive mode without any interaction. The AIC was -1294. The best model was put under diagnostics to find out whether the built linear model was adequate. Figure 8

illustrates the diagnostic plot, wherein no evidence was found on any trend in the distribution of residual against the fitted value. The Q-Q plot also proved that the response had a perfect normal distribution, which was quite anticipated because of the box-cox transformation. Table 3 shows the ANOVA on stiffness as influenced by the aforesaid factors. The settling time was found to be insignificantly influencing the stiffness of the mat. This was so because the medium having more of fibrous content was not affected by the settling time.

Figure 9a to c shows the Tukey plots relevant to the three influencing factors on the stiffness of the mat. It may be noted that none of the paired differences between treatments of each factor contains zero in its range and hence each pair statistically was proven to be significantly different. On analyzing the mean values of the stiffness as plotted between the ripping stress (mat stiffness) and the settling time (Figure 10a), the presence of base layer (corrugated sheet) improved the mat strength from about 115 to 155 g cm^{-2} proven to be

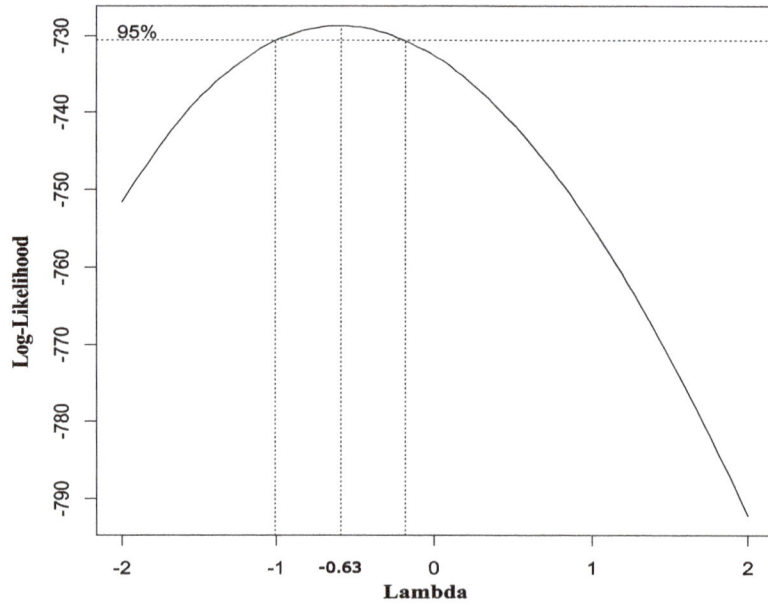

Figure 7. Determination of Box-Cox transformation parameter.

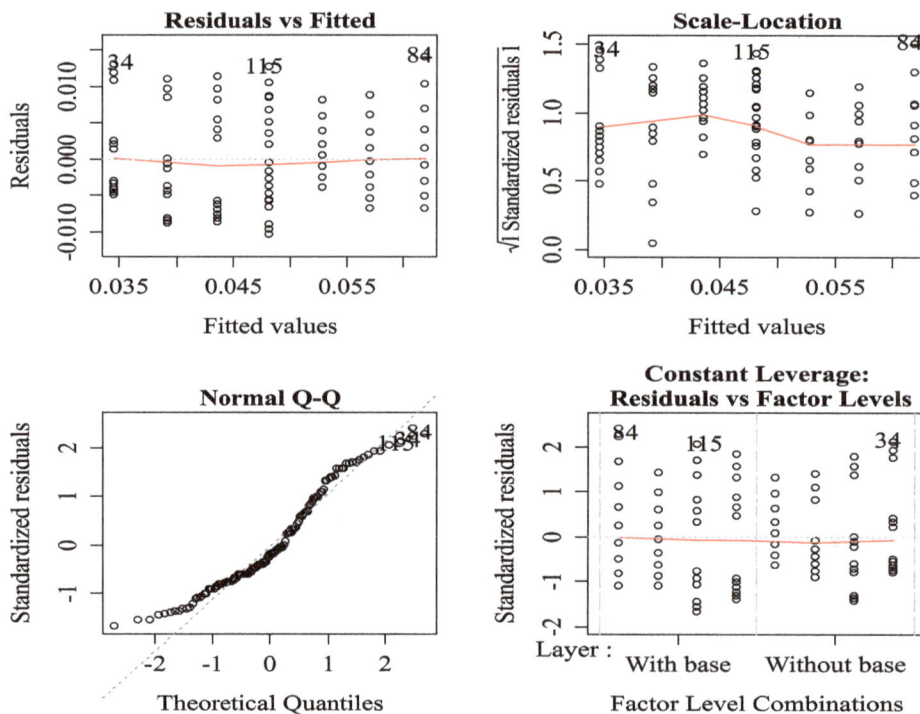

Figure 8. Diagnostics on model of mat stiffness affected by growth medium without nursery.

significant from the ANOVA. Though the corrugated sheet is thoroughly soaked for 14 days, the mat stiffness is higher.

Figure 10b depicts the influence of the mixing ratio on the stiffness. The equal proportion of soil and sieved/fibrous coirpith had caused an increase of mat strength by 20 g cm^{-2} over that of a 2:1 soil and sieved /fibrous coirpith (Table 3). This significantly higher mat strength was due to the presence of more organic matter by way of more soil in the medium causing better binding, thereby better stiffness. As for the influence of the type of medium, the fibrous coirpith media had about 170 g cm^{-2} of mat strength compared to 105 g cm^{-2} as exhibited by medium having sieved coirpith. The longer fibrous

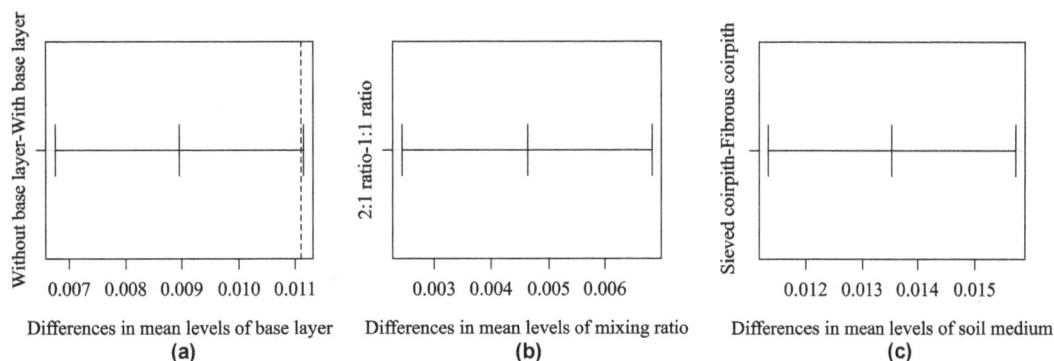

Figure 9. Tukey's mean comparison on effect of media characteristics on mat stiffness.*All differences at 95 % family-wise confidence level.

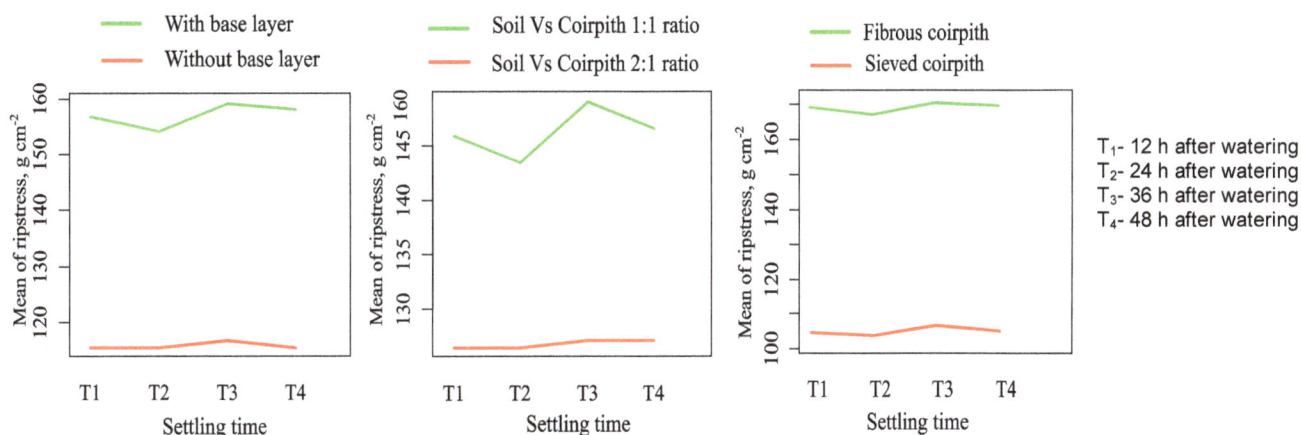

Figure 10. Influence of characteristics of growth media on mat stiffness.

Table 3. ANOVA on mat stiffness as influenced by media characteristics.

	Df	Sum squared	Mean Squared	F value	Pr(>F)
Layer	1	0.002563	0.002563	65.39	4.73e-13***
Medium	1	0.005861	0.005861	149.54	< 2e-16***
Mix ratio	1	0.000687	0.000687	17.52	5.34e-05***
Residuals	124	0.004860	0.000039		

Significance level: ***, 0; **, 0.001; *, 0.01.

content in the fibrous coirpith had interwining structure which had caused the significantly higher mat strength. The response curves in all the above graphs (Figure 10) happened to be flat exhibiting the fact that the settling time did have insignificance on the mat stiffness. A listing of model and the results obtained from factorial analysis is given subsequently as a sample factorial analysis.

From this analysis, it was concluded that the mat stiffness would be maximum for the media with fibrous coirpith, a media mixture of soil and coirpith at 1:1 ratio and with a corrugated sheet base layer followed by the same mixture at 2:1 ratio. The 2:1 ratio itself gave the

sufficient mat stiffness suitable for SRI transplanter and hence optimized this ratio.

The optimized soil medium for growth and for stiffness were judiciously combined and studied. Vermi soil contains field soil, vermicompost and fibrous coirpith in the proportion of 2:1:1. FYM soil contains field soil, FYM and fibrous coirpith in the ratio of 2:1:1 and a base layer of corrugated sheet were incorporated in the above two soil media to stiffen the mat further. For these optimized soil mediums plant health and picking performance of transplanter were studied. There were no noticeable differences were observed between the vermin soil and FYM soil. Since

FYM soil is easily available as well as cheaper in cost, it was optimized for the SRI transplanter spaced mat nursery. This spaced mat nursery was suitable for loading into the seedling box of the transplanter for SRI method of transplanting.

ACKNOWLEDGEMENT

The authors express their sincere thanks to the Council of Scientific and Industrial Research, Government of India, New Delhi and Tamil Nadu Agricultural University, Coimbatore - 3 for their financial assistance and permission to conduct the study, respectively. The authors also greatly acknowledge the help rendered by pear scientists and other technical supporting staff.

REFERENCES

Ahamed AS, Ravi V (2006). National Symposium on System of rice Intensification present status - future prospects, 17-18th November, Abstract. 27, Hyderabad, Andhra Pradesh, India.

Ahmed SKZ, Ravisankar N, Srivastava RC (2008). Participatory evaluation of SRI in Tsunami affected villages of south Andaman, 3rd National symposium on SRI at Tamil Nadu Agricultural University, Coimbatore, pp. 47-50.

Dewangan KN, Thomas EV, Ghosh BC (2005). Performance evaluation of a laboratory model rice technology. Agric. Eng. Today 29:38-45.

Hameed KA, Jaber FA (2007). System of Rice Intensification in IRAQ during 2007. Al-Mishkhab Rice Research Station (MRRS) Najaf-Iraq. http://ciifad.cornell.edu/sri/countries/iraq/iraqhameedrpt1207.pdf

Junsripibul W (1988). Effect of soaking-drying on seed viability, vigor and storability in RD 23 and KDML 105 rice. AGRIS Categories: Plant physiology-growth and development; Seed production. P. 81. http://agris.fao.org/agris.

Nghiep HV, Gaur A (2005) Efficacy of seed treatment in improving seed quality in rice (Oryza sativa L.). Omonrice 13:42-51.

Sahay CS, Satapathy KK, Agrawal KN, Mishra AK (2002). Evaluation of self propelled rice transplanter in valley and terraced lands of north eastern hilly region. Agric. Eng. Today 26:1-10.

Sharma PK, Masand SS (2008). Evaluation of system of rice intensification in a high rainfall area of North Western Himalayas. Oryza 45:206-211.

Vijayakumar M, Sundarsingh SD, Prabhakaran NK, Thiyagarajan T M (2004). Effect of SRI practices on the yield attributes yield and water productivity of rice (Oryza sativa L). Acta Agronomica Hungarica 52:399-408.

3

Eco-farming models and their regional differentiations of Shaanxi, China

Jiang Che[1], Yun Cheng Liao[1] , Xiao Hui Ding[2] and Yan Liu[1]

[1]College of Agriculture, Northwest A&F University, Yang ling, 712100, China.
[2]College of Resources and Environment, Northwest A&F University, Yang ling, 712100, China.

By statistical analysis and cluster analysis, the study investigated eco-farming models and their regional distributions of twelve counties of Shaanxi Province. It categorized the eco-farming models into seven basic models and analyzed the distributions of the basic models of the different counties. Its results showed that of the sampled counties of Shaanxi, the average number of the eco-farming models was 6.2 and the average coverage degree of the models was 88.1%; and the proportion of the crop planting model was the highest and the proportion of the courtyard based model was the lowest. According to its cluster analysis results on the proportions of the different eco-farming models, the study divided the twelve sampled counties under four groups, and analyzed and revealed the distribution characteristics of the natural resources and environments of the groups, put forward the priorities for the different regions to practice the different eco-farming models, and provided research methods on eco-farming models and their distribution of other regions.

Key words: Shaanxi, eco-farming model, regional differentiation pattern.

INTRODUCTION

Shaanxi province is commonly divided into Central, South and North Shaanxi and this division is done in terms of its natural geographical conditions. The study separately screened four counties of the three regions as its research targets, which were the District and countries of Yangling, Meixian, Dali and Chang'an of Central Shaanxi, the counties of Mianxian, Chenggu, Xixiang and Ningqiang of South Shaanxi and the counties of Jingbian, Luochuan, Suide and Zichang of North Shaanxi. The study carried out on-site surveys of these countries and collected the data of the counties from relevant year books concerned with them. The study conducted its processing and analysis of the data thus obtained to probe into the countries. It is well-known that to practice eco-farming has become one efficient approach and one inevitable choice for sustainable agricultural development,

and eco-farming models has been the essence and concrete reflection for the different regions to practice eco-farming (Krishna., 2011; Kurosh and Saeid, 2010), and thus to investigate eco-farming models and their distribution characteristics of different regions is helpful to orienting regional eco-farming development, helpful to comparing the limiting factors for different regions to practice eco-farming and solve their main problems while their practicing eco-farming, helpful for similar regions to learn how to practicing eco-farming, so as to promote eco-farming development (Liu and Jiao, 2002) . The two methods that the study adopted and the statistical analysis and cluster analysis that the study carried out have different research targets and the research results obtained by them are different as well; the study carried out its statistic analysis to probe into the basic situations of the different countries concerned and considering that statistic analysis is the basis of cluster analysis, cluster analysis could be carried out to categorize similar regions and thus to set up corresponding categorization patterns and as a result the research results thus obtained will be of very great practicality (Wackernagel et al., 2002). Although there are related researches in eco-farming models, they are not deep enough and their analysis are not penetrative enough as well, so that it is necessary to do further research on eco-farming models.

THEORY

Main eco-farming models and their regional classification of Shaanxi

Eco-farming models

The classification of eco-farming models is crucial to studying, promoting and expanding eco-farming as well as setting up development and assessment standards of eco-farming. Because most of eco-farming models are generalized from practical experiences, many scientific workers summarize eco-farming models by regional example enumeration or practical classification. In the recent years, some scholars try to systematically sort, sum up and classify typical eco-farming models. Qi (1992) proposed that depending on the scales or administrative ranks of the regions that they cover, eco-farming models be classified into cities, countries, townships(towns), villages and farms and households of eco-farming; depending on their natural, social and economic conditions of the regions where they are practiced, they be classified into plain-based eco-farming, mountainous eco-farming, hilly eco-farming, aquatic eco-farming, grassland-based eco-farming, courtyard-based eco-farming and littoral eco-farming and urban eco-farming; depending on their products, eco-farming models be classified into single-product and multi-product models; depending on their farming types, eco-farming models be classified into crop-planting models, forest plus fruit tree models, animal-raising models, farm model, business models; depending on their resource exploiting modes, eco-farming models be classified into multi-layer resource-exploiting models, integrated resource-exploiting models, recyclable resource exploiting models, self-cleaning resource models and bio-ring-added resource exploiting models (models with bio-rings added to or removed from their biological chains and symbiotic model). Li (2000) put forward what simplified the aforementioned classifications, holding that eco-farming could be classified according to the scales and natural, social and economic conditions plus main products, or three kinds of classification standards of main industries of the regions where the models were practiced, that is, according to the scales or administrative ranks of the

regions where eco-farming was adopted, expanded and practiced, eco-farming models could be classified into into cities, countries and townships, villages and households of eco-farming; according to the natural, geographic, social and economic conditions of the regions where they were practiced, eco-farming models could be classified into plain based models, mountainous models, hilly models, aquatic models, grassland based models, courtyard-based models, littoral models and urban plus suburb models; according to the major products or major industries of the regions where they were adopted, eco-farming models could be classified into single product-or single industry-dominated models, or at least two or three product- or industry-dominated integrated models. Li (2008) divided eco-farming models of China into four groups, multi-layer substance-exploiting models, symbiotic models, resource-exploiting models and environment-controlling models, tourism attraction models according to its agricultural development characteristics, social and economic developments and resources status quo. In their study of the Standard Systems and Important Technical Standards of Eco-farming. Qiu and Ren (2008) adopted the classification method developed by Li (2008). Currently, the understanding on the classifications of eco-farming still need to be unified. The basic types and their intensions of eco-farming models are shown in Table 1.

The study put forward that the classification of eco-farming models was one basic classification and other classifications of eco-farming models carried out in light of their purposes were not excluded.

The afore-described nuclear classifications of eco-farming models showed that there are seven basic eco-farming groups, that is, seven basic models. Of course, this modification is not absolute and in production practices, different basic eco-farming models mutually penetrate into and interact with one another, so that basic eco-farming models can be extended to form an integrated model by adding such non-farming industries as processing and tourism to them.

Classifications of main eco-farming models of Shaanxi

Eco-farming models of Central, South and North Shaanxi were classified on the basis of on-site survey data and their statistical analysis results of the regional distributions of the main eco-farming models of the three regions and according to the requirements of basic eco-farming models (Bicknell et al., 1998). The classifiers of the classification, taking the form of region+serial number+ type, are shown in Table 1. For example, YL1-Y1 stands for a crop planting dominated eco-farming model, which is practiced in Yang ling, that is, YL, and whose serial number is 1. Table 2 presents main eco-farming models and their basic models of Yang ling.

Table 1. Basic eco-farming model.

Model code	Basic eco-farming model	Citing (basic intension)
Y_1	Crop planting model	Being aimed at improving farmland environments and properly arranging crop-planting modes on farmlands
Y_2	Forest plus fruit tree model	Being aimed at promoting ecological forests and fruit trees and improving land use efficiencies
Y_3	Animal-raising model	Being aimed at promoting ecological raising dominated by aquatic culture and animal husbandry
Y_4	Courtyard based model	Being aimed at promoting courtyard economies
Y_5	Business model	Being aimed at properly processing agro-products and producing organic agro-products
Y_6	Environment-conserving model	Being aimed at improving eco-environments and environment qualities
Y_7	Tourism attraction model	Emphasizing on conserving natural landscapes and promoting sustainable tourism

Table 2. Main eco-farming models and their basic types of Yang ling.

Eco-farming model	Basic model
The model of "Straw composting and incorporation into soil"	$YL1-Y_1$
The model of "fruit tree and grass intercropping"	$YL2-Y_2$
The model of "pig raising-biogas generation –fruit tree planting"	$YL3-Y_3$
The model of safe latrine and waste stacking and composting	$YL4-Y_4$
The model of processing byproduct and urban organic waste recycling	$YL5-Y_5$
The model of countryside tree planting and polluted land rehabilitation	$YL6-Y_6$
The model of developing the city of Agricultural Sciences and Technology Yang ling as a tourism attraction	$YL7-Y_7$

METHODS

Proportions and coverage degrees of the eco-farming models

Depending on main eco-farming models and their basic models of the different regions of Shaanxi (Bastianoni et al., 2001), the study calculated the percentages of the different eco-farming models of the different regions, or the proportions of them, that is, the proportion of the number of one individual eco-farming model of one region to the total number of all the eco-farming models of the region (y_{ij}), whose computation formula is as follows:

$$y_{ij} = \frac{\text{Number of the eco farming model of j of the Region of i}}{\text{Number of all the eco farming models of the Region of i}} \times 100\% \qquad (1)$$

In which $i=1,2,3,...,11$ which stand for the different regions; $j=1,2,...,7$, which stands separately for the crop planting model, the forest plus fruit tree model, the animal-raising model, the courtyard based model, the business model, the environment-conserving model and the tourism attraction model.

Therefore, the value of y_{ij} indicates how important the different models are to some extent. In addition, the study calculated the coverage degrees of the different models (C_i), which was the percentage of the number of the models practiced in one region to the total number of the models of the region and the total model number of the study was seven. C_i was calculated by the following

formula:

$$C_i = \frac{\text{Number of the models practiced by the Region of i}}{7} \times 100\% \qquad (2)$$

Similarities of the eco-farming models by cluster analysis

As described before, the establishments of eco-farming models and their basic groups of different regions are closely related to the local social, economic and environmental conditions of the regions (Coleman et al., 1992). Thus, the proportions of different eco-farming models (y_{ij}) of different regions can be used as the original cluster analysis variables to find out similar regions for different eco-farming models as well as the social and economic or environmental characters of these regions, so that a scientific foundation for promoting and adopting and practicing eco-farming can be provided (Tilley and Swank, 2003). Here, the study adopted the un-weighted pair group method with arithmetic means (UPGMA) to conduct its cluster analysis (Brown and Buranakarn, 2003). In which:

$$x_{ij} = \frac{y_{ij} - \overline{y}_j}{S_{ij}} \qquad (3)$$

Table 3. Percentages (proportions) and coverage degrees of the different eco-farming models of the different regions.

Region	Crop planting	Forest plus fruit tree type	Animal-raising types	Courtyard based type	Business types	Environment -conserving type	Tourism attraction type	Number of models	Coverage degree (%)
Yang ling	13.2	26.4	26.8	10.6	15.5	2.5	7.8	7	100
Mei xian	28	30.5	10.2	5.3	14.5	0	5.2	6	85.7
Chan gan	20.4	14.8	15.3	11.5	12	2.3	20.5	7	100
Da li	18.6	29.2	27.6	9.8	18	0	15	6	85.7
Mian xian	30.1	10.2	12.5	0	15	10.5	13.5	6	85.7
Cheng gu	27.6	8.4	9.8	0	20.8	9	10	6	85.7
Xi xiang	21.4	20	15.9	0	14.8	12.9	8.8	6	85.7
Ning qiang	40.8	18.9	11	10.5	0	5.7	6.9	6	85.7
Luochuang	20.5	32.1	5.3	0	10.4	15.1	6.9	6	85.7
Jing bian	30.6	18.9	6.5	0	25.1	19.3	5.1	6	85.7
Sui de	33.5	29.2	7.1	5.1	0	17.2	4.2	6	85.7
Zi chang	36.8	27.9	6.2	6.2	0	19.1	9.8	6	85.7
Average	26.8	22.8	12.8	4.9	12.2	9.5	9.5	6.2	88.1

of which i and j mean the same as afore mentioned;

$$\overline{y}_j = \frac{1}{11}\sum_{i=1}^{12} y_{ij}$$

(4)

of which y_{ij} is the average of the proportions of the eco-farming model of j; and

$$S_{ij} = \sqrt{\frac{1}{12-1}\sum_{i=1}^{12}\left(y_{ij}-\overline{y}_j\right)^2}$$

(5)

of which , S_{ij} is the standard deviation of the proportions of the eco-farming model of j.

And then, the distance between two regions could be represented with the Minkowski distance of $d_{ij}(q)$, which was calculated by the following formula:

$$d_{ij}(q) = \left[\sum_{k=1}^{m}\left|x_{ik} - x_{jk}\right|^{q}\right]^{1/q}$$

(6)

Here, the study adopted Euclidean distance to represent the distance coefficient between two regions concerned (De Koeije et al., 1987,), that is, q=2. Then,

$$d_{ij}(2) = \sqrt{\sum_{k=1}^{7}\left(x_{ik} - x_{jk}\right)^2}$$

(7)

Then the cluster distance was calculated as follows:

$$D_{pq}^2 = \frac{1}{n_p n_q}\sum_{x_i \in G_p}\sum_{x_j \in G_q} d_{ij}^2$$

(8)

The recurrence formula was as follows:

$$D_{ir}^2 = \frac{n_p}{n_r}D_{ip}^2 + \frac{n_p}{n_r}D_{iq}^2$$

(9)

In which np and nq separately stand for the numbers of regions that involve Gp and Gq, and nr stands for the number of the regions that involve Gr , a new group formed from Gp and Gq by merging them, and that nr= np+nq .

RESULTS

Proportions and coverage degrees of the different eco-farming models of the different clusters of the different regions

So far, the study obtained the percentages (Proportions) and coverage degrees of the different eco-farming models of the different clusters of the different regions (Table 3).

From Table 3 we can see that the minimum proportions of the different eco-farming models of the different regions, that is, the regions where the minimum proportions appeared, showed that the proportion of the crop planting model of Yang ling was 13.2%, the proportion of the forest plus fruit tree model of Chengu was 8.4%, the proportion of the animal-raising model of Luochuan was 5.3%, Mianxian no longer practiced the courtyard based model, Ningqian no longer practiced the business model, Meixian no longer practiced the environment-conserving model, and the proportion of the tourism attraction model of Suide was 4.2%. The minimum proportions of the different eco-farming models revealed that Yangling, located in Central Shaanxi, had an well developed economy and as result its traditional crop planting model shrunk; Chenggu, situated in an plain area, had a weak capacity to practice the forest plus fruit tree model because of its climatic constraints; Luchuan as a traditional big apple producer did not pay much attention to its animal raising so that it did not widely practice the animal raising model; Located in South Shaanxi, Mianxian had did not widely adopted the courtyard-based model because the County had poor transportation and information accesses and that the model was a newly emerged one. However, the authors of the paper held that Mianxian still had too low a proportion of the courtyard based model and that the county had the capacity and necessity to widely practice the model. Ningqiang as a mountainous county poorly practiced the business model, but the authors of the paper considered that Ningqiang had its own advantages, such as medicinal herbs, which could be fully exploited while it practiced the business model. Meixian, located in Central Shaanxi, an economically well developed region, generally paid more attention to its economic development but ignored its environment conservation. Suide as a remote county had poor transportation accesses, and thus it was constrained to practice the tourism attraction model, but the county had quite a few tourism attractions, so the country needed to strength its practicing the tourism attraction model. It follows that some eco-farming models are not preferentially practiced on priority by all the regions and no doubt this was because of the local conditions of these regions, but these model needed to be improved and their practicing need to be strengthened.

From Table 3 it could be found that, the maximum proportions of the different eco-farming models of the different regions in the table, that is, the regions where the maximum proportions appeared, showed that the proportion of the crop planting model of Ningqiang was 40.8% , the proportion of the forest plus fruit tree model of Luochuan was 32.1%, the proportion of the animal-raising model of Dali was 27.6%, the proportion of the courtyard based model of Chang'an was 11.5%, the proportions of the business model and the environment-conserving model of Jingbian were separately 25.1 and 19.3%, and the proportion of the tourism attraction model of Chang'an was 20.5%. the rationale for these were that: Niqiang was a remote county with abundant water resources so it had a high proportion of the traditional crop planting model, but because its proportion of the model was too high, the county needed to strength practicing the other models. Luochuan had such geographical advantages as high elevation above sea level so that it extensively practiced the forest plus fruit tree model. Dali was located in the economically well developed area and thus its newly emerging animal raising model expanded widely. Chang'an was a metropolitan suburb county, easy to get access to new ideas and information, and thus the County practices the courtyard based model extensively. Jingbian had abundant reserves of petroleum and natural gas and thus the county had many related enterprises, with consequence that it extensively practiced the business model. Nonetheless, Jingbian still practiced the environment-conserving model widely, and this was a

welcoming phenomenon, which indicates that enterprises traditionally thought as polluters (Donald et al., 2002) could be completely transformed into non-pollution ones. Chang'an widely practiced the tourism attraction model, because the County, as aforementioned, had well developed transportation systems as well rich tourism resources. The maximum proportions of the different eco-farming models clearly indicated what eco-farming models the different regions should mainly adopt and practice.

The coverage degrees of the different eco-farming models showed that Yangling and Chang'an, for instance, had the highest coverage degrees which amounted to 100% , and the reason for this was that the both of them are located in economically well developed areas and have favorable natural conditions and thus they had such high coverage degrees; and the coverage degrees of the different ecological models of the different regions no doubt depend on the social and economic conditions and actual eco-environments of the regions, however, they cannot be separated from the administration of the local governments of the regions (Antle and Mcaucking, 1993).

According to the sampling survey on the different regions of Shaanxi, the average number of the eco-farming models of the sampled counties is 6.2 and the average coverage degree of the models of the counties was 88.1%. The crop planting model had the highest proportion (26.8%), followed by the forest plus fruit tree model (with a proportion of 22.8%), which indicated that these two models still were the key eco-farming models, and the courtyard based model had the lowest proportion of only 4.9%, so that Shaanxi mainly practiced the traditional crop planting model, but the province also widely practiced the newly emerging forest plus fruit tree model, which indicates that the province was right in its orientation for eco-farming construction, but considering that its courtyard based model has a proportion of 4.9%, the province did not practice the models in a balanced manner, thereby needing to promote its adopting some models.

Average coverage degrees of the eco-farming models of the different clusters and the average proportions of the eco-farming models obtained by cluster analysis

By programming and computer operation, results obtained by cluster analysis are given in Figure 1. Considering its realities, the study carried out its classification with the distance coefficient D^2 =17 as its classification threshold, thus obtaining the four clusters shown in Table 4.

The first cluster included Yangling, Chang'an and Dli, the second cluster included Meixian and Luchuan, the third cluster included Mianxian, Chenggu, Xixiang and

Dendrogram using Average Linkage (Between Groups)

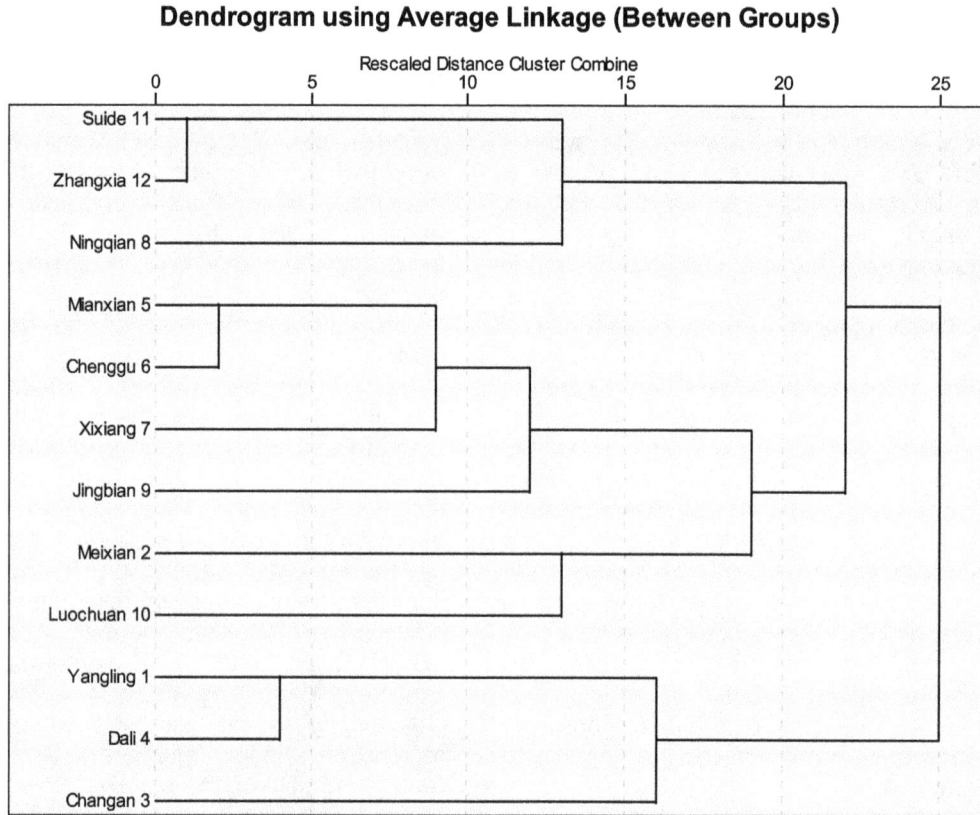

Figure 1. Clustering analysis.

Table 4. Four clusters obtained by cluster analysis.

Case	4 Clusters
1:Yangling	1
2: Meixian	2
3: Changan	1
4: Dali	1
5: Mianxian	3
6: Chenggu	3
7: Xixiang	3
8: Ningqiang	4
9: Jingbian	3
10: Luochuan	2
11: Suide	4
12: Zichang	4

Jingbian, and the fourth cluster included Ningqiang, Suide and Zichang.

So far the average proportions and the average coverage degrees of the different eco-farming models of the different clusters were obtained (Table 5).

From Table 3 we can find that, the first cluster covered three counties and the coverage degrees of its eco-farming models were the highest, averaging 95.2%; and of the three counties, the coverage degrees of the eco-farming models of two counties reached 100%, the coverage degree of the forest plus fruit tree model was the highest, amounting to 23.5%, and the coverage degree of the environment-conserving model was the lowest, amounting to only 1.6%. it can be seen that all the three counties were located in economically well developed regions and had favorable natural conditions so that they could provide favorable material conditions for practicing the different eco-farming models, but they had too low a coverage degree of the environment-conserving model, which was a common problem of environment conservation ignorance frequently occurring in newly emerging regions with well developed economies (Wackernagel et al., 2011), so that they needed to strengthen its environment conservation. Generally speaking, the new eco-farming models of the regions belonging to the first cluster expanded quickly but the environment conserving model expanded slowly.

The second cluster covered two and the coverage degree of its eco-farming model were 85.7%; and of the counties, the coverage degree of the forest plus fruit tree model was the highest, amounting to 31.3%, and the coverage degree of the courtyard based model was the

Table 5. Average coverage degrees and the average proportions of the eco-farming models of the different clusters.

Cluster	Coverage (%)	Crop planting	Forest plus fruit tree type	Animal-raising types	Courtyard based type	Business types	Environment -conserving type	Tourism attraction type
The first cluster	95.2	17.4	23.5	23.2	10.6	15.2	1.6	14.4
The second cluster	85.7	24.3	31.3	7.8	2.7	12.5	7.6	6.1
The third cluster	85.7	27.4	14.4	11.2	0	14.4	12.9	9.4
The fourth cluster	85.7	37	25.3	8.1	7.3	0	14	7

lowest, amounting to 2.7%. However, the two counties were located in two regions and the reason for this was that Meixian was near a mountain although economically well developed and Luochuan had a quickly growing apple industry presenting a bright economic development future although it was located in an economically underdeveloped mountainous area. In the meantime, Meixian depended on kiwi production for its economic development and thus the two counties widely adopted the forest plus fruit tree model, but they expanded the new Courtyard based model slowly so that they had a slightly low coverage of the model.

The third cluster covered four counties and the coverage degree of its eco-farming models was 85.7%. Among the coverage degrees of the eco-farming models of the four counties, the coverage degree of the crop planting model was the highest, amounting to 27.4% and the coverage degree of the courtyard based model was the lowest, equal to zero percent. Of the four counties, three counties, located in south Shaanxi, had moderate transportation access and moderate economic development and the other one, located in north Shaanxi, a economically poorly developed region, but Jingbian had its own unique advantage, that is, rich reserves of petroleum and natural gas, so that the county saw a quick economic development, which was comparable to those of the different counties of South Shaanxi. These regions with poorly developed economies were characterized by the dominance of the traditional models and slow expansion of the new models, but they expanded the newly emerging courtyard based model too slowly so that they needed to accelerate the expansion.

The fourth cluster totally covered three counties and the coverage degree of its eco-farming models was 85.7%. Among the coverage degrees of the cluster, the coverage degree of the crop planting model was the highest, amounting to 37% and the coverage degree of the business model was the lowest, equal to zero percent. Of the three counties, Suide and Zichang are located in North Shaanxi, a region with both poor transportation access and backward economies (Jin Lian et al., 2010), and the other county, Ningqiang, is located in South Shaanxi, a region with a relatively good economy, but it is a mountainous county. Because of their relatively backward economies and poor

transportation accesses, the regions belonging to the cluster had a high coverage degree of the traditional crop planting model and say a slow expansion of the newly emerging business model, but because of their highland locations, they expanded the forest plus fruit tree model rapidly and because of the attention of China on environment conservation, they expanded the environment conserving model quickly (Larsen, 2010). In general, their expansions of the different models were imbalanced.

It can be seen from the above analysis that of the different regions of Shaanxi, the coverage degree of the different eco-farming models was 88.1% and this indicated that the eco-farming models of the regions of Shaanxi were rich and diverse and that the regions knew what their orientations towards the eco-farming models were and what main problems they needed to solve. It can be seen from the cluster analysis of the study that the regions belonging to the first and second clusters had higher proportions of the newly emerging forest plus fruit tree model and the business model and relatively lower proportions of the traditional models than the regions belonging to the other clusters and this just right revealed local natural conditions and economic developments of the regions. But in the meantime, it also revealed some problems. Regions with poorly developed economies had too high proportions of the traditional models and too low proportions of the newly emerging models (Gordon, 1991). For example, the regions belonging to the third and fourth clusters had too high a proportion of the traditional crop planting model and too low proportions of the business model and the forest plus fruit tree model; and this was doubtlessly related to the local characteristics and economic conditions of the regions, but it also indicted unbalanced expansion of the eco-farming models and the human factors resulting in slow expansion of the newly emerging models (Gonway, 1991).

DISCUSSION

The study investigated the eco-farming models and their regional distributions of Shaanxi, concluding that the models could be divided into four groups, and summarizing some characteristics and patterns of the models that of the regions with well developed economies and favorable natural conditions, the proportions of the traditional models was lower and the proportions of the newly emerging models were increasing, meanwhile, the proportions of the different models were balanced; and of the regions with relatively backward economies and relatively poor natural conditions, the situations were opposite that the proportions of the traditional models was high and the proportions of the newly emerging models were lower, meanwhile, the expansions of the different models were hardly balanced. By statistical analysis and cluster analysis, the study accurately found out what problems there existed, what caused them and how they could be solved where the eco-farming models were practiced and expanded, so that it provided concrete and highly operable research methods for researching on solving similar problems as well as thinking approaches for eco-farming development and environment management.

In the future, backward regions should be promoted to expand and practiced the newly emerging models by improving their transportation and information accesses and human qualities and in the meantime, the governments should increase their investment and guidance to create the conditions for these backward regions to practice and expand the newly emerging models. Regions with relatively well developed economies should step up their investments for environment conservation while promoting their economic development

Meanwhile, where same types of problems are investigated in the future, attention should be paid first to the rationale and integrity for screening eco-farming models and then to developing the index system for the screened eco-farming models, so as to deepen the research on the problems.

ACKNOWLEDGEMENTS

The study was sponsored by the Cultivation Fund of the Key Scientific and Technological Innovation Project of the Ministry of Education of China (NO706054) and the New Century Excellent Talents in University (NCET-07-0700).

REFERENCES

Antle JM, Mcaucking T (1993). Technological innovation, agricultural Productivity, and environmental quality. Agricultural and Environmental Resource Economies. Oxford University Press, New York. pp. 175-220

Bicknell KB, Ball RJ, Cullen R, Bigsby HR (1998). New methodology For the ecological footprint with an application to the New Zealand economy. Ecol. Econ. 27:149-160.

Bastianoni S, Marchettini N, Panzieri M, Tiezzi E (2001). Sustainability assessment of a farm in the Chianti area (Italy). J. Cleaner Prod. 9:365-373.

Brown MT, Buranakarn V (2003). Emergy indices and ratios for sustainable Material cycles and recycle options. Resour. Conserv. Recycl. 38:1-22.

Coleman D, Codum EP, Crossley DA (1992). Soil biology, Soil ecology And global change. Biol. Fert. Soils 14(2):104-111.

Donald PF, Pisano G, Rayment MD, Pain DJ (2002). The Common Agricultural Poliey, EU enlargement and the conservation of Europe's farmland birds. Agric. Ecosyst. Environ. 89:167-182.

Jin L, Wang YP, Liu XL (2010). Summary of the Developmental Pattern of Special Agriculture [J]. J. Anhui Agric. Sci. 11:146-153. (in Chinese with English abstract).

Krishna BKC (2011). Modeling and measuring the economic success of farming families using remote sensing and GIS: An example from mountains of Nepal. J. Geogr. Reg. Plann. Vol. 4(7):4401-416.

Kurosh R-M, Saeid S (2010). Agricultural specialists' intention toward precision agriculture technologies: Integrating innovation characteristics to technology acceptance model. Afr. J. Agric. Res. 5:1191-1195.

Liu YF, Jiao LM (2002). The Application of BP Networks to Land Suitability Evaluation[J]. Geo-Spatial Inform. Sci. 1:55-61.

Li XP (2000). Theoretical Basis and Research Developments of Chinese Ecological Agriculture [J]. Res. Agric. Modernization. 6:33-47.

Li JC (2008). Categories of eco-agriculture models in China [J]. Chinese J. Eco-Agric. 5:16-20.

Larsen LB (2010). Strategic Implication of Environmental Reporting [J]. Cororate Environ. Strategy 7:276-279.

Qi XS (1992). An Approach on the Establishment of the Synthetical Index System of Eco-Agriculture Assessment [J]. J. Ningbo Univ. 2:25-27.

Qiu JJ, Ren (2008). Elementary study on the framework of standard system for eco-agriculture in China [J]. Chinese J. Eco-Agric. 5:55-61. (in Chinese with English abstract.

Tilley DR, Swank WT (2003). Emergy-based environmental systems assessment of a multi-purpose temperate mixed—forest watershed of the southern appalachian Mountains, USA. J. Environ. Manage. 69:213-227.

Wackernagel M, Schulz NB, Deumling D, Linares AC, Jenkins M, Kapos V, Monfreda C, Loh J, Myers N, Norgaard R, Randers J(2002). Tracking the ecological Overshoot of the human, economy[J]. Washington, DC, USA. 99(14):9266-9267.

Wackernagel M, Schulz NB, Deumling D, Linares AC, Jenkins M, Kapos V, Monfreda C, Loh J, Myers N, Norgaard R, Randers J. (2011).Taking the ecological overshoot of the humane economy [J]. Washington, DC, USA. 99(14):9266-9267.

4

Peroxidase and polyphenol oxidase activity on the yield of grafted and ungrafted cucumber plants

Douglas Seijum Kohatsu[1]*, Valdir Zucareli[1], Wilian Polaco Brambilla[2], Elizabeth Orika Ono[3], Tiago Roque Benetoli da Silva[1] and João Domingos Rodrigues[3]

[1]Professores Doutores, UEM/CCA/ DCA Agronomia Umuarama – PR, Brazil.
[2]Pós graduando em Agronomia, Unesp - Botucatu-SP, Brazil.
[3]Professores Doutores, Departamento de Botânica, Instituto de Biociências, Unesp - Botucatu-SP, Brazil.

The first experiment was carried out under green house and involved nine treatments: 'Tsuyoi' cucumber, 'Shelper' squash and 'Green-stripped cushaw squash' ungrafted plants and 'Tsuyoi' cucumber plants grafted onto 'Shelper' squash and 'Green-stripped cushaw squash (lower, mid and upper region of the recommended and non-recommended rootstock, respectively). After grafting, plant tissue samples were collected 1, 4, 7, 10 and 13 days after grafting for analysis of peroxidase (EC 1.11.1.7) and polyphenol oxidase (EC 1.10.3.1) activity. In the second experiment, yield and number of marketable fruits were evaluated. The differences in peroxidase activity at the rootstock region and in polyphenol activity at the region between the scion and the rootstock seem to be determining factors for a successful grafting process, increasing the yield and the number of marketable fruits.

Key words: Cucumis sativus, antioxidant enzymes, yield, Japanese cucumber.

INTRODUCTION

In Brazil, some vegetable producers have intensively cultivated cucumber plants in protected environments since the 1980s. Although this activity has contributed to increasing production and quality, it has resulted in problems related to incidence of diseases caused by soilborne fungi and infestation of nematodes, as well as higher salinity levels in the soil, which are impractical for some cultures and lead farmers to use new production systems, adopting as a short-term solution in some cases grafting onto resistant materials (Da Hora, 2006). Currently, a large part of Japanese cucumber cultivation in protected environment already uses cucumber grafted onto squash (Cardoso, 2010).

The grafting activity has widely expanded, improving the yield by increasing the capacity of nutrient uptake by the soil, preventing infection by soilborne pathogens and increasing the tolerance to low temperatures, salinity and soaking of the soil (Martinez-Balesta et al., 2010).

The incompatibility mechanism is not yet fully understood and there are a large number of studies aimed at understanding the mechanism of grafting development. Most reports, however, are related to biochemical responses and peroxidase activity, which occur at the initial phase in response to grafting, as well as to the consequences of these events for future responses associated with the plant development (Pina and Errea, 2005).

Peroxidase activity and phenol levels are important for the scion and the rootstock joining and may thus influence the stress response of incompatibility in the grafting process (Rodrigues et al., 2002). According to Santamour (1992), to allow the vascular system to work in the grafting region, peroxidases must be similar in both the scion and the rootstock, so that lignin is produced. In plants with similar peroxidases, incompatibility is rare. Incompatibility between the scion and the rootstock

---

*Corresponding author. E-mail: dskohatsu2@uem.br.

Table 1. Activity of the enzyme peroxidase in samples from ungrafted 'Tsuyoi' cucumber, 'Shelper' squash, Green-stripped cushaw squash and grafted plants in samples collected from three different regions of the stem, at 1, 4, 7, 10 and 13 days after grafting (nmol min^{-1} purpurogallin mg protein^{-1}).

Treatment	Days				
	1	4	7	10	13
	nmol min^{-1} purpurogallin mg protein^{-1}				
Cucumber hybrid Tsuyoi ungrafted	7.076^c	6.498^c	1.461^b	2.662^b	1.171^d
Squash hybrid Shelper ungrafted	5.543^c	5.883^c	4.424^b	3.818^b	4.249^c
Green-stripped cushaw squash ungrafted	3.563^c	6.889^c	3.365^b	7.630^b	3.091^c
Region below the grafting (Tsuyoi onto Shelper)	20.435^b	31.813^b	28.665^a	17.650^a	1.500^d
Grafting region(Tsuyoi onto Shelper)	43.297^a	40.773^a	37.433^a	19.430^a	15.293^a
Region above the bud (Tsuyoi onto Shelper)	13.991^b	6.225^c	5.116^b	5.787^b	2.193^d
Region below the grafting (Tsuyoi onto GSCS)	4.340^c	5.599^c	2.784^b	4.870^b	4.457^c
Grafting region (Tsuyoi onto GSCS)	42.616^a	31.152^b	25.054^a	23.002^a	22.980^a
Region below the grafting (Tsuyoi onto GSCS)	4.968^c	2.881^c	2.226^b	3.697^b	3.198^c
VC (%)	34.18	41.29	66.67	40.87	17.24
F value	26.6*	21.8*	11.7*	15.4*	182.0*

*Means followed by same letter in the same column do not differ significantly from each other according to Scott-Knott test at 5% probability. Collections for enzymatic analyzes: Samples for enzymatic analyzes were collected from the grafting region; from 1 cm above the grafting region; and from 1 cm below the grafting region for grafted plants. For ungrafted plants, samples were collected from heights similar to those of grafted plants. Such plant material was frozen in liquid nitrogen and stored at -80°C. Enzyme extract was obtained according to the method described by Ekler et al. (1993) and the activity of peroxidase (POD - EC 1.11.1.7) was determined according to the methodology described by Teisseire and Guy (2000). Experimental design was completely randomized with nine treatments and four replicates of five plants. Treatments consisted of the following stem samples: 1. Cucumber hybrid Tsuyoi – ungrafted; 2. Squash hybrid Shelper – ungrafted; 3. Green-stripped cushaw squash (GSCS) – ungrafted; 4. Region below the grafting (Tsuyoi grafted onto 'Shelper'); 5. Grafting region (Tsuyoi grafted onto 'Shelper'); 6. Region above the bud (Tsuyoi grafted onto 'Shelper'); 7. Region below the grafting (Tsuyoi grafted onto GSCS); 8. Grafting region (Tsuyoi grafted onto GSCS); 9. Region below the grafting (Tsuyoi grafted onto GSCS).

may manifest in future responses, as noted by Pina and Errea (2005), consuming research time due to the large number of tested rootstocks. Therefore, the aim of this study was to verify whether the enzymatic difference between the scion and the rootstock, before and after the healing process, can be used as a rapid mechanism to verify incompatibility, reducing thus experimentation time in the field.

RESULTS AND DISCUSSION

For ungrafted plants, peroxidase activity was not greater than 8,000 nmol purpurogallin min^{-1} μg protein^{-1} (Table 1) during the assessment period; it was five-fold lower compared to the grafting region on the day following the process. Some authors have suggested that the different peroxidase activity between the scion and the rootstock could be the limiting factor for compatibility. This hypothesis, however, cannot be accepted in this experiment since there was no statistical difference among ungrafted plants until the tenth day of analysis.

Nevertheless, peroxidase seems to have more influence after grafting, ranging in activity between the rootstocks used in this experiment. After grafting, the site of contact between the scion and the rootstock showed high activity of this enzyme, probably due to healing,

typical of injured tissues that are lignified. However, increased POD activity at this site can be caused by the stress of grafting, since this enzyme is considered a stress marker, which could only be elucidated with specific analysis of lignin peroxidase.

POD activity was low below the grafting joining region (1 cm) throughout the experimental period, regardless of the used rootstock, maintaining thus the activity equal to that before the grafting. However, below the grafting joining region (1 cm), POD activity was high in the recommended rootstock; this activity was similar to that at the grafting joining region, not statistically differing, until the tenth day of analysis, which is the period for the grafting establishment. The non-recommended rootstock had low POD activity in the lower region compared to the grafting joining region; this activity was similar to that in ungrafted plants.

At the grafting region, initially, there was no statistical difference for POD activity between the used rootstocks. However, this activity decreased during the establishment of grafts, so that the amplitude was approximately 17,550 nmol purpurogallin min^{-1} mg protein^{-1} in cucumber plants grafted onto the non-recommended rootstock and 5,900 nmol purpurogallin min^{-1} mg protein^{-1} in those grafted onto the recommended rootstock. This lower amplitude of peroxidase activity for the recommended rootstock is probably due to hydrogen peroxide supply below the

Table 2. Activity of the enzyme polyphenol oxidase in samples from ungrafted 'Tsuyoi' cucumber, 'Shelper' squash, Green-stripped cushaw squash and grafted plants in samples collected from three different regions of the stem, at 1, 4, 7, 10 and 13 days after grafting (nmol of purpurogallin min^{-1} µg of protein^{-1}).

Treatment	Days				
	1	4	7	10	13
	nmol purpurogallin min^{-1} µg protein^{-1}				
Cucumber Hybrid Tsuyoi ungrafted	3.658^b	2.517^a	3.343^a	1.562^b	1.250^c
Squash hybrid Shelper ungrafted	4.058^b	3.238^a	1.155^b	2.392^b	3.447^a
Green-stripped cushaw squash ungrafted	2.290^c	2.080^b	1.788^b	3.717^a	2.089^b
Region below the grafting (Tsuyoi onto Shelper)	3.820^b	3.178^a	1.756^b	3.776^a	1.792^b
Grafting region (Tsuyoi in Shelper)	1.506^c	1.347^b	1.200^b	710^b	1.341^c
Region above the bud (Tsuyoi onto Shelper)	5.105^a	2.937^a	3.749^a	1.792^b	2.058^b
Region below the grafting (Tsuyoi onto GSCS)	2.717^c	1.659^b	1.814^b	3.445^a	2.461^b
Grafting region (Tsuyoi onto GSCS)	2.248^c	2.554^a	2.181^b	1.329^b	2.048^b
Region below the grafting (Tsuyoi onto GSCS)	2.691^c	2.213^b	1.907^b	4.248^a	2.091^b
VC (%)	20.68	23.74	36.77	29.34	19.63
F value	12.0*	5.1*	5.3*	11.6*	10.1*

*Means followed by same letter in the same column do not differ significantly from each other according to Scott-Knott test at 5% probability. Collections for enzymatic analyzes: Samples for enzymatic analyzes were collected from the grafting region; from 1 cm above the grafting region; and from 1 cm below the grafting region for grafted plants. For ungrafted plants, samples were collected from heights similar to those of grafted plants. Such plant material was frozen in liquid nitrogen and stored at -80°C. Enzyme extract was obtained according to the method described by Ekler et al. (1993) and polyphenol oxidase activity (POL - EC 1.10.3.1) was determined based on the method described by Kar and Mishra (1976). Treatments consisted of the following stem samples: 1. Cucumber hybrid Tsuyoi – ungrafted; 2. Squash hybrid Shelper – ungrafted; 3. Green-stripped cushaw squash (GSCS) – ungrafted; 4. Region below the grafting (Tsuyoi grafted onto 'Shelper'); 5. Grafting region (Tsuyoi grafted onto 'Shelper'); 6. Region above the bud (Tsuyoi grafted onto 'Shelper'); 7. Region below the grafting (Tsuyoi grafted onto GSCS); 8. Grafting region (Tsuyoi grafted onto GSCS); 9. Region below the grafting (Tsuyoi grafted onto GSCS)

grafting region, which showed high peroxidase activity, providing a better connection of vessels.

The high POD activity at the grafting joining region in 'Green-stripped cushaw squash' on the day after grafting may be attributed more to the stress caused to that region than to the lignification process itself, since it did not persist during the days after collection, as shown for the recommended rootstock. This study suggests that the involvement of this enzyme in the different enzymatic activity between the rootstock and the scion is not before grafting but after grafting, that is, the smaller the difference in activity between the rootstock and the grafting region, the better the connection of vessels; however, this fact can only be confirmed by verifying the plant anatomy.

Peroxidases have been studied in apricot plants, indicating that compatible cultivars have higher activity of this enzyme than incompatible ones and such incompatibility could be provided by inefficient lignifications at the grafting joining region (Quesada and Macheix, 1984).

Cucumber plants grafted onto 'Shelper' squash had a reduction in POD activity during the evaluation period, which remained high until the seventh day of analysis, when it started deeply decreasing, maintaining a similar activity until the end of the experiment. According to Da Hora (2006), who studied grafting onto cucumber plants, the perfect connection of vessels occurs at 8 days after the procedure.

Therefore, the reduced peroxidase activity in grafting onto 'Shelper' squash around the eighth day after grafting would result from the complete lignification process. Even with such a reduction, the values remained higher compared to those at 1 cm above and below the grafting joining region, probably due to the high concentration of auxin, responsible for the tissue differentiation.

As shown in Table 2, there was statistical difference among the three regions analyzed for grafting onto 'Shelper' squash, so that the grafting joining region showed significantly lower POL activity compared to the lower and the upper region.

Such a lower activity of POL enzyme at the grafting joining region is suggested to be a result of lower levels of phenolic compounds due to the use of this substrate by the enzyme peroxidase in the lignifications process.

The higher activity at the grafting joining region on the day after grafting is a consequence of increased levels of phenolic compounds in regions which have undergone some type of injury, as in the grafting process.

After the reduction in POL activity in plants grafted onto 'Shelper' squash over the experiment, there was an increase on the 13th day, probably due to the levels of polyphenols or monophenols, responsible for the degradation of auxin and the inhibition of this process, respectively.

POL activity was not significantly different among the three regions analyzed for grafting onto Green-stripped cushaw squash, except for the eleventh day after

Table 3. Average number of marketable fruits (fruits per plant) and yield (kg per plant) of 'Tsuyoi' cucumber plants grafted onto 'Shelper' squash or ' Green-stripped cushaw squash' and ungrafted plants, São Manuel-Paraná, Brazil. 2010.

Treatment	Fruits marketable per plant	Yield (kg per plant)
Cucumber 'Tsuyoi' - ungrafted	15.0[b]	2.430[a]
Grafting onto 'Shelper' squash	21.8[a]	3.261[a]
Grafting onto GSCS	12.4[c]	1.980[b]
VC (%)	20.68	23.74
F value	12.0*	5.1*

*Means followed by same letter in the same column do not differ significantly from each other according to Scott-Knott test at 5% probability. Fruits were harvested when their length was between 20 and 22 cm at every two days during 8 weeks. Yield: All fruits harvested during the whole collection period divided by the number of plants; the result was expressed as kg plant^{-1}. Number of marketable fruit: Fruits that remained after the removal of those showing internal length / external length rate inferior to 0.85; the result was expressed as marketable fruits per plant. The experimental design for cucumber production was in randomized blocks, consisting of three treatments: Ungrafted cucumber hybrid Tsuyoi, 'Tsuyoi' cucumber grafted onto 'Shelper' squash, and 'Tsuyoi' cucumber grafted onto Green-stripped cushaw squash, with 10 replicates of 3 plants per plot.

grafting. Enzyme activity in the non-recommended rootstock was different from that in the rootstock recommended for the culture, both for peroxidase and for polyphenol oxidase. Both enzymes may be closely related in the process of compatibility between the rootstock and the scion.

Therefore, higher activity of POD, directly responsible for the process of vessel lignifications, and lower activity of POL, which uses phenolic compounds essential for this process, would provide a more rapid and efficient grafting process, changing thus the transport of nutrients, photoassimilates, plant hormones and consequently plant vigor.

Grafting onto 'Shelper' squash increased the number of marketable fruits and the yield, while plants grafted onto Green-stripped cushaw squash were negatively influenced by these parameters, showing lower yield and fewer marketable fruits, which demonstrates that there may be levels of compatibility between the different rootstocks used in this experiment.

The rootstock 'Shelper' provided 31.2 and 43.2% increase in the number of marketable fruits and 35.5 and 39.5% in the yield, relative to ungrafted plants and cucumber grafted onto Green-stripped cushaw squash, respectively (Table 3). According to Khryanin (2007), cytokinin can induce feminization in various plant species. Zhou et al. (2007) studied cucumber grafted onto *Cucurbita ficifolia* and found that grafted plants had twice the level of cytokinin at a temperature without chilling; when the temperature was reduced, cytokinin level in grafted plants was up to 33-fold higher than that in ungrafted plants. These data support the hypothesis that the rootstock 'Shelper' may have increased cytokinin level, providing a larger number of fruits, as observed in this study, particularly because this experiment was partially performed during the cold period.

According to Ciobotari et al. (2010), incompatibility may have negative consequences on the growth and the development of plants. In studies of grafting in cucumbers,

data related to yield are contradictory; depending on the experiment a higher or a lower yield is obtained, which explains the difference between the rootstocks in this experiment. Thus, successful grafting is closely related to the used rootstock.

Conclusion

The difference in enzyme activity between the rootstock and the grafting joining region during the healing process influences the performance of grafted plants and could be used as a rapid mechanism to verify incompatibility.

REFERENCES

Cardoso AII (2010). Enxertia em pepinos mostra vantagens. Campo. Negócios. Uberlândia 52:70-71.

Ciobotari G. Brinza M, Morariu A, Gradinariu G (2010). Graft Incompatibility Influence on Assimilating Pigments and Soluble Sugars Amount of some Pear (*Pyrus sativa*) Cultivars. Notulae Botanicae Horti Agrobotanici Cluj-Napoca. 38:187-192.

Da Hora RC (2006). Avaliação de pepineiro enxertado em diferentes ambientes. Tese (Doutorado Agronomia/Horticultura) – Faculdade de Ciências Agronômicas, Universidade Estadual Paulista, Botucatu, São Paulo. p. 69.

Ekler Z, Dutka F, Stephenson GR (1993). Safener effects on acetochlor toxicity, uptake, metabolism and glutathione S-transferase activity in maize. Weed Res. 33:311-318.

Kar M, Mishra D (1976). Catalase, peroxidase, and polyphenoloxidase activities during rice leaf senescence. Plant Physiol. 57:315-319.

Khryanin VN (2007). Evolution of the pathways of sex differentiation in plants. Russian J. Plant Physiol. 54:845-852.

Martinez-Ballesta MC, Alcaraz-López C, Muries B, Mota-Cadenas C, Carvajal M (2010). Physiological aspects of rootstock-scion interactions. Sci. HortiC. 127:112-118.

Pina A, Errea P (2005). A review of new advances in mechanism of graft compatibility incompatibility. Scientia Horticulturae 106:1-11.

Quesada MP, Macheix JJ (1984). Caractérisation d'une peroxydase implique, spécifiquement dans la lignification en relation avec l'incompatibilité au greffage chez l'abricotier. Physiologie Végétale 22(5):533-540.

Rodrigues AC, Diniz AC, Fachinello JC, Silva JB, Carvalho Faria JL(2002). Peroxidase e fenóis totais em tecidos de porta enxertos de

Prunus sp. nos períodos de crescimento vegetativo e de dormência. Ciência Rural, Santa Maria 32(4):559-564.

Santamour JFS (1992). Predicting graft incompatibility in woody plants: combined proceedings International Plant Propagators Society. New York: International. Soc. Hortic. Sci. 42:131-134.

Teisseire H; Guy V (2000). Copper-induced changes in antioxidant enzymes activities in fronds of duckweed (*Lemna minor*). Plant Sci. 153:65-72.

Zhou Y, Huang, L, Zhang Y, Shi, K, Yu J, Nogués S (2007). Chill-induced decrease in capacity of RuBP carboxylation and associated H_2O_2 accumulation in cucumber leaves are alleviated by grafting onto figleaf gourd. Annals Bot. 100:839-849.

5

An analysis of hybrid sterility in rice (*Oryza sativa* L.) using genetically diverse germplasm under temperate ecosystem

S. Najeeb, M. Ashraf Ahangar and S. H. Dar

SHER-I-KASHMIR University of Agricultural Sciences and Technology, Kashmir Mountain Research Centre for Field Crops Khudwani, Anantnag, 192102, Ashmir, India.

Sixty cross combinations consisting of 15 intervarietal *indica*, 15 intervarietal *japonica* and 15 crosses each of inter-subspecific and three-way crosses were generated through different mating patterns in *kharif* 2008 and evaluated for hybrid sterility under two different agroecologies of Kashmir during *kharif* 2009. The estimates of pollen and spikelet sterility were very high in *indica /japonica* crosses followed by *indica/indica* and *japonica/japonica* crosses. The magnitude of pollen and spikelet sterilities were observed high in $L_1 \times T_1$ followed by $L_4 \times T_3$, $L_1 \times T_3$, $L_5 \times T_3$, $L_3 \times T_3$ and so on all belonging to inter-subspecific group. The mean estimates of pollen and spikelet sterilities of different crossing blocks were very high when compared to their corresponding parental mean. Only inter-subspecific cross $L_1 \times T_2$ yielded pollen and spikelet sterilities of less than 50% and another three crosses revealed a range between 50 and 70% for these traits. Five crosses were grouped as highly sterile by depicting sterility estimates greater than 70%. The mean estimates of pollen sterility and spikelet sterility for inter-subspecific crosses got reduced to the level of 30.80 and 25.75% respectively through three-way crosses by deploying wide compatibility cultivar (WCV) (Dular) as bridging parent. The estimated mean percentage overcome was 36.62% for pollen sterility and 63.92% for spikelet sterility. Such kind of modest attempt has widened the genetic variability by combining the genetic background across the two sub species and tailors the new genotypes. This in turn has paved way for the concept of ideotype breeding for unique cold temperate agroecologies.

Key words: Hybrid sterility, inter-subspecific crosses, intra-subspecific crosses, rice, wide compatibility cultivar (WCV).

INTRODUCTION

The strong hybrid vigor in the F_1s between *indica* and *japonica* subspecies of Asian cultivated rice (*Oryza sativa* L.) has attracted a large amount of research interest, with the hope for developing hybrid rice by making use of such heterosis (Yang et al., 1962; Yuan, 1994). However, hybrid sterility frequently occurs in such inter-subspecific crosses (Ikehashi, 1982), the fertility of indica-japonica hybrids varies widely from fully fertile to almost completely sterile, with the majority of such hybrids showing significantly reduced fertility (Oka, 1988; Liu et al., 1996; Zhang et al., 1997). Hybrid sterility is a major form of post zygotic reproductive isolation that restricts gene flow between populations and inter-sub specific hybrids are usually sterile (Jiang et al., 2012). Wide compatibility cultivars (WCVs) produce fertile hybrids when crossed to both *indica* and *japonica* cultivars. The discovery of WCVs brought hope by breaking the fertility barriers between *indica* and *japonica* subspecies and provided possibility for exploiting the very strong heterosis demonstrated in cross between the two subspecies. Consequently, there has been considerable interest in understanding the mechanism underlying wide compatibility and hybrid sterility. Under Kashmir valley- a unique temperate climate, *indica* sub-species are grown at an altitude range of 1350 to 1900 m amsl, whereas, *japonica* sub-species find their habitat under foot hills at an altitude of 1950 to 2300 m amsl. *Indica* types have higher yielding ability, wide adaptability and good grain quality, while as *japonica* types possess high degree of cold tolerance, early maturity and fertilizer responsiveness. The objective to combine the traits across the two subspecies is to develop the varieties with traits most suitable for the two agroecologies of Kashmir to raise productivity, quality and resilience to biotic and abiotic stresses. However, many attempts were made in this direction but major bottleneck in this effort was the factor of hybrid sterility. A modest effort in this direction was to assess and estimate the magnitude of hybrid sterility in intra- and inter-subspecific crosses *viz-a-viz* to determine the level of overcome by deploying wide compatibility cultivar (WCV) as bridging parent through three-way crosses and to get the most suitable recombinants for two different agroecologies of Kashmir.

MATERIALS AND METHODS

The materials for the present study comprised of twelve genetically diverse cultivars of rice (*O. sativa* L.) belonging to two sub-specific groups viz. *indica* and *japonica*. The six typical *japonica* types [Koshihikari (P_7), K-332 (P_8), GS-503 (P_9), GS-504 (P_{10}), Kohsar (P_{11}) and K-508 (P_{12})] and six typical *indica* cultivars including one wide compatibility cultivar (WCV) [Jhelum (P_1), SK-382 (P_2), Shalimar Rice-1 (P_3), China-1039 (P_4), Chenab (P_5) and Dular (P_6) (WCV)] were selected out of the germplasm collection maintained at Rice Research and Regional Station, Khudwani of Sher-e-Kashmir University of Agricultural Sciences and Technology of Kashmir.

Fifteen intra-sub specific hybrids were created from each half-diallel-mating (Griffing, 1956) using six parental lines each of *indica* and *japonica* rices in *Kharif* 2008. In the same season five *indica* genotypes as lines Jhelum (L_1), SK-382 (L_2), SR-1 (L_3), China-1039 (L_4) and Chenab (L_5) were crossed with three typical *japonica* cultivars as testers Koshihikari (T_1), K-332 (T_2) and Kohsar (T_3) in a line x tester fashion (Kempthorne, 1957) to generate 15 inter-subspecific (*indica*/*japonica*) hybrids. Five intervarietal single crosses using *indica* parents each with Dular (WCV) as one of the parent were mated to three typical *japonica* genotypes viz., Koshikikari, K-332 and Kohsar to generate 15 three-way crosses.

A total of sixty cross combinations generated (viz.15 *indica* x *indica*, 15 *japonica* x *japonica*, 15 *indica* x *japonica*, and 15 three-way F_1s) were evaluated at two research stations namely, Rice Research and Regional Station, Khudwani (1650 m amsl) and the High Altitude Rice Research Sub-Station, Larnoo (2285 m amsl)

during *Kharif* 2009. Each progeny line was evaluated in two-row of 3 m length in a randomized complete block design (RCBD) with three replications. Twenty-eight days old seedlings with single plant/hill were planted at inter- and intra- row spacing of 20 and 15 cm, respectively. Recommended crop management practices were followed.

Pollen sterility and spikelet sterility are the best indices of hybrid sterility (Zhang and Lu, 1993). Hence hybrid sterility was studied in terms of these two aspects. Pollen sterility was observed under a light microscope using an iodine potassium iodide (I_2KI) [0.1%] staining method (Anonymous, 1996). Samples for pollen were collected from at least 10 florets from individual plants at the heading stage (6th growth stage). At least three random microscopic fields were used to count non-sterile pollen grains (dark stained round) and sterile pollen grains (unstained withered, unstained spherical, or partially stained round). Pollen sterility was calculated as ratio of sterile pollen to the total number of pollen grains. Similarly main/primary panicles of five randomly selected plants were bagged before heading stage. The harvested panicles were observed for filled and empty spikelets and mean spikelet sterility was calculated. The mean of the pooled data over the locations for pollen and spikelet sterility were subjected to statistical analysis for interpretation of results. The programme R-Software was used for estimating the variance analysis.

RESULTS AND DISCUSSION

Analysis of variance for pollen and spikelet sterilities revealed significant mean squares for all the genotypes on the whole and as different groups of crosses; however, higher estimates were observed for inter-subspecific crossing block. It was observed that variance due to inter-subspecific crosses was much more than the total variance contributed by all the groups of crosses together. Similarly variances due to interactions of crossing blocks with respect to locations were found significant to highly significant for these traits (Table 1). Estimate of interactions mean square again were observed high for *indica* intra-subspecific crosses followed by inter-subspecific group of crosses. The estimates of pollen and spikelet sterilities over locations of each genotype when crossed to its own or other subspecies and averaged over all possible combinations are listed in the Table 2. Similarly it also depicts pollen and spikelet sterilities of cross progenies involving single cross (one parent as WCV) with *japonica* parents averaged over two locations. Generally it was observed that F_1 progenies of *indica* genotypes exhibited relatively high sterility when crossed with *japonica* group of genotypes. Similar kind of trend was observed in F_1s of *japonica* genotypes when crossed with *indica* genotypes in line x tester mating design, however, reverse was the case for F_1s of intra-subspecific groups (*indica* x *indica* and *japonica* x *japonica*). The estimates of both kinds of sterilities were found non-significant in case of intra-subspecific (*indica* x *indica*) F_1s, however, low but significant estimates were observed for intra-subspecific (*japonica* x *japonica*) F_1s. The estimates of hybrid sterility (pollen and spikelet sterilities) of all single crosses with WCV (Dular) as one of the parent exhibited moderate

Table 1. Analysis of variance of various crossing blocks of rice for pollen and spikelet sterilities.

Source	df		Mean squares	F-test
	Main	Partitioned		
Location	1		11.04	4.92*
Block (loc)	2		5.5	2.45
Lines	59		5.6	2.5**
Among groups		3	42.17	4.82**
Within i x i		14	52.97	3.25**
Within j x j		14	10.78	2.45*
Within i x j		14	140.6	9.19**
Within three-way		14	8.28	2.24**
Lines x Location	59		2.24	
Loc x among groups		3	8.75	4.91**
Loc x within i x i		14	16.30	9.15**
Loc x within j x j		14	4.40	2.47*
Loc x within i x j		14	15.30	8.59**
Loc x within three-way		14	3.70	2.07*
Lines x Block (loc)	118		1.78	1.78

*,** indicate significance at 5 and 1% level, respectively.

expression. All but one single cross exhibited moderate levels of expression for one or other kind of sterility, whereas, their corresponding three-way crosses followed similar trend but the values were non-significant. The higher and significant estimates of pollen and spikelet sterilities in inter-subspecific crosses and their low and non-significant expression in three-way crosses suggest the intervention of WCV.

The intensity of pollen and spikelet sterility varied greatly among and within crossing blocks [*indica* x *indica*, *japonica* x *japonica* and *indica* x *japonica*] from highly sterile to fully fertile. Barring a few exceptions, the magnitude of hybrid sterility of intra-subspecific (*indica* x *indica* and *japonica* x *japonica*) crosses, were much lower than those recorded for inter-subspecific (*indica* x *japonica*) crosses (Table 3). Further high mean pollen sterility of 58.26% was observed for *indica* x *japonica* crosses ranging between 8.3 and 96.3 % followed by relatively medium range (9.27 and 85.40%) and mean (30.25%) for intra-subspecific (*indica* x *indica*) crosses. The spikelet sterility did not perform differently and observed the similar trend with higher mean (76.51%) and wide range (31.80 and 96.60%) for inter-sub specific crosses. The higher estimates of hybrid sterility in inter-subspecific (*indica* x *japonica*) crosses were reported by number of workers (Oka, 1964; Liu et al., 1996, 1997). Among intra-sub specific crosses the high value of pollen sterility (85.4%) was observed on cross combination P_2 x P_6 (*indica* x *indica*). The lowest mean pollen sterility (10.97%) was recorded on intra-subspecific (*japonica* x *japonica*) crosses, besides, a constricted range (6.99 to 19.6%). Further spikelet sterility of intra-subspecific (*indica* x *indica*) cross progenies revealed a mean estimate of 27.32%. The intra-subspecific (*Japonica* x

Japonica) crosses recorded a lower mean value for the spikelet sterility (16.78%) and relatively a narrower range (6.50 to 31.74%) in comparison to range of intra-subspecific (*indica* x *indica*) cross progenies (13.11 to 74.00%).

The magnitude of pollen and spikelet sterilities were observed high on cross combination L_1 x T_1 (95.85 and 75.81%) followed by L_4 x T_3 (95.3 and 93.05%), L_1 x T_3 (92.03 and 86.08%), L_5 x T_2 (92.60 and 91.50%) and L_3 x T_1 (91.66 and 84.20%) all belonging to inter-subspecific crossing block. The only cross combination P_2 x P_6 belonging to intervarietal (*indica* x *indica*) group depicted the higher estimates for pollen and spikelet sterilities, that is, 85.4 and 74.0%, respectively. The mean pollen sterility estimates for intra-subspecific groups, that is, *indica* x *indica* (30.25%) and *japonica* x *japonica* (10.97%) crosses were about thrice and twice when compared to their corresponding parental mean values of 9.39 and 5.72%, respectively. The mean spikelet sterility of intra-sub specific (*indica* x *indica*) crosses (27.32%) was almost fourfold than the corresponding mean parental value (7.7%), whereas, for intra-sub specific group (*japonica* x *japonica*) the estimate for said sterility (16.78%) was just half of the corresponding parental mean (33.32%). High mean spikelet sterility of *japonica* parents than expected was owing to higher value of the same recorded on two *japonica* parents GS-503 and GS-504. The reason was due to their late heading which coincided with early commencement of cold weather as a result of untimely snowfall on adjoining mountains of High Altitude Rice Research Sub-Station, Larnoo (2200 m amsl)-one of the locations. This infact caused the cold stress that increased the rate of spikelet sterility for these parents.

Table 2. Pollen and spikelet sterilities (±SD) in the F₁s of each parent of both subspecies averaged over two locations.

Variety	Sub species	Crossed to all varieties	Indica	Japonica	Selfing
Jehlum	I	45.78±8.12**(34.82±7.12)**	26.27±14.10(16.38±10.42)	78.31±15.13** (64.56±15.86)**	7.35(6.59)
SK-382	I	36.12±8.21** (42.23±8.18)**	34.49±21.22(33.53±19.64)	38.84±9.22** (74.74±12.28)**	9.55(7.26)
SR-1	I	33.56±6.22** (45.43±12.12)**	22.28±14.20(26.81±16.24)	52.36±18.02** (79.78±18.92)**	12.20(10.12)
China 1039	I	41.2±4.04** (42.34±11.22**)	23.27±14.78(20.85±15.18)	71.13±15.01** (78.14±28.95)**	8.30(6.91)
Chenab	I	35.34±5.28** (47.56±12.37)**	27.75±16.21(24.89±17.12)	50.66±14.15** (82.59±9.23)**	12.65(9.28)
Dular	I^w	39.45±25.86** (40.86±26.20)**	39.45±25.86(40.86±26.20)	-	6.29(6.46)
Koshihikari	J	36.90±9.95** (45.91±14.39**)	13.87±3.62** (69.71±18.46)**	62.66±17.17** (22.04±4.18)**	6.80(51.40)
K-332	J	30.87±9.50** (74.70±20.80)**	47.87±10.07** (77.32±20.11)**	13.87±3.60** (22.86±6.08)**	4.88(11.88)
GS-503	J	8.88±1.81** (16.57±4.59)**	-	8.88±1.81** (16.57±4.59)**	7.49(55.29)
GS-504	J	9.15±1.91** (15.04±5.10)**	-	9.15±1.91** (15.04±5.10)**	6.66(60.18)
Kohsar	J	38.59±10.24** (41.52±14.24)**	64.25±14.28** (82.51±18.57)**	12.92±4.65** (16.53±5.16)**	3.86(8.62)
K-508	J	9.86±3.03** (15.19±4.52)**	-	9.86±3.03** (15.19±4.52)**	4.65(9.57)
Jhelum x Dular	(I x D)	24.7±3.02**(15.85±8.82)	-	16.51±2.32** (16.89±3.18)**	-
SK-382 x Dular	(I x D)	65.4±12.56**(74.0±20.45**)	-	31.58±8.90** (38.99±8.40)**	-
SR-1 x Dular	(I x D)	28.0±17.81(59.87±25.6*)	-	24.46±7.01** (25.16±4.06**)	-
China-1039 x Dular	(I x D)	26.8±15.38(15.17±8.90)	-	44.23±13.62** (25.64±3.70)**	-
Chenab x Dular	(I x D)	32.72±4.89** (39.42±20.7*)	-	37.25±12.32** (22.09±4.19)**	-

Figures in parentheses denote spikelet sterilities; I=*indica* sub species, j=*japonica* sub species, I^w=wcv with *indica* background and D=*Dular* (WCV). *, **, significant at 5 and 1% level respectively.

One inter-subspecific cross L_1 x T_2 yielded pollen and spikelet sterilities of less than 50% and another three crosses showed sterility estimates in between 50 and 70%. Five crosses were highly sterile (pollen and spikelet sterility-spikelet sterilities were also reported by earlier workers on the same phenomenon. The mean pollen and spikelet sterilities of inter-sub specific (*indica x japonica*) crosses viz. 58.26 and 76.51% were manifold when compared to mean of 8 parents (5 lines and 3 testers) for the same traits, that is, 18.19 and 14% respectively. It was also observed that for intra-sub specific (*indica x indica*) crosses and in three-way crosses, pollen sterility showed positive and significant association with spikelet sterility, however, no such kind of association was observed for other two crossing blocks. These results suggest that the two phenomena can occur independently or can happen together and are responsible for low seed setting (grain yield). When pollen and spikelet sterilities were correlated with number of grains harvested per panicle (Data not shown) it was highly significant and negative suggesting that both these aspects are accountable for yield, but in a negative fashion. Song et al. (2005) also reported that the pollen and embryo sac sterility contributed almost equally to fertility. However, Zhang and Lu (1993) and He et al. (1994) pointed out that the hybrid sterility within Asian cultivated rice mainly resulted from pollen abortion, whereas, reduced affinity between the uniting gametes is also an important cause for sterility (Liu et al., 2004). Kubo et al. (2011) reported that in inter-sub specific crosses the hybrid male sterility gene *S24* caused the selective abortion of male gametes carrying the *japonica* allele (*S24-j*) via an allelic interaction in the heterozygous hybrids. Yan et al. (2003) demonstrated that inter-subspecific hybrid sterility mainly resulted from pollen abortion with the pollen deteriorating at various stages, and the correlation between pollen and spikelet sterility was highly significant in typical *indica–japonica* crosses.

The inter-subspecific test crosses L_1 x T_2, L_2 x T_2, L_2 x T_3, L_3 x T_2, L_5 x T_1 and L_5 x T_3 showed pollen sterility of just 25%, however, very high spikelet sterility suggesting that the semi sterility might be due to female gamete abortion. Since25% sterility means 75% fertile pollen grains and the proportion was quite enough to set the

Table 3. Relative comparison of pollen and spikelet sterility averaged over two locations for *indica x indica*, *japonica x japonica* and *indica x japonica* crosses.

Indica x Indica	Pollen sterility (%)	Spikelet sterility (%)	Japonica x Japonica	Pollen sterility (%)	Spikelet sterility (%)	Indica x Japonica	Pollen sterility (%)	Spikelet sterility (%)
Jhelum x SK-382 (P_1 x P_2)	25.12	14.35	Koshihikari x K-332 (P_7 x P_8)	15.21	32.07	Jhelum x Koshihikari (L_1 x T_1)	95.85	75.81
Jhelum x SR-1 (P_1 x P_3)	30.49	13.11	Koshihikari x GS-503 (P_7 x P_9)	8.19	11.91	Jhelum x K-332 (L_1 x T_2)	47.05	31.80
Jhelum x Ch 1039 (P_1 x P_4)	14.36	24.11	Koshihikari x GS-504 (P_7 x P_{10})	11.25	9.79	Jhelum x Kohsar (L_1 x T_3)	92.03	86.08
Jhelum x Chenab (P_1 x P_5)	37.13	17.5	Koshihikari x Kohsar (P_7 x P_{11})	14.09	25.05	SK-382 x Koshihikari (L_2 x T_1)	62.06	60.58
Jhelum x Dular (P_1 x P_6)	24.27	15.85	Koshihikari x K-508 (P_7 x P_{12})	6.99	31.74	SK-382 x K-332 (L_2 x T_2)	17.25	81.09
SK-382 x SR-1 (P_2 x P_3)	16.9	19.82	K-332 x GS-503 (P_8 x P_9)	11.83	9.79	SK-382 x Kohsar (L_2 x T_3)	37.21	82.56
SK-382 x Ch-1039 (P_2 x P_4)	16.9	27.16	K-332 x GS-504 (P_8 x P_{10})	10.77	10.80	SR-1 x Koshihikari (L_3 x T_1)	91.66	84.20
SK-382 x Chenab (P_2 x P_5)	28.15	32.33	K-332 x Kohsar (P_8 x P_{11})	19.6	10.84	SR-1 x K-332 (L_3 x T_2)	20.79	85.64
SK-382 x Dular (P_2 x P_6)	85.4	74.00	K-332 x K-508 (P_8 x P_{12})	11.98	10.75	SR-1 x Kohsar (L_3 x T_3)	44.65	69.52
SR-1 x Ch-1039 (P_3 x P_4)	26.74	21.99	GS-503 x GS-504 (P_9 x P_{10})	7.33	25.29	Ch-1039 x Koshihikari (L_4 x T_1)	55.43	44.77
SR-1 x Chenab (P_3 x P_5)	9.27	19.29	GS-503 x Kohsar (P_9 x P_{11})	7.72	26.55	Ch-1039 x K-332 (L_4 x T_2)	61.67	96.60
SR-1 x Dular (P_3 x P_6)	28.00	59.87	GS-503 x K-508 (P_9 x P_{12})	9.37	9.35	Ch-1039 x Kohsar (L_4 x T_3)	96.3	93.05
Ch-1039 x Chenab (P_4 x P_5)	31.51	15.91	GS-504 x Kohsar (P_{10} x P_{11})	9.35	13.75	Chenab x Koshihikari L_5 x T_1)	8.3	83.22
Ch-1039 x Dular (P_4 x P_6)	26.87	15.17	GS-504 x K-508 (P_{10} x P_{12})	7.08	17.63	Chenab x K-332 (L_5 x T_2)	92.60	91.50
Chenab x Dular (P_5 x P_6)	32.72	39.42	Kohsar x K-508 (P_{11} x P_{12})	13.88	6.50	Chenab x Kohsar (L_5 x T_3)	51.10	81.35
Mean	30.25	27.32		10.97	16.78		58.26	76.51
Range	9.27-85.40	13.11-74.00		6.99-19.60	6.50-32.07		8.3-96.3	31.80-96.60
CD at 5% level	4.30	1.216		4.044	8.632		7.97	7.22
Parental mean	9.39	7.7		5.72	33.32		-0.15	0.17
Correlation coefficient (r)			0.68 **					

grains, however, the empty grains obtained could be solely attributed to spikelet sterility. Similar sort of suggestions came from Ikehashi and Araki (1986) who reported that hybrid sterility involved egg killer and induced abortion of megaspore. Female gamete abortion in *indica japonica* hybrids was also observed by Chen et al. (2008). Other possible reason is that the actual fertility (germinability) of the pollen produced by the *indica x japonica* hybrids would be much lower than what is observed with I_2KI staining. Because of the drastic fall in night temperature pollen fertility observed with I_2KI staining did not coincided with actual pollen germination. Liu et al.

(1992) reported that low germinability of morphologically normal pollen (<10%) could be a sort of factor for sterility in *indica x japonica* hybrids.

Further cold stress experienced under high altitude testing ecology (2285 m amsl) at late heading stage (last week of August) is suggested a possible reason for varied expression of hybrid sterility within and more particularly among various crossing blocks particularly for inter-subspecific crosses generated for the study. This is supposed a major probability for recording differential expression of panicle fertility or seed setting in the present experiment. Qi et al. (1993) and Liu et al. (2004) proposed that the differential

expressions of hybrid sterility were in fact determined by temperatures in inter-subspecific hybrids which are highly prone to low temperature.

The other objective of the study was to estimate overcome in hybrid sterility by deploying WCVs as the bridging parent. It was observed that the estimates of mean pollen sterility (58.26%) and spikelet sterility (76.51%) for inter-subspecific crosses got reduced to the level of 30.80 and 25.75% through three way-crosses with mean percentage overcome of 36.62 and 63.92%, respectively (Table 4). The highest estimate for pollen sterility was recorded on inter-subspecific

Table 4. Comparative estimates of pollen sterility and spikelet sterility (%) averaged over two locations for *indica x japonica* (*i x j*) crosses and their corresponding three-way crosses using WCV (*Dular*).

S/N	Cross combination (i x j crosses)	Pollen/spikelet sterility (%)	Cross combination (Three-way crosses)	Pollen/spikelet sterility (%)	Percentage overcome in sterility
1	Jhelum x Koshihikari L_1 x T_1	95.85 (75.81)	(Jhelum x Dular) x Koshihikari (L_1 x D) x T_1	19.18 (15.44)	79.98 (79.63)
2	Jhelum x K-332 L_1 x T_2	47.05 (31.81)	(Jhelum x Dular) x K-332 (L_1 x D) x T_2	14.90 (21.69)	68.33 (31.69)
3	Jhelum x Kohsar L_1 x T_3	92.03 (86.08)	(Jhelum x Dular) x Kohsar (L_1 x D) x T_3	15.47 (13.56)	83.19 (84.25)
4	SK-382 x Koshihikari L_2 x T_1	62.06 (60.58)	(SK-382 x Dular) x Koshihikari (L_2 x D) x T_1	47.49 (40.39)	23.47 (30.67)
5	SK-382 x K-332 L_2 x T_2	17.25 (81.09)	(SK-382 x Dular) x K-332 (L_2 x D) x T_2	25.55 (47.35)	-48.11 (33.28)
6	SK-382 x Kohsar L_2 x T_3	37.21 (82.56)	(SK-382 x Dular) x Kohsar (L_2 x D) x T_3	21.71 (20.89)	41.65 (74.70)
7	SR-1 x Koshihikari L_3 x T_1	91.66 (84.20)	(SR-1 x Dular) x Koshihikari (L_3 x D) x T_1	39.26 (21.45)	57.17 (74.52)
8	SR-1 x K-332 L_3 x T_2	20.79 (85.64)	(SR-1 x Dular) x K-332 (L_3 x D) x T_2	26.43 (30.69)	-27.13 (64.16)
9	SR-1 x Kohsar L_3 x T_3	44.65 (69.52)	(SR-1 x Dular) x Kohsar (L_3 x D) x T_3	7.69 (23.34)	82.78 (66.43)
10	China-1039 x Koshihikari L_4 x T_1	55.43 (44.77)	(China-1039 x Dular) x Koshihikari (L_4 x D) x T_1	6.41 (20.41)	88.43 (54.41)
11	China-1039 x K-332 L_4 x T_2	61.67 (96.60)	(China-1039 x Dular) x K-332 (L_4 x D) x T_2	93.88 (43.74)	-52.23 (54.72)
12	China-1039 x Kohsar L_4 x T_3	96.3 (93.05)	(China-1039 x Dular) x Kohsa (L_4 x D) x T_3	32.40 (12.79)	66.35 (86.25)
13	Chenab x Koshihikari L_5 x T_1	8.3 (83.22)	(Chenab x Dular) x Koshihikari (L_5 x D) x T_1	7.53 (17.27)	9.28 (79.25)
14	Chenab x K-332 L_5 x T_2	92.60 (91.05)	(Chenab x Dular) x K-332 (L_5 x D) x T_2	91.40 (37.29)	1.30 (59.24)
15	Chenab x Kohsar L_5 x T_3	51.10 (81.35)	(Chenab x Dular) x Kohsar (L_5 x D) x T_3	12.84 (11.73)	74.87 (85.58)
	Mean	58.26 (76.51)		30.80 (25.75)	36.62 (63.92)
	Range	8.3-96.3 (31.80-96.60)		6.41-93.88 (11.73-54.10)	
	CD at 5 % level	7.97		7.65	
	Correlation coefficient (r)			0.64**	

Figures in parenthesis denote spikelet sterility]: D=Dular; L and T denote lines and testers, respectively.

cross L_1 × T1 (95.85%) followed by L_5 × T3 (96.3%), L_1 × T3 (92.03%) and L4 × T3 (91.66%). Among three-way crosses that recorded low estimates for pollen sterility were (L_3 × D) × T_3 (7.69%), (L_5 × D) × T_1 (7.53%) and (L_4 × D) × T_1 (6.41%), whereas, larger estimates among three-way crosses were noticed on (L_4 × D) × T_2 (93.88%) followed by (L_5 × D) × T_2 (91.40%). The three-way cross (L_4 × D) × T_1 recorded highest overcome in pollen sterility (88.43%) followed by (L_1 × D) × T_3 (83.19%) and (L_3 × D) × T_3 (82.78%) and so on. The reduction was noticed in negative direction on three three-way crosses namely (L_2 ×

D) × T_2 (-48.11%), (L_3 × D) × T_1 (-27.13%) and (L_4 × D) × T_2 (-52.23%) meaning that WCV could not overcome this type of sterility even to the smallest degree. Regarding the spikelet sterility, the three-way crosses narrowed down the range from a low 11.73 % to a high 54.10 % with mean reduction of 63.92% against the corresponding inter-subspecific single crosses. The magnitude of spikelet sterility was recorded highest for single cross L_4 × T_2 (96.60%) followed by L_4 × T_3 (93.05%) and L_5 × T_3 (91.50%). The estimates of spikelet sterility were relatively lower for L_1 × T_2 and L_4 × T_1 with magnitude of 31.80 and 44.77%,

respectively. The sterility got reduced by a big margin of 86.25% in three-way cross (L_4 × D) × T_3 against corresponding single cross L_4 × T_3. The high reduction for the trait was also observed for three-way crosses (L_5 × D) × T_3 (85.58%) and (L_1 × D) × T_3 (84.25%) when compared with corresponding single crosses L_5 × T_3 and L_1 × T_3, respectively. Yang et al. (2012) showed that wide compatibility gene $S_5{}^n$ can overcome the embryo-sac sterility between *indica–japonica* hybrids caused by the S_5 locus located on chromosome 6. Embryo-sac fertility was more than 93% in the $S_5{}^n$ gene–harboring hybrids, whereas embryo-sac fertility was relatively low in control hybrids between typical *indica* and *japonica* cultivars without the S_5^n gene, suggesting that S_5^n can overcome the sterility between *indica–japonica* hybrids.

The *hybrid sterility* analysis showed a considerable variation in F_1s with the same WCV (Dular) in different *indica* and *japonica* backgrounds. Differential expression under different backgrounds suggests the inadequate nature or inability of wide compatibility gene (WCG) from Dular source only, although both kinds of sterilities got reduced to an appreciable extent in some cross combinations. The reason of deploying only one WCV in the present experiment was due to non-availability of such established lines in the germplasm bank at the Regional Rice Research Station, Khudwani together with scanty information regarding the germplasm pertaining to this phenomenon. The other possible reason for this variable sterility is that additional genes might be involved that modify hybrid fertility in the presence of wide compatible gene WCG, since a large number of genes exist that affect hybrid fertility in one way or another. Epistasis or non-allelic interactions may also be involved for differential degrees of expression in different backgrounds. Kumar and Chakraborty (2000) observed different degree of expression of WCGs and suggested the set of modifier gene(s) and epistasis. Kinoshita (1995) emphasized the role of additional genes modifying hybrid fertility in the presence of WCG in one way or another, whereas, Ikehashi and Wan (1996) proposed other loci with minor effect in addition to S5 in improving hybrid fertility. Kubo and Yoshimura (2005) and Kubo et al. (2008) suggested the role of epistatic interactions for differential expression of hybrid semi sterility.

The present study demonstrates differential expression of hybrid sterility among and within crossing blocks in terms of both pollen and spikelet sterility. The findings give us clue about the genetic relatedness regarding the germplasm available at the research station and explore high level of genetic distinctness between *indica* and *japonica* genotypes followed by within *indica* and close relatedness within *japonica* types. Since this was the modest and first successful attempt under cold temperate conditions of Kashmir, although few attempts earlier were made but they failed to make any conclusion. Now the genetically distinct parents identified will be crossed *inter se* followed by pedigree method of selection under different agroecologies to throw suitable promising segregants. Besides ideotype breeding/super rice breeding can be initiated by combining desirable traits from *indica* genotypes (desirable grain shape, texture and quality) with *japonica* genotypes (lodging resistance, early maturity and cold tolerance) and reciprocally through combination breeding. Under the present investigation only one WCV was attempted, hence several more need to be assessed so that many improved WC cultivars with *indica* and *japonica* backgrounds usable as cultivars and parents in hybrid breeding could be developed for unique temperate

conditions of Kashmir and equivalent ecologies worldwide. The new materials developed/derived will definitely overcome the yield barriers and will prove useful at an altitude range of 1500 to 2300 m amsl unique rice grown temperate agro-ecologies.

REFERENCES

Anonymous (1996). Standard evaluation system for rice. INGER, Genetic Resource Centre, 4th ed. IARI, Manila, Philippine.

Chen J, Ding J, Ouyang Y, Du H, Yang J, Cheng K, Zhao J, Qiu S, Zhang X, Yao J (2008). A triallelic system of S5 is a major regulator of the reproductive barrier and compatibility of *Indica–japonica* hybrids in rice. Proc. Natl. Acad. Sci. 105:11436-11441.

Griffing B (1956). A generalized treatment of the use of diallel crosses in quantitative inheritance. Heredity 10:31-50.

He GH, Zhang JK, Yin GD, Yang ZL (1994). Genetic fertility of F1 between *indica* and *japonica*. Chin. J. Rice Sci. 8:177-180.

Ikehashi H (1982) Prospects for overcoming barrier in utilization of *indica- japonica* crosses in rice breeding. Oryza 19:69-77.

Ikehashi H, Araki H (1986). Genetics of F_1 sterility in remote crosses of rice. Rice Genetics IRRI. pp. 119-130.

Ikehashi H, Wan J (1996). Differentiation of alleles at seven loci for hybrid sterility in cultivated rice (*Oryza sativa* L.). Rice Genetics III, IRRI. pp. 404-408.

Jiang Y, Zhao X, Cheng K, Du H, Ouyang Y, Chen J, Qiu S, Huang J, Jiang Y, Jiang L, Ding J, Wang J, Xu C, Li X, Zhang Q (2012). A killer-protector system regulates both hybrid sterility and segregation distortion in rice. Science pp. 1336-1340.

Kubo T, Yoshimura A (2005). Epistasis underlying female fertility detected in hybrid breakdown in a *japonica/indica* cross of rice (*Oryza sativa* L.). Theoret. Appl. Genet. 110:346-355.

Kubo T, Yamagata Y, Eguchi M, Yoshimura A (2008). A novel epistatic interaction at two loci causing male sterility in an inter-subspecific cross of rice (*Oryza sativa* L.). Gene Genet. Syst. 83:443-453.

Kubo T, Yoshimura A, Kurata N (2011). Hybrid male sterility in rice is due to epistatic interactions with a pollen killer locus. Genetics 189:1083-1092.

Kempthorne O (1957). An introduction to genetic studies. John Wiley and Sons, New York.

Kinoshita T (1995). Report of the committee on gene symbolization, nomenclature and linkage groups. Rice Genet. Newslett. 12:24-93.

Kumar S, Chakrabarti SN (2000). Identification of improved source of wide compatibility gene and sterility gene(s) in inter-subspecific crosses of rice. Indian J. Agric. Sci. 70:446-449.

Liu KD, Wang J, Li HB, Xu CG, Liu AM (1997). A genomewide analysis of wide compatibility in rice and the precise location of the S5 locus in the molecular map. Theor. Appl. Genet. 95:809–814.

Liu A, Zhang Q, Li H (1992). Location of a gene for wide compatibility in the RFLP linkage map. Rice Genet. Newslett. 9:134-136.

Liu KD, Zhou ZQ, Zu CG, Zhang Q, Maroof MA (1996). An analysis of hybrid sterility in rice using a diallel cross of 21 parents involving *indica*, *japonica* and wide compatibility cultivars. Euphytica 90:257-280.

Liu HY, Xu CG, Zhang Q (2004). Male and female gamete abortion, and reduced affinity between the uniting gametes as the causes for sterility in an *Indica/Japonica* hybrid in rice. Sex Plant Reprod. 17:55-62.

Oka HI (1964). Consideration on the genetic basis of intervarietal sterility in *Oryza sativa* L. Rice Genetics and Cytogenetics (Ed. R.F. Chandler). pp. 158-174.

Oka HI (1988). Origin of cultivated rice. Elsevier Japan Science Society Press, Tokyo. pp. 181-209.

Qi ZB, Chai YT, Li BJ (1993). A study of the factor influencing *indica-japonica* hybrid fertility. Guangdong Agric. Sci. 2:2-6.

Song X, Qiu SQ, Yu CG, Li XH, Zhang Q (2005). Genetic dissection of embryo sac fertility, pollen fertility, and their contributions to spikelet fertility of inter-subspecific hybrids in rice. Theor. Appl. Genet. 110:205-211.

Yang SR, Shen XY, Gu VYL, Cao DJ (1962). The report of rice breeding by *indica-japonica* crosses. Acta Agron. Sinica 1:97-102.

Yang Y, Hong L, Fei T, Qasim S, Xiong C, Wang L, Li J, Li X, Lu Y (2012). Wide-compatibility gene $s_5{}^n$ exploited by functional molecular markers and its effect on fertility of intersubspecific rice hybrids. Crop Sci. pp. 669-675.

Yuan LP (1994). Increasing yield potentials in rice by exploitation of heterosis. In : Hybrid Rice Technology. "New Development and Future Prospects". [Ed. S.S. Virmani], IRRI, Phillipines. pp. 1-6.

Zhang GQ, Lu YG (1993). Genetic studie aematoxylin staining for determining pollen fertility of *Indica/Japonica* F_1 hybrids and study on its gamete fertility. Rice Genet. Newslett. 13:125-126.

Zhang GQ, Liu KD, Yang GP, Saghai M, Xu CG, Zhou ZQ (1997). Molecular marker diversity and hybrid sterility in *Indica-japonica* rice crosses. Theor. Appl. Genet 95:112-118.

6

Genetic studies of morpho-physiological traits of maize (*Zea mays* L.) seedling

Qurban Ali[1], Muhammad Ahsan[1], Muhammad Hammad Nadeem Tahir[1] and Shahzad Maqsood Ahmed Basra[2]

[1]Department of Plant Breeding and Genetics, University of Agriculture, Faisalabad, Pakistan.
[2]Department of Crop Physiology, University of Agriculture, Faisalabad, Pakistan.

The proposed study was carried out in the glasshouse of the Department of Plant Breeding and Genetics, University of Agriculture Faisalabad, Pakistan to evaluate eighty maize genotypes for seedling traits during the crop season in February 2011. The higher value of genotypic variance was found for chlorophyll contents, photosynthetic rate and sub-stomatal CO_2 concentration. The higher values of heritability and genetic advance were also recorded for chlorophyll contents, photosynthetic rate and sub-stomatal CO_2 concentration for both after 14 and 28 days of seedling emergence. It was concluded that the selection of higher yielding genotypes can be made and may also be used as potential genotypes for higher grain and fodder yields.

Key words: *Zea mays*, genotypic variance, heritability, genetic advance, Pakistan.

INTRODUCTION

Maize (*Zea mays* L.) is world's leading cereal food crop with added importance for countries like Pakistan where rapidly increasing population has already outstripped the available food supplies. Maize is third important cereal in Pakistan after wheat and rice. Maize accounts for 5.67% of the value of agriculture output. It accounts for 950 thousands hectares of total cropped area in Pakistan with annual production of 3487 thousands tons (Anonymous, 2010). Maize is dual purpose crop such as food for human and feed for livestock and also used as industrial raw material for the manufacture of different products. It has highest crude protein 9.9% at early and at full bloom stages which decreases to 7% at milk stage and to 6% at maturity. Maize has highly nutritive value as it contains starch 72%, protein 10%, oil 4.80%, fiber 9.50%, sugar 3.0%, ash 1.70%, endosperm 82%, embryo 12%, bran testa 5% and tip cap 1% (Chaudhary, 1983; Bureau of

Chemistry, U.S., 2010). Pakistan have livestock population of 154.7 million heads which produce approximately 43.562 million tons of milk, 1.601 million tons of beef and 0.590 million tons of mutton (Anonymous, 2009-2010). Livestock sector contributes approximately 53.2% of the agriculture value added and 11.4% to national GDP of Pakistan (Anonymous, 2009-2010). Green fodder is the most valuable and cheapest source of food for livestock. It is rich source of cellulose 35 to 40%, hemicelluloses 25.28%, fat 0.30%, crude fiber 28.70%, ADF 37.22%, NDF 70.85% dry matter 40.6%, ash 4%, carbohydrates 48.86%, moisture 9.22%, ether extract 2.84% and crude proteins 11% (Chaudhary, 1983; Bureau of Chemistry, U.S., 2010). With quality nutritional fodder, milk production can be increased up to 100% (Maurice et al., 1985). The present study was conducted to evaluate maize accessions for morpho-physiological seedling traits.

MATERIALS AND METHODS

The present study was carried out in the glasshouse of the Department of Plant Breeding and Genetics, University of Agriculture, Faisalabad to evaluate the maize genotypes for seedling traits during the crop season in February 2011. The experimental material was consisting of 80 accessions including ten check varieties namely: F-121, F-128, F-150, F-142, F-151, F-118, F-117, F-130, F-140, F-143, F-113, F-111, F-114, F-136, F-122, F-134, F-147, F-105, F-148, F-146, B-303, B-316, B-306, B-303, B-313, B-314, B-305, B-321, B-326, B-308, B-304, B-312, EV-344, EV-343, EV-310, POP/209, EV-342, EV-347, F-96, EV-324, EV-335, EV-323, EV-334, EV-330, EV-329, EV-338, EV-340, E-349, E-352, E-341, E-351, E-322, E-346, E-336, BF-337 BF-248, BF-212, BF-236, BF-238, F-98, B-96, F-135, VB-06, B-121, B-15, B-11, Sh-213, Sh-139, SWL-2002, Sawan-3, Pak-Afgoee, Gold Isalamabad, Islamabad W, VB-51, EV-1097, EV-7004Q, Raka-Poshi, BS-2 and POP/2007). The seeds were sown in iron trays filled with sand following a randomized complete block design (RCBD) with three replications. The seed were sown at the depth of 2.5 cm. Twenty seedlings of each accession were established in each replication. The data of 5 plants was recorded for Chlorophyll contents, Photosynthetic rate, Stomata conductance, Leaf temperature, Transpiration rate, Sub-stomata CO_2 concentration, fresh root weight and fresh shoot weight after 14 and 28 days of sowing with the help of Chlorophyll Meter and IRGA (Infrared Gas Analyzer). The data was analyzed statistically using analysis of variance technique and Duncan Multiple Range (DMR) test at 1% significance level was used to compare the treatments means (Steel et al., 1997).

RESULTS AND DISCUSSION

Chlorophyll contents (mg g^{-1} fr. wt.)

It persuaded from Table 1a and b that highly significant differences were found among genotypes. The higher mean for chlorophyll contents (4.384±0.022 mg g^{-1} fr. wt.) was recorded for 28 days old seedlings of genotypes while lower for 14 days old seedlings. Table 1 indicated that the genotypic variance, phenotypic variance and environmental variance were 23.068, 23.07 and 0.002, respectively while genotypic coefficient of variance, phenotypic coefficient of variance and environmental coefficient of variance were 110.31, 110.32 and 0.98%, respectively. From Table 1b it is cleared that genotypic, phenotypic and environmental variances were 22.81, 22.81 and 0.0003, respectively while genotypic coefficient of variance, phenotypic coefficient of variance and environmental coefficient of variance were 108.84, 108.84 and 0.42%, respectively. Table 1a and b also indicated that highest estimates of heritability (100%, 100%) and genetic advance were 154.43 and 152.52% for 14 and 28 days old seedlings respectively. The higher values of heritability and genetic advance indicated that selection for chlorophyll contents of seedling may be effective for selecting higher yielding maize genotypes (Moulin et al., 2009; Ali et al., 2011b).

Figure 1a and 1b indicated that the genotypes Raka-Poshi and EV-344 showed higher chlorophyll contents for 14 and 28 days, respectively while lowest chlorophyll contents was found for F-146 and B-316 followed by B-327 and F-134 for 14 and 28 days, respectively. The higher values of chlorophyll contents indicated that the photosynthesis is enhanced due to which the root and shoot fresh/dry weights are increased that increased plant productivity (Moulin et al., 2009).

Photosynthetic rate (µg CO$_2$ s^{-1})

It persuaded from Table 1a and b that highly significant differences were found among genotypes. The higher mean for photosynthesis rate (7.517±0.084 µg CO$_2$ s^{-1}) was recorded for 14 days old seedlings of genotypes while lower for 28 days old seedlings. It is indicated from Table 1 that the genotypic variance, phenotypic variance and environmental variance were 35.14, 35.14 and 0.002, respectively while genotypic coefficient of variance, phenotypic coefficient of variance and environmental coefficient of variance were 78.86, 78.86 and 0.57%, respectively. From Table 1a and b it is cleared that genotypic, phenotypic and environmental variances were 36.23, 36.24 and 0.013, respectively while genotypic coefficient of variance, phenotypic coefficient of variance and environmental coefficient of variance were 80.96, 80.98 and 1.55%, respectively.

Table 1a and b also indicated that highest estimates of heritability (100%, 100%) and genetic advance were 110.34 and 110.90% for 14 and 28 days old seedlings respectively. The higher values of heritability and genetic advance indicated that selection for photosynthesis rate of seedling may be effective for selecting higher yielding maize genotypes. It is cleared from the graphs (Figure 2) that the genotypes F-146 and E-341 showed higher photosynthesis rate, followed by EV-343 and BS-2 for 14 and 28 days respectively while lowest photosynthesis rate was found for B-305 and B-313 for 14 and 28 days, respectively. The values of photosynthetic rat indicated that the production and accumulation of food reserves are increased that leads towards the improvement in the crop yield and productivity to increase grain and fodder production in maize (Grzesiak et al., 2007; Wang et al., 2007; Moulin et al., 2009).

Stomatal conductance (mmol m^{-2} s^{-1})

It persuaded from Table 1a and b that highly significant differences were found among genotypes. The higher mean for stomatal conductance (0.376±0.040 mmol m^{-2} s^{-1}) was recorded for 28 days old seedlings of genotypes while lower for 14 days old seedlings. It is indicated from Table 1a that the genotypic variance, phenotypic variance and environmental variance were 0.014, 0.017 and

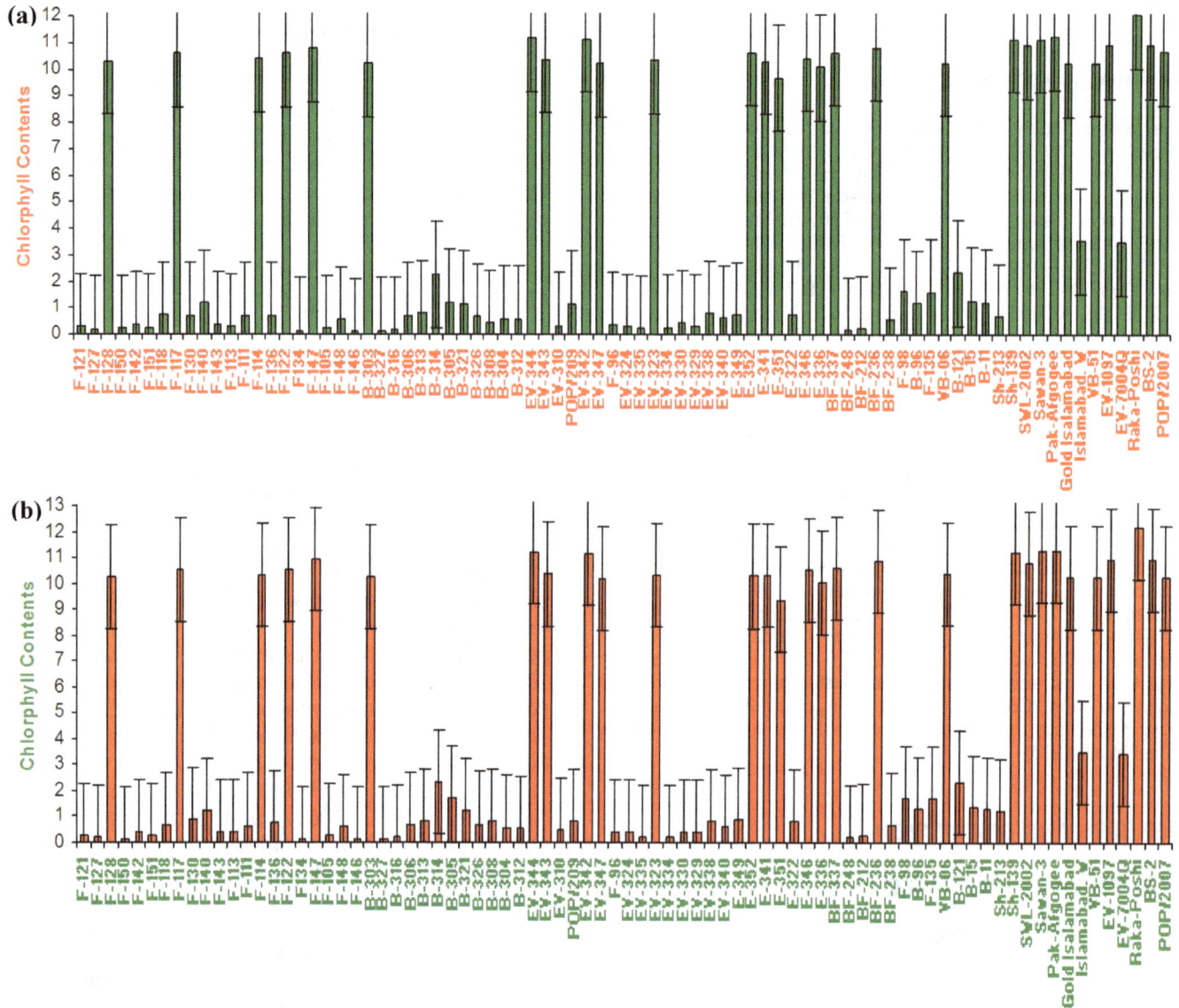

Figure 1. Chlorophyll contents (mg g^{-1} fr. wt.) of (a) 14 days old seedlings; (b) 28 days old seedlings.

Table 1a. Genetic components of 14 days old maize seedlings.

Traits	S.S	G.M±S.E	GV	GCV%	PV	PCV %	EV	ECV %	h$^2_{bs}$ %	S.E h$^2_{bs}$	GA %
Chl.C	3645.098**	4.354±0.045	23.068	110.31	23.07	110.32	0.002	0.98	100	0.023	154.43
PR	5551.91**	7.517±0.084	35.14	78.86	35.14	78.86	0.002	0.57	100	0.019	110.34
SC	2.75*	0.325±0.054	0.014	36.90	0.017	40.59	0.003	16.92	82.60	0.912	46.95
LT	557.149**	33.22±0.114	3.513	5.64	3.526	5.65	0.013	0.34	99.60	0.059	7.88
TR	22.482*	0.682±0.031	0.141	55.09	0.142	55.31	0.001	4.59	99.20	0.296	76.81
SSCC	7179115**	257.75±3.494	45539.15	82.79	45551.36	82.80	12.21	1.36	100	0.0005	115.89
FRW	3.905*	0.715±0.156	0.024	21.82	0.025	21.99	0.0004	2.75	98.40	0.712	30.30
FSW	4.757*	0.814±0.015	0.029	21.24	0.03	21.32	0.0002	1.85	99.2	0.643	78.99

** = Significance at 5% level, * = significance at 1% level; S.S = Sum of Squares, G.M. = grand mean, S.E. = standard error, GV = genotypic variance, GCV = genotypic coefficient of variance, PV = phenotypic variance, PCV = phenotypic coefficient of variance, EV = environmental variance, ECV = environmental coefficient of variance, h$^2_{bs}$ = broad sense heritability, S.E h$^2_{bs}$ = standard error for broad sense heritability, GA = genetic advance, Chl.C = chlorophyll contents, PR = photosynthetic rate, SC = stomatal conductance, LT = Leaf temperature, TR = transpiration rate, SSCC = sub-stomatal CO_2 concentration, FRW = fresh root weight, FSW = fresh shoot weight.

Table 1b. Genetic components of 28 days old maize seedlings.

Traits	S.S	G.M ±S.E	GV	GCV %	PV	PCV %	EV	ECV %	h²bs%	S.E h²bs	GA%
Chl.C	3604.142**	4.384±0.022	22.81	108.94	22.81	108.94	0.0003	0.42	100	0.023	152.52
PR	5725.664**	7.434±0.114	36.23	80.96	36.24	80.98	0.013	1.55	100	0.018	110.90
SC	3.37*	0.376±0.040	0.019	37.33	0.021	38.84	0.002	10.75	92.3	0.789	50.21
LT	533.671**	33.058±0.143	3.357	5.54	3.378	5.56	0.020	0.43	99.40	0.061	7.73
TR	21.808*	0.682±0.013	0.138	54.44	0.138	54.47	0.0002	1.95	99.90	0.299	76.17
SSCC	7074406**	255.438±2.739	44772.89	82.84	44774.72	82.84	1.832	0.53	100	0.0005	115.96
FRW	16.263**	1.07±0.014	0.103	29.96	0.103	29.98	0.0002	1.31	99.80	0.347	41.89
FSW	152.391**	2.53±0.051	0.962	38.77	0.965	38.82	0.0026	2.00	99.70	0.113	54.20

** = Significance at 5% level, * = significance at 1% level; S.S = Sum of Squares, G.M = grand mean, S.E = standard error, GV = genotypic variance, GCV = genotypic coefficient of variance, PV = phenotypic variance, PCV = phenotypic coefficient of variance, EV = environmental variance, ECV = environmental coefficient of variance, h^2_{bs} = broad sense heritability, S.E h^2_{bs} = standard error for broad sense heritability, GA = genetic advance, Chl.C = chlorophyll contents, PR = photosynthetic rate, SC = stomatal conductance, LT = Leaf temperature, TR = transpiration rate, SSCC = sub-stomatal CO_2 concentration, FRW = fresh root weight, FSW = fresh shoot weight.

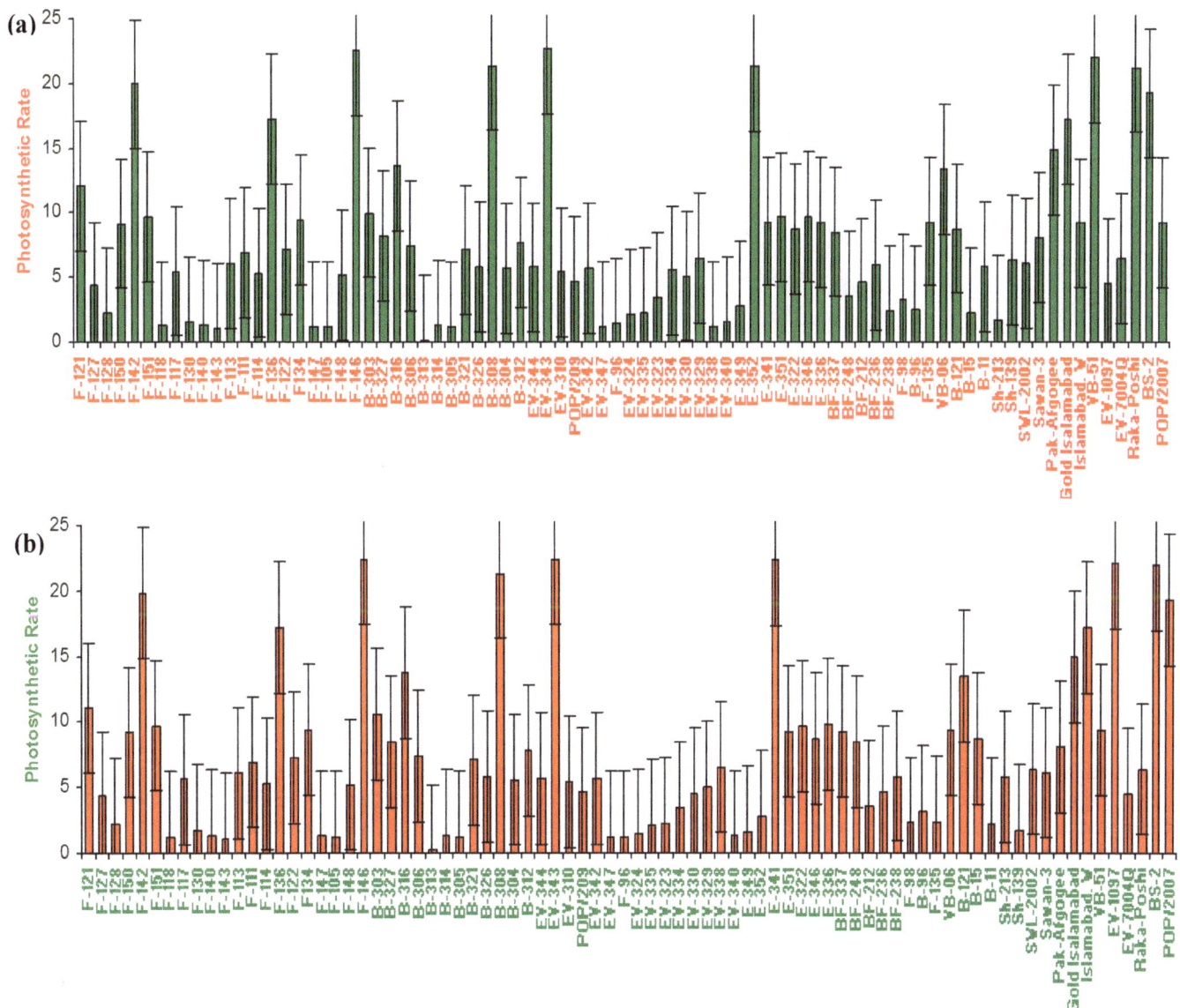

Figure 2. Photosynthesis rate ($\mu g\ CO_2\ s^{-1}$) of (a) 14 days old seedlings; (b) 28 days old seedlings.

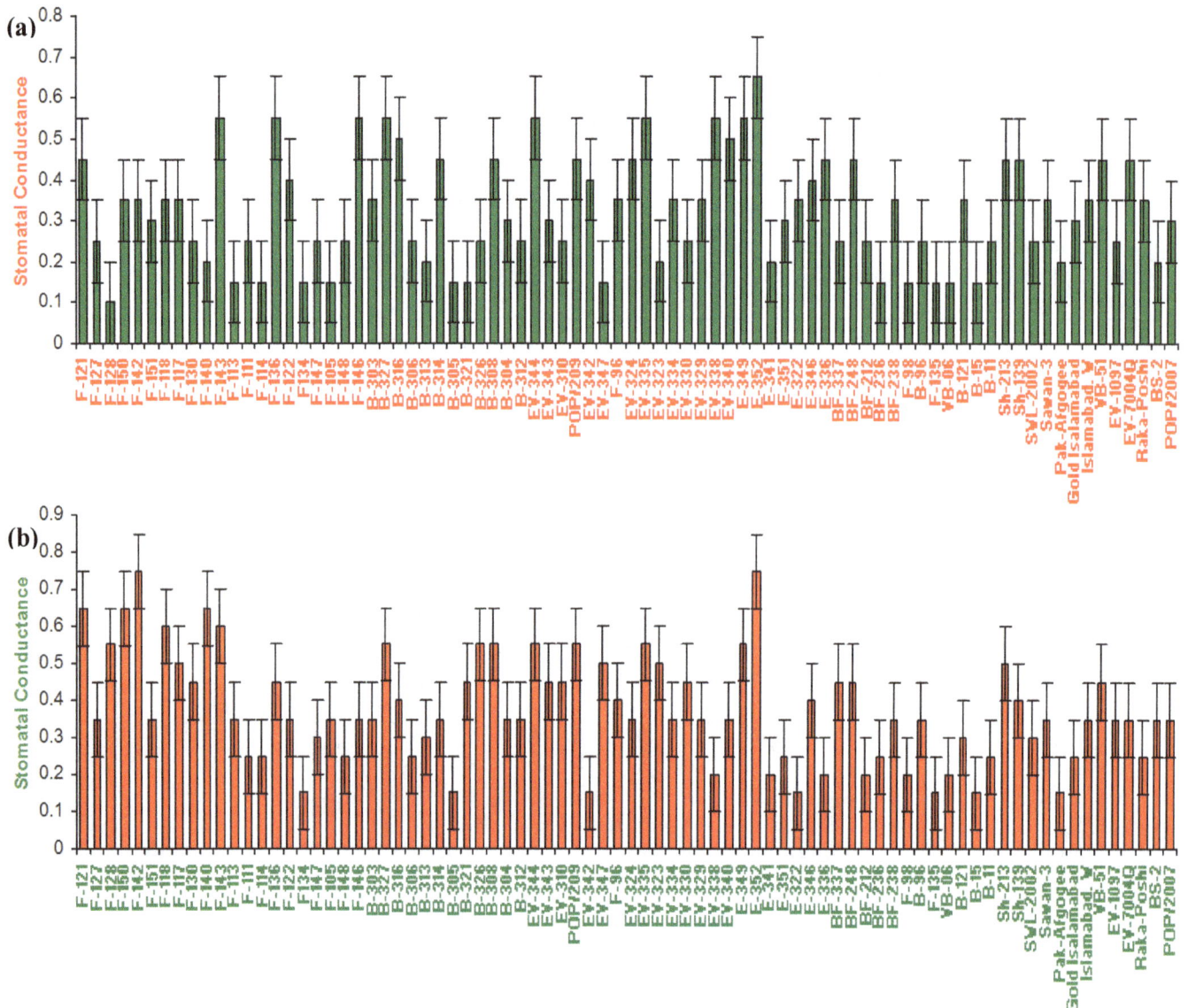

Figure 3. Stomata conductance (mmol m^{-2} s^{-1}) of (a) 14 days old seedlings; (b) 28 days old seedlings.

0.003, respectively while genotypic coefficient of variance, phenotypic coefficient of variance and environmental coefficient of variance were 36.90, 40.59 and 16.92%, respectively. From Table 1a and b it is cleared that genotypic, phenotypic and environmental variances were 0.019, 0.021 and 0.002, respectively while genotypic coefficient of variance, phenotypic coefficient of variance and environmental coefficient of variance were 37.33, 38.84 and 10.75%, respectively. Table 1a and b also indicated that highest estimates of heritability (82.60%, 92.30%) while lower genetic advance were 46.95 and 50.21% for 14 and 28 days old seedlings respectively. The higher values of heritability indicated that selection for stomatal conductance of seedling may be effective for selecting higher yielding maize genotypes. The increased stomatal conductance

caused to enhance the CO_2 absorption due to which photosynthetic rate increased that leads towards increase in grain and fodder yield in maize (Grzesiak et al., 2007; Wang et al., 2007; Moulin et al., 2009; Ahsan et al., 2011). It is cleared from the graphs (Figure 3) that the genotypes E-352 and BS-2 showed higher stomatal conductance, followed by E-349 and F-142 for 14 and 28 days respectively while lowest stomatal conductance was found for F-128 and B-305 followed by F-105 and F-134 for 14 and 28 days, respectively.

Leaf temperature (°C)

It persuaded from Table 1a and b that highly significant differences were found among genotypes. The higher

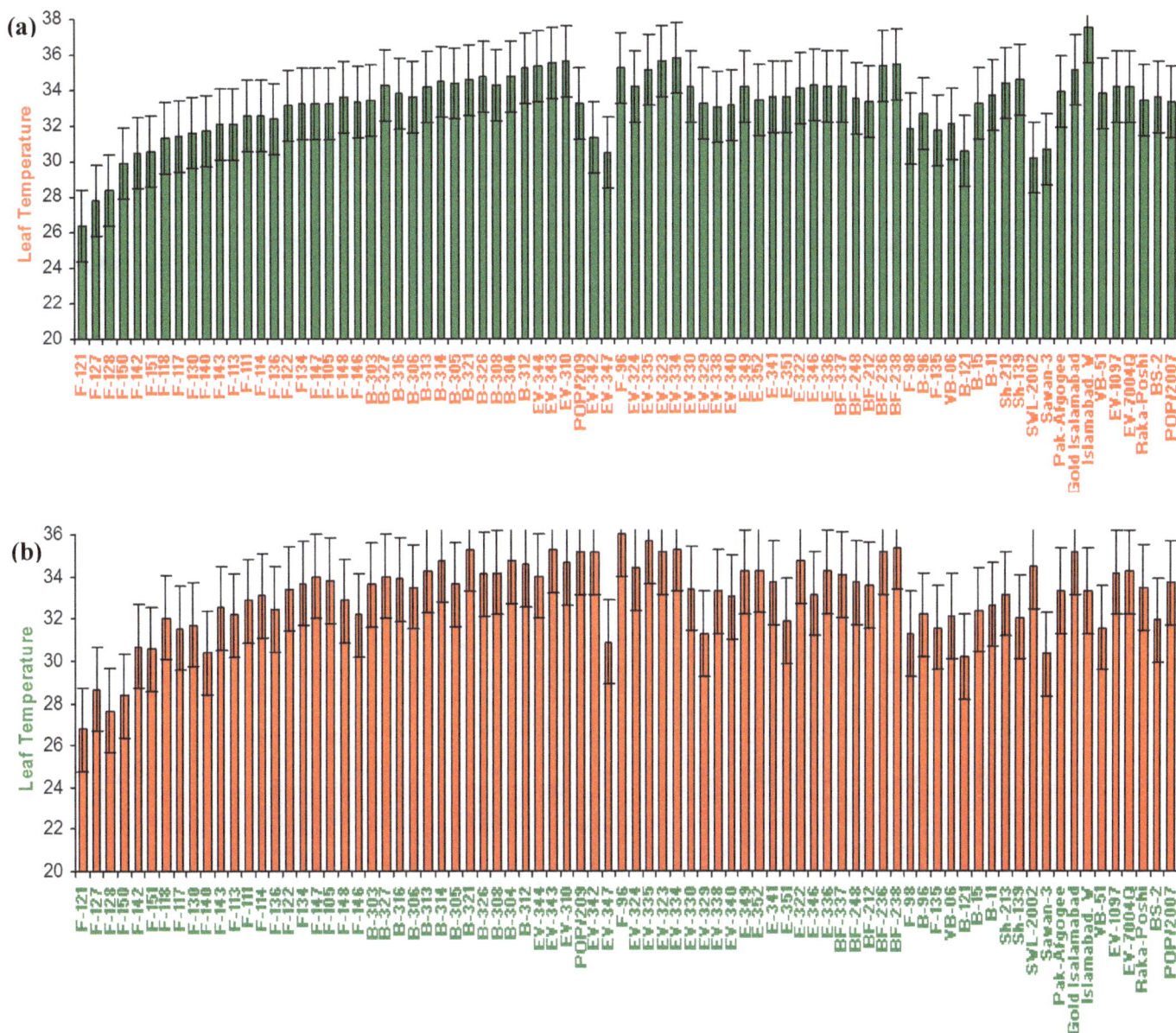

Figure 4. Leaf temperature (°C) of (a) 14 days old seedlings; (b) 28 days old seedlings.

mean for leaf temperature (33.22±0.114°C) was recorded for 14 days old seedlings of genotypes while lower for 28 days old seedlings. It is indicated from Table 1a that the genotypic variance, phenotypic variance and environmental variance were 3.513, 3.526 and 0.013, respectively while genotypic coefficient of variance, phenotypic coefficient of variance and environmental coefficient of variance were 5.64, 5.65 and 0.34%, respectively. From Table 1b it is cleared that genotypic, phenotypic and environmental variances were 3.357, 3.378 and 0.020, respectively while genotypic coefficient of variance, phenotypic coefficient of variance and environmental coefficient of variance were 5.54, 5.56 and 0.43%, respectively. Table 1a and b also indicated that

highest estimates of heritability (99.60%, 99.40%) and genetic advance were 7.88 and 7.73% for 14 and 28 days old seedlings respectively. The higher values of heritability and genetic advance indicated that selection for leaf temperature of seedling may be effective for selecting higher yielding maize genotypes. The change in leaf temperature caused change in stomatal conductance, photosynthetic and transpiration rate (Grzesiak et al., 2007; Wang et al., 2007; Moulin et al., 2009; Ahsan et al., 2011; Ali et al., 2011b). It is cleared from the graphs (Figure 4) that the genotypes Islamabad W and F-96 showed higher leaf temperature, for 14 and 28 days respectively while lowest leaf temperature was found for F-121 for 14 and 28 days respectively.

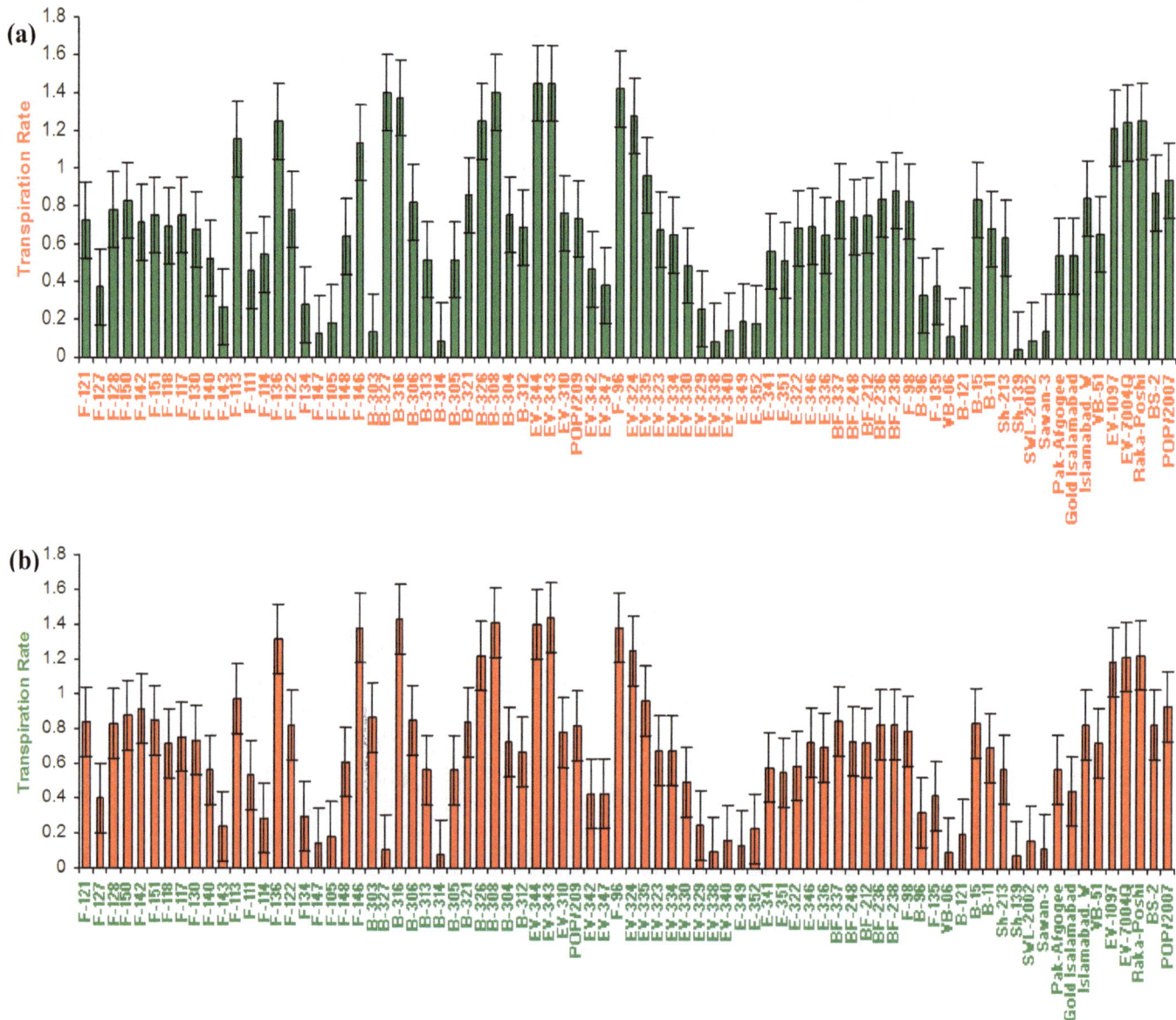

Figure 5. Transpiration rate (mm day^{-1}) of (a) 14 days old seedlings; (b) 28 days old seedlings.

Transpiration rate (mm day^{-1})

It persuaded from Table 1a and b that highly significant differences were found among genotypes. The higher mean for transpiration rate (0.682±0.031 mm day^{-1}) was recorded for 14 days old seedlings of genotypes while lower for 28 days old seedlings. It is indicated from Table 1a that the genotypic variance, phenotypic variance and environmental variance were 0.141, 0.142 and 0.001, respectively while genotypic coefficient of variance, phenotypic coefficient of variance and environmental coefficient of variance were 55.09, 55.31 and 4.59%, respectively. From Table 1b it is cleared that genotypic, phenotypic and environmental variances were 0.138, 0.138 and 0.0002, respectively while genotypic coefficient

of variance, phenotypic coefficient of variance and environmental coefficient of variance were 54.44, 54.47 and 1.95%, respectively. Table 1a and 1b also indicated that highest estimates of heritability (99.20%, 99.90%) and moderate genetic advance were 76.81 and 76.17 for 14 and 28 days old seedlings respectively. The higher values of heritability and moderate genetic advance indicated that selection for transpiration rate of seedling may be effective for selecting higher yielding maize genotypes. Higher transpiration rate causes changes in stomatal conductance and photosynthetic rate (Grzesiak et al., 2007; Wang et al., 2007; Moulin et al., 2009; Ahsan et al., 2011). It is cleared from the graphs (Figure 5) that the genotypes EV-343 and B-316 showed higher transpiration rate, followed by EV-344 and EV-343 for 14

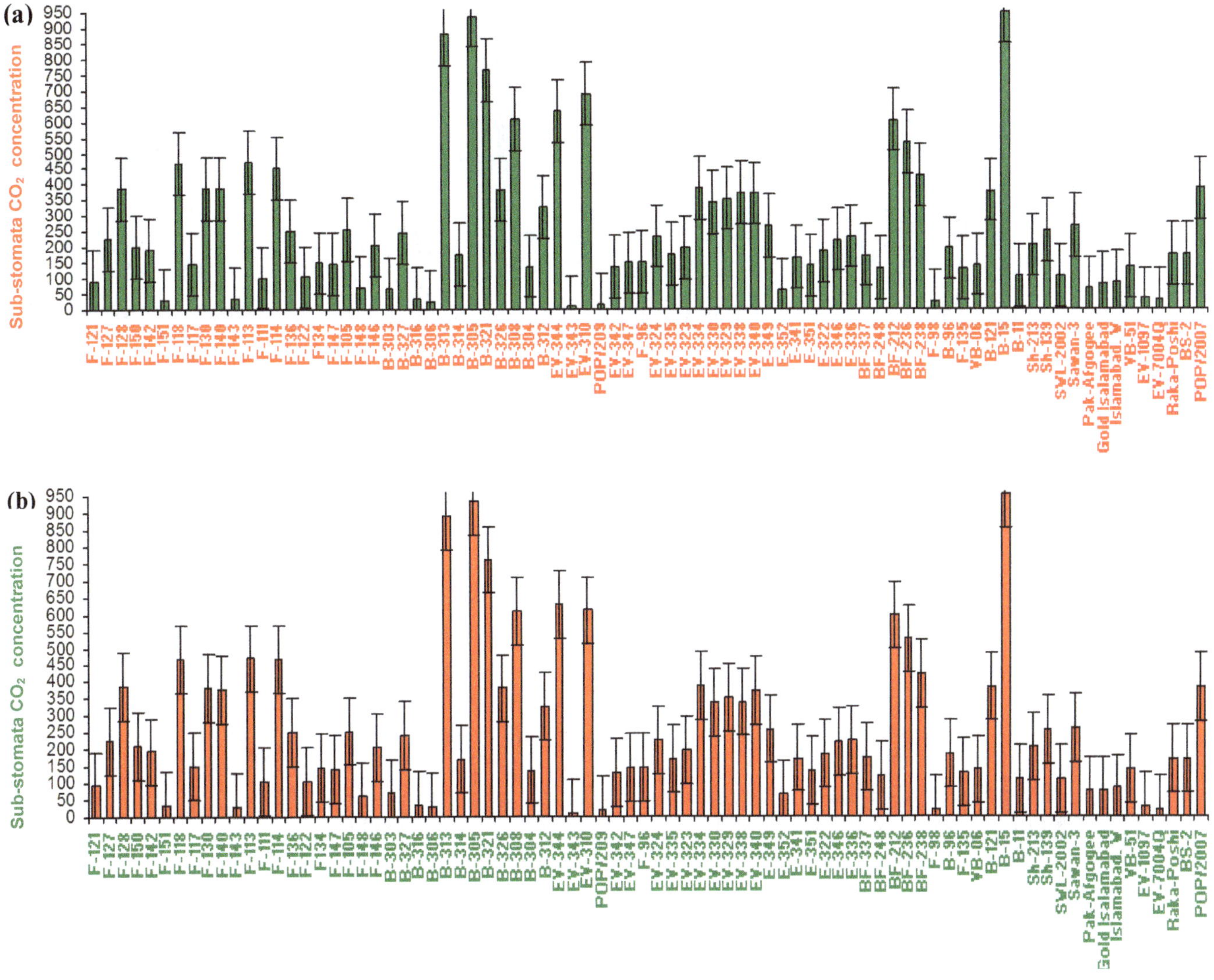

Figure 6. Sub-stomata CO_2 concentration ($\mu mol\ mol^{-1}\ CO_2$) of (a) 14 days old seedlings; (b) 28 days old seedlings.

and 28 days, respectively while lowest transpiration rate was found for B-314 and Sh-139 for 14 and 28 days, respectively.

Sub-stomata CO_2 concentration ($\mu mol\ mol^{-1}\ CO_2$)

It persuaded from Table 1a and b that highly significant differences were found among genotypes. The higher mean for Sub-stomata CO_2 concentration (257.75±3.494 $\mu mol\ mol^{-1}\ CO_2$) was recorded for 14 days old seedlings of genotypes while lower for 28 days old seedlings. It is indicated from Table 1 that the genotypic variance, phenotypic variance and environmental variance were 45539.15, 45551.36 and 12.21, respectively while genotypic coefficient of variance, phenotypic coefficient of variance and environmental coefficient of variance were 82.79, 82.80 and 1.36%, respectively. From Table 1b it is

cleared that genotypic, phenotypic and environmental variances were 44772.89, 44774.72 and 1.832, respectively while genotypic coefficient of variance, phenotypic coefficient of variance and environmental coefficient of variance were 82.84, 82.84 and 0.53%, respectively. Table 1a and b also indicated that highest estimates of heritability (100%, 100%) and genetic advance were 115.89 and 115.96% for 14 and 28 days old seedlings respectively. The higher values of heritability and genetic advance indicated that selection for Sub-stomata CO_2 concentration of seedling may be effective for selecting higher yielding maize genotypes. Higher Sub-stomata CO_2 concentration increased CO_2 absorption that caused an increase in the stomatal conductance, photosynthetic and transpiration rate (Grzesiak et al., 2007; Wang et al., 2007). It is cleared from Figure 6 that the genotypes B-15 showed higher while EV-343 showed lowest Sub-stomata CO_2

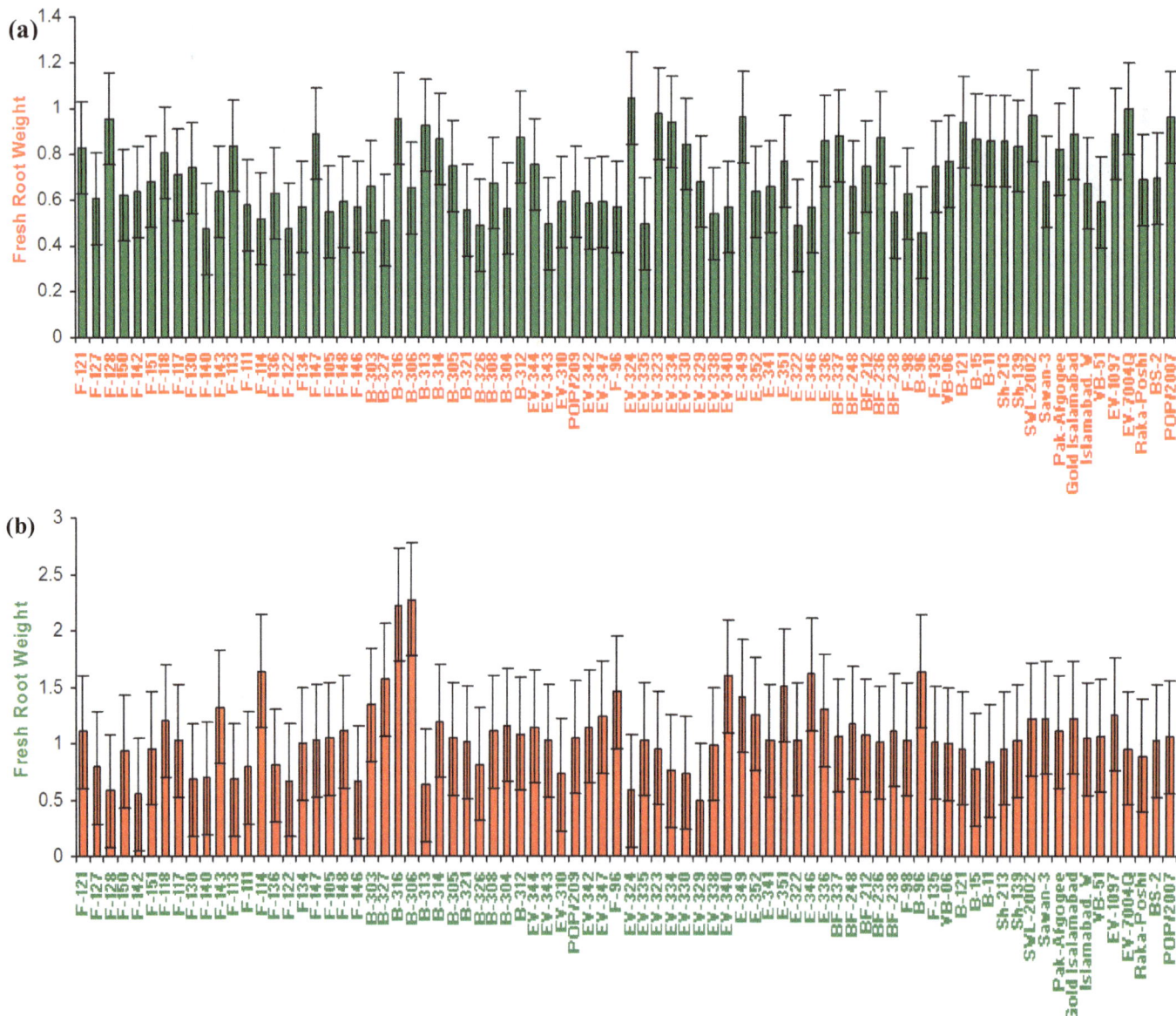

Figure 7. Fresh root weight (g) of (a) 14 days old seedlings; (b) 28 days old seedlings.

concentration for 14 and 28 days.

Fresh root weight (g)

It persuaded from Table 1a and b that highly significant differences were found among genotypes. The higher mean for fresh root weight (1.07±0.014 g) was recorded for 28 days old seedlings of genotypes while lower for 14 days old seedlings. It is indicated from Table 1 that the genotypic variance, phenotypic variance and environmental variance were 0.024, 0.025 and 0.0004, respectively while genotypic coefficient of variance, phenotypic coefficient of variance and environmental coefficient of variance were 21.82, 21.99 and 2.75%, respectively. From

Table 1b it is cleared that genotypic, phenotypic and environmental variances were 0.103, 0.103 and 0.0002, respectively while genotypic coefficient of variance, phenotypic coefficient of variance and environmental coefficient of variance were 29.96, 29.98 and 1.31%, respectively. Table 1a and b also indicated that highest estimates of heritability (98.40%, 99.80%) and low and moderate values of genetic advance were 30.30 and 41.89% for 14 and 28 days old seedlings, respectively. The higher and moderate values of heritability and genetic advance indicated that selection for fresh root weight of seedling may be less effective for selecting higher yielding maize genotypes (Mehdi and Ahsan, 2000a; Ali et al., 2011a, b). It is cleared from the graphs (Figure 7) that the genotypes EV-324 and B-306 showed

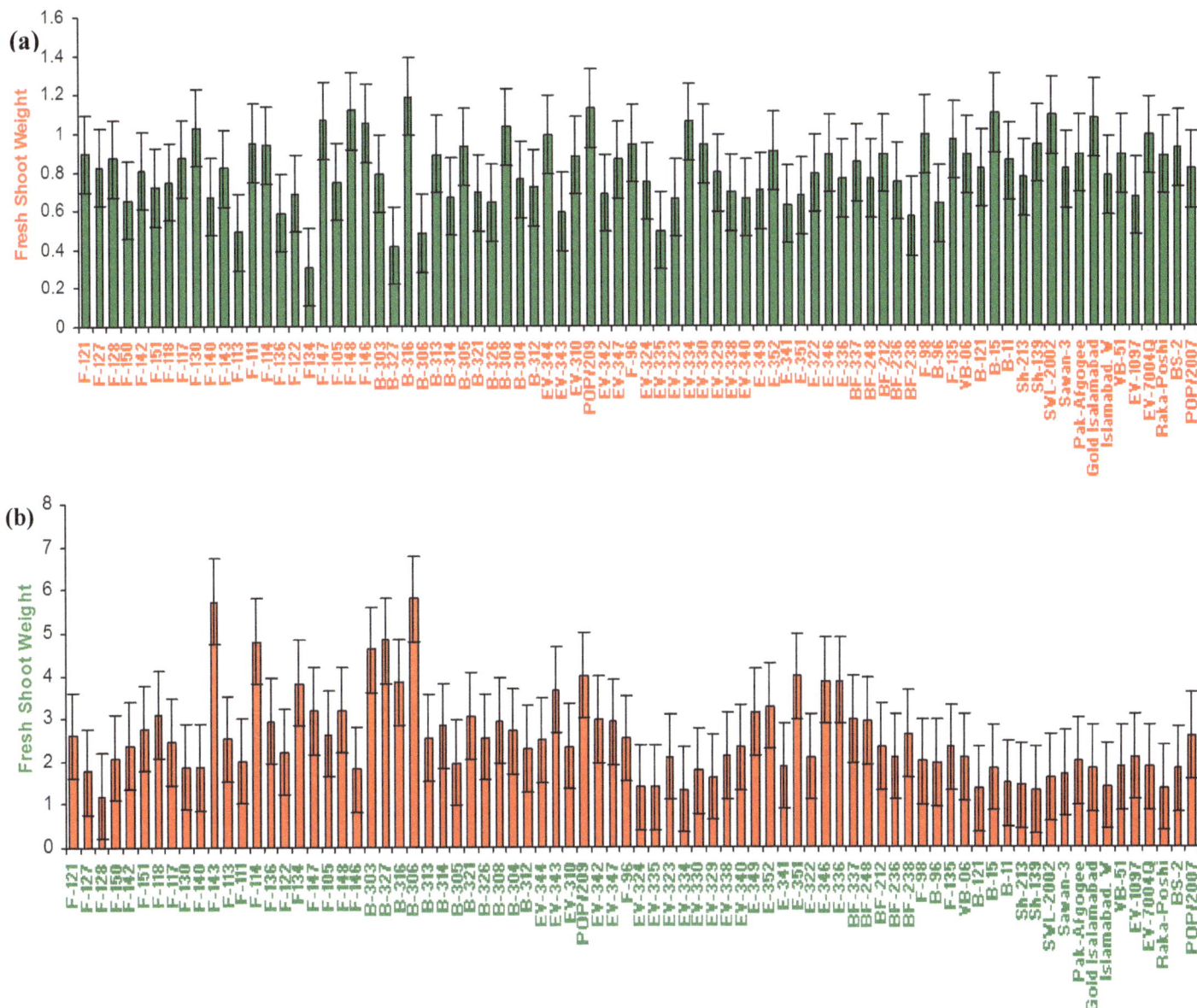

Figure 8. Fresh shoot weight (g) of (a) 14 days old seedlings; (b) 28 days old seedlings.

higher fresh root weight, followed by EV-7004Q and B-316 for 14 and 28 days respectively while lowest fresh root weight was found for F-140 and F-142 followed by B-96 and EV-329 for 14 and 28 days, respectively.

Fresh shoot weight (g)

It persuaded from Table 1a and b that highly significant differences were found among genotypes. The higher mean for fresh shoot weight (2.53±0.051 g) was recorded for 28 days old seedlings of genotypes while lower for 14 days old seedlings. It is indicated from Table 1b that the genotypic variance, phenotypic variance and environmental

variance were 0.029, 0.03 and 0.0002, respectively while genotypic coefficient of variance, phenotypic coefficient of variance (PCV) and environmental coefficient of variance were 21.24, 21.32 and 1.85%, respectively. From Table 1b it is cleared that genotypic, phenotypic and environmental variances were 0.962, 0.965 and 0.0026, respectively while genotypic coefficient of variance, phenotypic coefficient of variance and environmental coefficient of variance were 38.77, 38.82 and 2.00%, respectively. Table 1a and b also indicated that highest estimates of heritability (99.20%, 99.70%) and low and higher values of genetic advance were 78.99 and 54.20% for 14 and 28 days old seedlings respectively. The higher values of heritability and genetic advance indicated that selection for fresh shoot weight of seedling may be effective for selecting higher yielding maize genotypes (Hussain et al., 1995; Akhtar, 2002; Mehdi and Ahsan, 2000a, b; Ali et al., 2011a, b). It is cleared from the graphs (Figure 8) that the genotypes B-316 and B-306 showed higher fresh shoot weight, followed by POP/209 and F-143 for 14 and 28 days respectively while lowest fresh shoot weight was found for B-327 and EV-324 followed by F-134 and F-128 for 14 and 28 days, respectively.

REFERENCES

Ali Q, Ahsan M (2011). Estimation of Variability and correlation analysis for quantitative traits in chickpea (Cicer arietinum L.). IJAVMS 5(2):194-200.

Ali Q, Elahi M, Ahsan M, Tahir MHN, Basra SMA (2011). Genetic evaluation of maize (Zea mays L.) genotypes at seedling stage under moisture stress. IJAVMS 5(2):184-193.

Anonymous (2008-2009). Economic Survey of Pakistan. Govt. of Pakistan, Finance and Economic Affairs Division, Islamabad.

Anonymous (2009-2010). Economic Survey of Pakistan. Govt. of Pakistan, Finance and Economic Affairs Division, Islamabad.

Bureau of Chemistry, U.S., Wiley, Harvey Washington (2010). Composition of maize (Indian corn), including the grain, meal, stalks, pith, fodder, and cobs. University of California Libraries, nrlf_ucb:GLAD-151223559.

Grzesiak MT, Rzepka A, Hura T, Hura K, Skoczowski A (2007). Changes in response to drought stress of triticale and maize genotypes differing in drought tolerance. Photosynth 45:280-287.

Hussain A, Muhammad D, Riaz S, Bhatti MB, Sartaj M (1995). Forage yield potential and quality differences among various sorghum genotypes under rainfed conditions. Sarhad J. Agric. 11(3):1127-1135.

Maurice EH, Robert FB, Darrel SM (1985). Forages. In: The Science of Grassland Agriculture. (4th ed.) Iowa State Univ. Press (Ames), Iowa, USA. pp. 413-421.

Mehdi SS, Ahsan M (2000a). Genetic coefficient of variation, relative expected genetic advance and inter-relationships in maize (Zea mays L.) for green fodder purposes at seedling stage. Pak. J. Bio. Sci. 11:1890-1891.

Mehdi SS, Ahsan M (2000b). Coefficient of variation, inter-relationship and heritability: Estimated for some seedling trails in Maize in recurrent selection cycle. Pak. J. Biol. Sci. 3:181-182.

Moulin S, Baret F, Bruguier N, Bataille C (2003). Assessing the vertical distribution of leaf Chlorophyll content in a maize crop. INRA - Unite Climat. Sol. ET Environ. (CSE) pp. 7803-7929.

Steel RGD, Torrie JH, Dicky DA (1997). Principles and procedures of Statistics. A Biometrical Approach 3rd Ed. McGraw Hill Book Co. Inc. New Yark. pp. 400-428.

Wang BQ, Li ZH, Duan LS, Zhai ZX (2007). Effect of coronatine on photosynthesis parameters and endogenous hormone contents in maize (Zea mays L.) seedling under drought stress. Plant Physiol. Comm. 43:269-272.

7

Uncertainty in seasonal climate projection over Tamil Nadu for 21st century

D. Rajalakshmi, R. Jagannathan and V. Geethalakshmi

Agronomy Agro Climate Research Centre, Tamil Nadu Agricultural University, Coimbatore – 641003, India.

Climate change is likely to impact every sector including agriculture. To understand the impact on agricultural production, future climate change projections are imperative, but these are uncertain. Quantifying uncertainties in the projection of future climate has been identified as critical research need in impact studies. So, a study was carried out at Agro Climate Research Centre, Tamil Nadu Agricultural University, Coimbatore, Tamil Nadu to quantify the uncertainty in seasonal climate under A1B scenario and the results suggested that solar radiation, wind speed and relative humidity had either no consistent increase or decrease in the PRECIS ensembles and RegCM4 regional climate models studied. Maximum temperature and minimum temperature had definite increase adding confidence to the range predicted. The information about rainfall was consistent for North East Monsoon (NEM), which showed an increasing trend.

Key words: Uncertainty, seasonal climate, A1B scenario, Tamil Nadu.

INTRODUCTION

Climate change becomes a major concern in countries where food production is a fundamental component of its economy. This is because climate change and agriculture are interrelated and they have significant effect on crop production and food security (Gahukar, 2009). Despite technical advances, weather and climate plays a key role in agricultural productivity. Hence, it is necessary for us to empathize the changing climate over a period of time to feed the growing population.

The future climate is uncertain and impossible to predict (Schenk and Lensink, 2007). To overcome such uncertainties, scenarios are used for future climate projections by employing different global and regional climate models around the world. These models have different types of uncertainties viz., unpredictability, structural and value uncertainty (IPCC, 2005). The Global Climate Models (GCMs) have coarse resolutions (Giorgi and Mearns, 1991), andhence Regional Climate Models (RCMs) with finer resolutions (IPCC, 2007) are employed

for regional climate studies. Future climate projections, used for decision making carry uncertainty. These uncertainties are characterized normally into two viz., aleatoric and epistemic uncertainties. Aleatoric uncertainty arises from randomness in computational predictions, which are irreducible. However, in many cases, the dominant uncertainties arise from lack of knowledge (particularly lack of knowledge of physical model parameters and imperfections in the mathematical models themselves). These are epistemic uncertainties, which can in principle be reduced (Oden et al., 2010).

A better understanding of the range of possible future change may be derived from estimates of full range of their possible outcomes (Webster and Sokolov, 2000). This range of possible outcomes can be obtained using ensemble technique, which demonstrated a significant success in climate simulations during the last decade (Broccoli et al., 2003; Murphy et al., 2004; Stainforth et al., 2005; Yoshimori et al., 2005). By using these climate

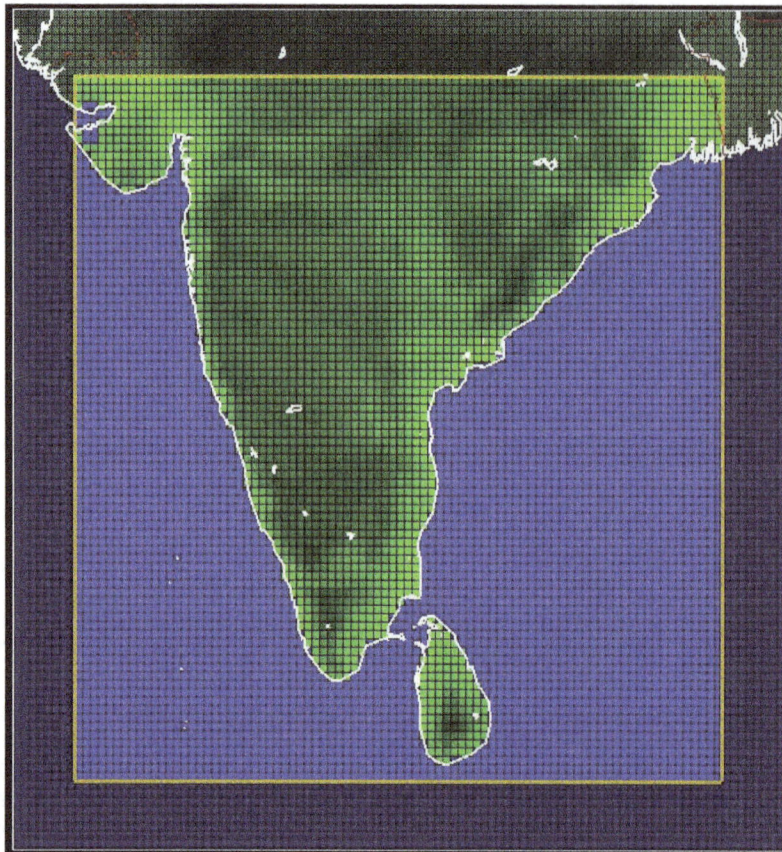

Figure 1. PRECIS domain.

uncertainty over a region can be predicted/obtained for further impact studies on different fields of science viz., agriculture and also to help stakeholders and policy makers to take decisions on adaptation and mitigation strategies in the face of many uncertainties about the future. In the present study seasonal climate projection uncertainty over Tamil Nadu State was assessed by using Providing Regional Climate for Impact Studies (PRECIS) and Regional Climate Model Version 4.0 (RegCM4) models.

MATERIALS AND METHODS

Regional climate models

The study employed two Regional Climate Models (RCMs), one was from the UK Met Office Hadley Centre's PRECIS and another one (RegCM4) was from the Abdus Salam International Centre for Theoretical Physics (ICTP), Italy.

PRECIS

The UK Met Office Hadley centre has provided boundary data for four simulations from a 17-member Perturbed Physics Ensemble (PPE), produced using HadCM3 under Quantifying Uncertainties in

Model Projections (QUMP) project. The PPE members designated as HadCM3Q0 to Q16 in which Q0 has the standard parameter setting while the remaining 16 were simultaneously perturbed with 29 of the atmospheric component parameters. These PPEs were formulated systematically to sample parameter uncertainties under the A1B scenario. The Hadley Centre after preliminary evaluation made a sub-selection (McSweeney and Jones, 2010) of four of its simulations viz., Q0, Q1, Q3 and Q16 and provided the Lateral Boundary Conditions (LBCs) to TNAU for running with PRECIS 1.9.2. The climate simulations were made for 129 years from 1970 to 2098 leaving the year 1970 for spin-up.

Open Software und System-Entwicklung (SuSE) was used as Linux operating system for the model as recommended by the Hadley centre (Wilson et al., 2008). The boundary data used in this study were:

HadCM3 Q0, Q1, Q3, Q16: A1B: Ensembles of SRES (Special Report on Emissions Scenarios) A1B scenario experiment (1970 - 2099).

PRECIS domain selection

The PRECIS model used rotated latitude map projection and was run with 0.22 × 0.22 degree resolution (~25 km). Domain selected covered 101 grids on EW direction and 104 grids on NS direction encompassing southern and central part of India. The extent of the boundary for the domain was 3.25 to 22.71°N latitude and 69.56°E to 89.81°E longitude (Figure 1).

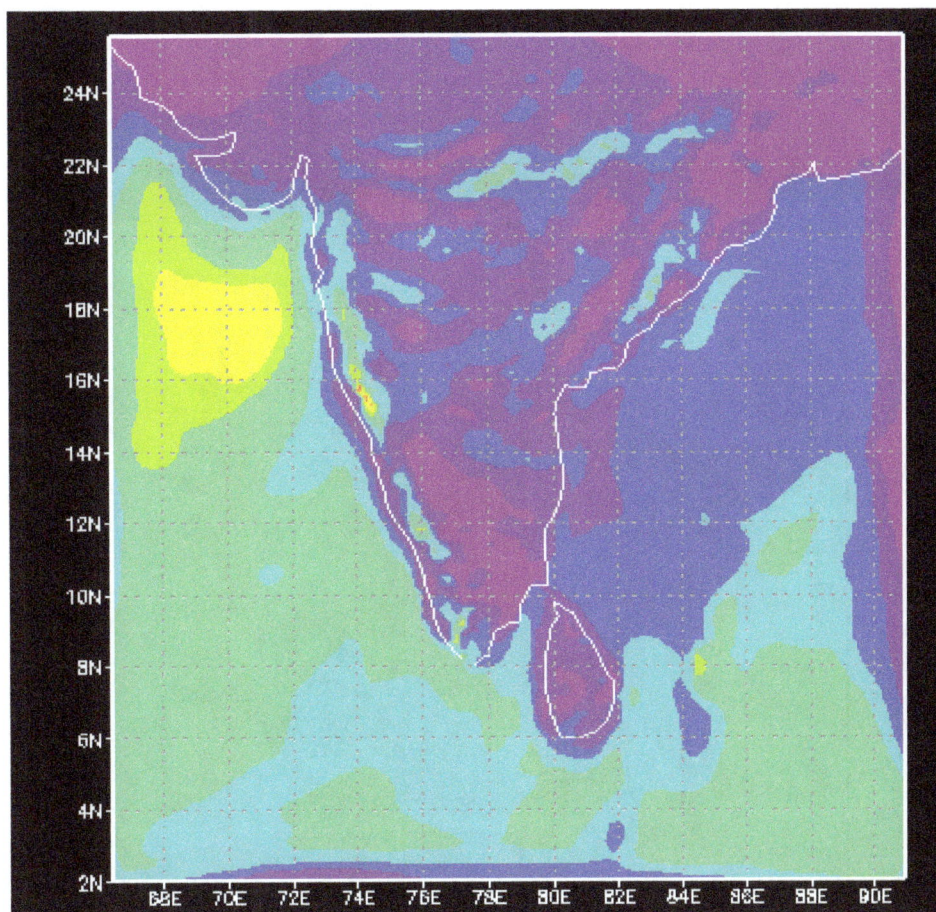

Figure 2. RegCM4 domain.

RegCM4

Regional Climate Model Version 4 released in 2010 (Elguindi et al., 2010) was used in the study. It is an open source RCM that can be used for climate simulation over different areas of interest. The 20th Century A1B ensemble was prepared from ECHAM5-MPIOM model of Max Plank Institute of Meteorology, Germany (hosted by ICTP as EH5OM) and its boundaries were downloaded and used to drive RegCM4 model. Simulations were also done with future climate using the same model's A1B scenario. The climate run started with 20th century boundaries from 1970 and continued upto 2100 for a total of 130 years.

RegCM4 domain selection

A domain covering most of the India was selected. The extent of the boundary for the domain was 2.00°N to 25.61°N latitude and 66.45°E to 90.9628°E longitude. The domain used was depicted in Figure 2. This covered 112 EW and 112 NS points. The horizontal resolution was 25 km.

Parameters retrieved

Both models generated large numbers of weather parameter as output from the simulations. However, only the solar radiation

(MJm^{-2}), maximum temperature (°C), minimum temperature (°C), rainfall (mm), wind speed (kmph) and relative humidity (%) were retrieved as these parameters were normally used for impact studies in agriculture.

Study area

The study area covered the State of Tamil Nadu in southern Peninsular India lies between 7.91°N to 13.65°N latitude and 76.17°E to 80.82°E longitude, an agriculturally predominant region and its climate favors majority of the crops cultivation. This covers 220 grid points in PRECIS and 218 grid points in RegCM4 models (Figure 3).

Study period

PRECIS was run for 128 years from 1971 to 2098 and RegCM4 was run for 130 years from 1971-2100. Data were retrieved to analyze the climate change in Tamil Nadu and as well for studying its impact.

Uncertainty

Uncertainty in seasonal climate projections over Tamil Nadu was

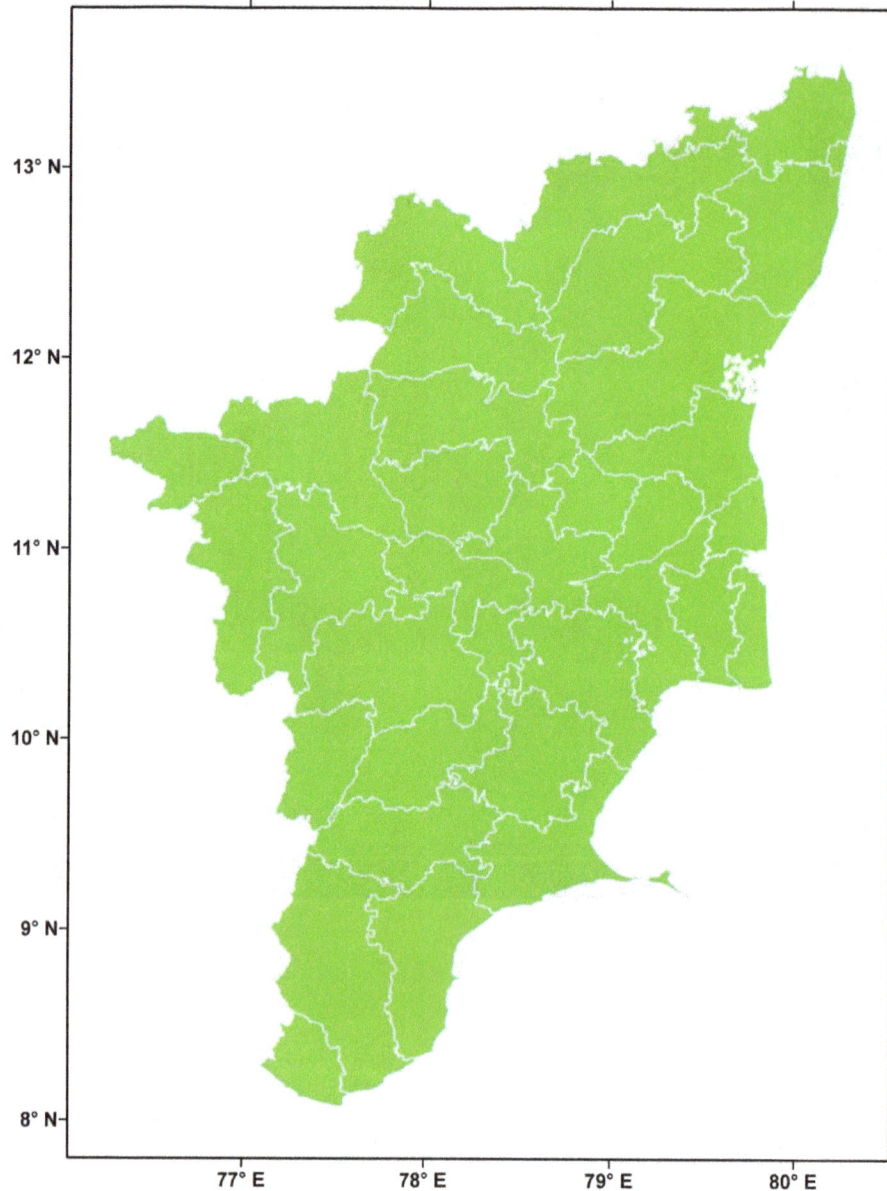

Figure 3. Tamilnadu.

carried out by converting daily data to seasonal output viz., winter, summer, south west monsoon and north east monsoon seasons for PRECIS ensembles and RegCM4 models by using perl programme. To find out the increase or decrease range at the end of the century, difference between 2091-2100 decade and base years (1971-2000) are considered.

RESULTS AND DISCUSSION

The RCM outputs of both the model were analyzed on season basis for all the six parameters to understand the uncertainty in these projections. Difference in seasonal averages during the end of the century over the base

period was arrived. These seasonal values were compared and the ranges of uncertainty arising out of these model comparisons are presented in this paper.

Uncertainty in intra season climate projections

Winter season

The period of two months, January and February is termed as winter season over Tamil Nadu. The Maximum temperature projected by the RCMs ranged from 2.54 to 3.96°C. In this Q16 (3.96°C) had the highest warming

Table 1. Uncertainty in winter climate projections at the end of 21st century over Tamilnadu.

Parameter/models	Q0	Q1	Q3	Q16	RegCM4
Solar radiation (MJm^{-2})	0.18	-0.02	-0.02	0.04	0.38
Maximum temperature (°C)	3.80	2.54	3.77	3.96	3.62
Minimum temperature (°C)	4.40	3.40	4.15	4.63	3.85
Rainfall (mm)	-6.95	-4.38	1.34	-29.22	-7.45
Wind speed (kmph)	0.06	0.06	-0.15	0.21	-0.57
Relative humidity (%)	-1.74	-0.66	-1.83	-2.02	0.70

Table 2. Uncertainty in summer climate projections at the end of 21st century over Tamilnadu.

Parameter/models	Q0	Q1	Q3	Q16	RegCM4
Solar radiation (MJm^{-2})	-0.17	-0.10	0.04	-0.49	0.41
Maximum temperature (°C)	3.62	2.50	3.96	3.98	3.58
Minimum temperature (°C)	4.34	3.07	4.33	5.04	3.86
Rainfall (mm)	3.97	-7.02	-27.11	-24.15	-30.14
Wind speed (kmph)	0.26	0.20	0.34	0.28	-0.58
Relative humidity (%)	-1.63	-0.87	-4.59	-0.73	-1.11

followed by Q0, Q3, RegCM4 and Q1 (2.54°C) respectively. Minimum temperature ranged from 3.4 to 4.63 while Q16 (4.63°C) had the highest value followed by Q0, Q3, RegCM4 and Q1 (3.40°C) respectively. The increase in range of minimum temperature was higher than the maximum temperature in the projections, similar to the projections as reported by Ramaraj et al. (2009) (Table 1).

Rainfall was predicted to decrease in RegCM4 and in all the PRECIS ensembles except Q3, which had an increment of 1.34 mm. Q16 had the highest decrement (-29.22 mm) followed by RegCM4, Q0 and Q1 (-4.38 mm). Solar radiation had declining trend in Q1 and Q3 but showed increasing trend in Q0, Q16 and RegCM4. RegCM4 had the highest increase of 0.38 MJm^{-2} followed by Q0 and Q16 (0.04 MJm^{-2}). The ensembles Q1 and Q3 had a decrement of (-)0.02 MJm^{-2}. The QUMP ensembles of PRECIS with varying initial conditions of Q1 and Q3 had same values showcasing some certainty of occurrence. Wind speed was predicted to increase for Q16 (0.21 Kmph) followed by Q0 and Q1 (0.06 Kmph). The ensemble Q3 and RegCM4 predicted a decline of (-)0.15 Kmph and (-)0.57 Kmph, respectively. Relative humidity was predicted to decrease by all the ensembles of PRECIS, where as RegCM4 predicted an increment of 0.70%. The ensemble Q16 (-2.02) had the highest decrement followed by Q3, Q0 and Q1 (0.66), respectively.

Summer season

The period of three months, March, April and May is termed

as summer season over Tamil Nadu. The Maximum temperature projected by the RCMs ranged from 2.5 to 3.98°C. In this Q16 (3.98°C) had the highest warming followed by Q3, Q0, RegCM4 and Q1, respectively. Minimum temperature ranged from 3.07 to 5.04°C while Q16 (5.04°C) had the highest value followed by Q0, Q3, RegCM4 and Q1, respectively. The range of increase in minimum temperature was higher than maximum temperature in all the ensembles. These ranges of temperature increase were in accordance with the findings of Geethalakshmi et al. (2011) for Cauvery Delta Zone and Lakshmanan et al. (2011) for Bhavani basin of Tamil Nadu (Table 2).

Rainfall was predicted to decrease in all the ensembles except Q0, which had an increment of 3.97 mm. RegCM4 had the highest decrement (-30.14 mm) followed by Q3, Q16 and Q1 (-7.02). Solar radiation had declining trend in Q0 and Q1and Q16 projection and had increasing trend in Q3 and RegCM4. RegCM4 had the highest increase of 0.41 MJm^{-2} followed by Q3 (0.04 MJm^{-2}). The ensembles Q16, Q1, Q0 had a decrement of -0.49, -0.17 and -0.10 MJm^{-2}, respectively. Wind speed was predicted to increase in Q3 (0.34 Kmph) followed by Q16, Q0 and Q1 (0.20). RegCM4 predicted a decline of (-)0.58 Kmph. Relative humidity was predicted to decrease in all the ensembles, were Q3 (-4.59) had the highest decrement followed by Q0, RegCM4, Q1 and Q16, respectively.

South West Monsoon (SWM)

The period of four months from June to September is termed as south west monsoon season over Tamil Nadu.

Table 3. Uncertainty in SWM climate projections at the end of 21st century over Tamilnadu.

Parameter/models	Q0	Q1	Q3	Q16	RegCM4
Solar radiation (MJm^{-2})	-0.06	-1.38	0.30	-1.20	-0.63
Maximum temperature (°C)	3.38	1.78	3.43	3.28	3.04
Minimum temperature (°C)	3.38	2.42	3.30	3.99	3.29
Rainfall (mm)	-18.59	-21.22	-33.77	28.94	113.12
Wind speed (kmph)	0.18	0.19	0.10	-0.26	0.17
Relative humidity (%)	-2.55	-0.59	-2.99	1.11	-0.42

The maximum temperature as projected by the RCMs found ranged from 1.78 to 3.43°C. In this Q3 (3.43°C) had the highest warming followed by Q0, Q16, RegCM4 and Q1 (1.78°C), respectively. Minimum temperature did range from 2.42 to 3.99 while Q16 (3.99°C) had the highest value followed by Q0, Q3, RegCM4 and Q1, respectively. The range of increase in minimum temperature was higher than maximum temperature in all the ensembles (Table 3). Rainfall was predicted to decrease in Q0, Q1 and Q3, while Q16 and RegCM4 predicted an increment in rainfall, which was similar to the findings of Rupa et al. (2006) for Indian region under A2 and B2 scenarios.

Solar radiation had shown declining trend in all the ensembles except Q3, which had an increment of 0.04 MJm^{-2}. The decrement was highest in Q1 (-1.38 MJm^{-2}) followed by Q16, RegCM4 and Q0 (-0.06 MJm^{-2}), respectively. Wind speed was predicted to increase in Q1 (0.19 Kmph) followed by Q0, RegCM4 and Q3 while Q16 predicted a decline of (-)0.26 Kmph. Relative humidity was predicted to get decrease by all the ensembles except Q16 (1.11). The Q3 (-2.99) had highest decrement followed by Q0, Q1 and RegCM4, respectively.

North East Monsoon (NEM)

The period of three months from October to December is termed as north east monsoon season over Tamil Nadu. The maximum temperature projected by the RCMs found ranged from 2.63 to 3.84°C. In this Q0 (3.84°C) had the highest warming followed by Q3 (3.17°C), Q16 (2.92°C), Q1 (2.80°C) and RegCM4 (2.63°C). Minimum temperature did range from 2.0 to 4.33°C while Q16 (4.33°C) had the highest value followed by Q3, Q0, RegCM4 and Q1 (2.0°C), respectively. The range of increase in minimum temperature was higher than the values of maximum temperature in all the ensembles (Table 4).

Invariably rainfall was predicted to increase in all the ensembles. RegCM4 (51.85 mm) had the highest increment followed by Q3, Q16, Q0 and Q1, respectively. This might be due to the intensification of rainfall in the

Indian region during the monsoon season, as a consequence of the anticipated increase in the greenhouse gas concentrations as reported by May (2002). Solar radiation had shown declining trend in Q3 (-0.09 MJm^{-2}), Q16 (-0.90 MJm^{-2}), RegCM4 (-1.22 MJm^{-2}), which might be due to solar dimming caused by increased aerosol concentrations and greenhouse gases as observed by Singh et al. (2009). Similar results were obtained by Rajalakshmi et al. (2012) for decadal mean projection over Cauvery Delta Zone and increasing trend in Q0 (0.90 MJm^{-2}) and Q1 (0.50 MJm^{-2}).

Wind speed was predicted to increase in Q0 (0.19 Kmph). Decrease in wind speed was predicted by Q16 (-0.18 Kmph) followed by Q3, Q1 and RegCM4 (-0.01), respectively. Relative humidity was projected to decrease in Q0 (-2.57), Q1 (-0.63) and Q3 (-0.10) while the same was projected to increase in Q16 (1.91) and RegCM4 (1.69).

Uncertainty in inter season climate projections

Overall comparison of all the four seasons revealed that maximum temperature over the seasons ranged from 3.43 to 3.98°C, in which summer season the highest temperature value of 3.98°C had followed by winter (3.96°C), NEM (3.84°C) and SWM (3.43°C) (Figure 4). In respect of minimum temperature, the variation was same as that of maximum temperature with highest minimum temperature projected during summer was 5.04°C followed by winter (4.63°C), NEM (4.33°C) and SWM (3.99°C) (Figure 5). Similar findings were also reported by Wiltshire et al. (2010) in which they analyzed AR4 multi-model ensemble as well as the RCM ensemble and found that there is confidence in a trend towards increasing temperature under the SRES A1B scenario. The increase in temperature as projected by the models would affect the crop production over Tamil Nadu. This was evident from the studies of Aggarwal and Mall (2002) for rice that a 2°C increase resulted in decrease in grain yield of rice. Another study by Lin et al. (2005) over China revealed that increase in temperature without carbon dioxide fertilization could reduce the rice, maize and wheat yields by up to 37% in the next 20 to 80 years.

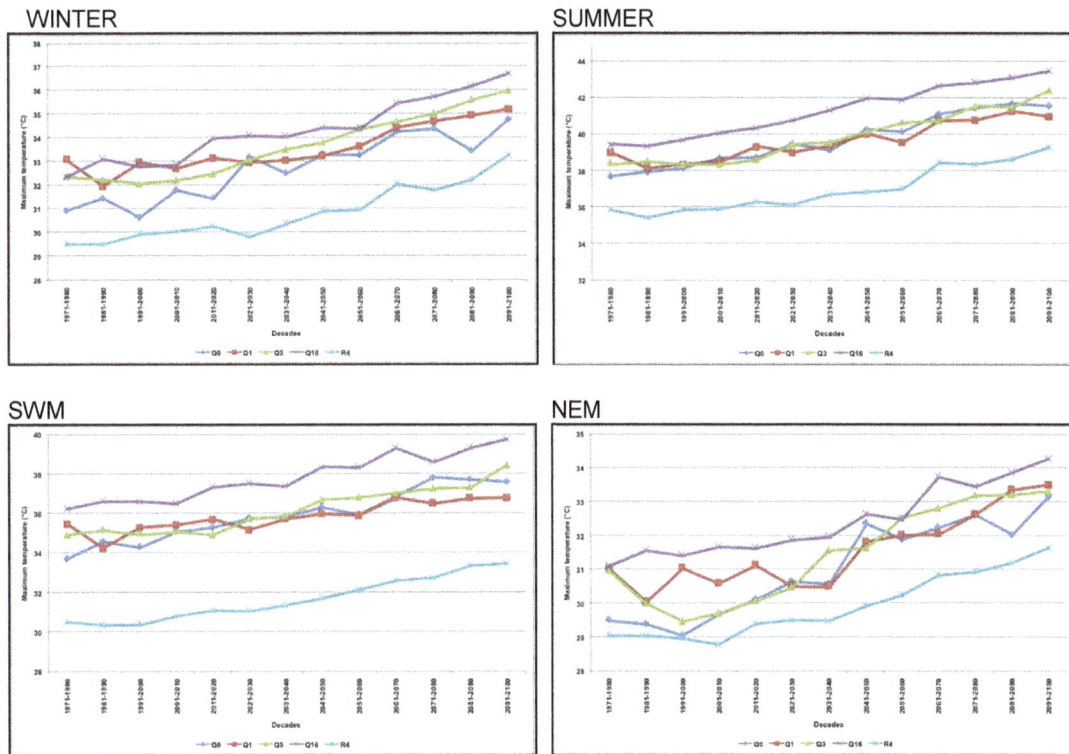

Figure 4. Maximum temperature projections at different time scales over Tamilnadu.

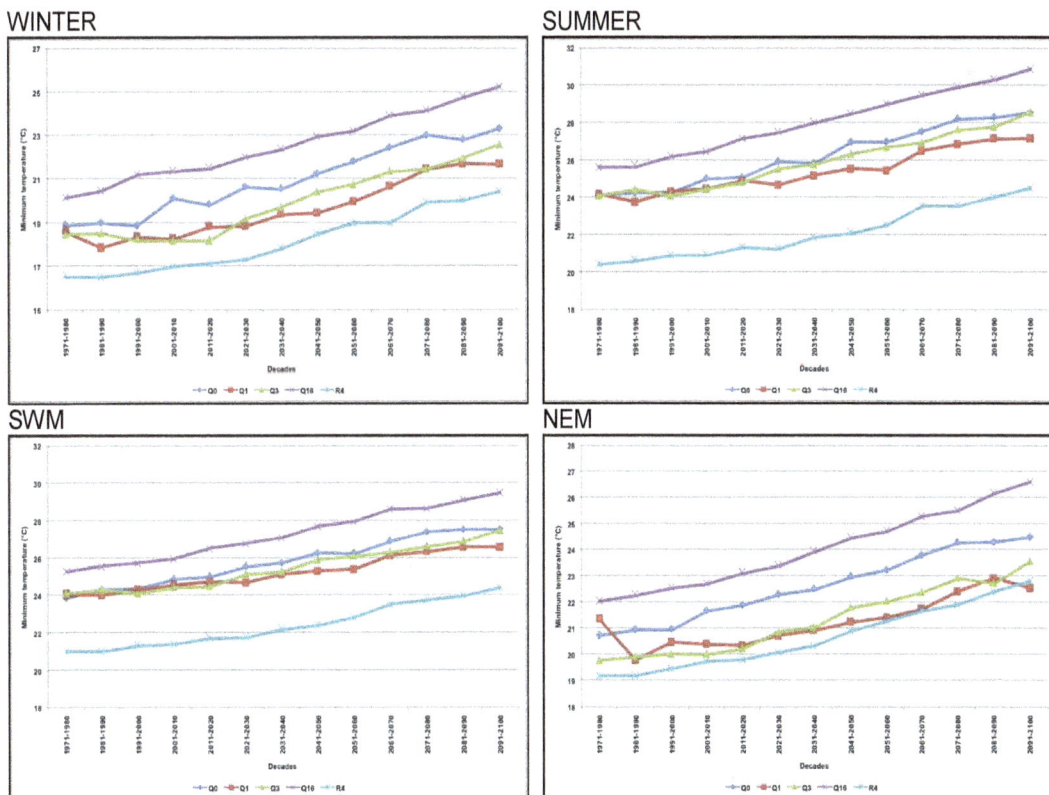

Figure 5. Minimum temperature projections at different time scales over Tamilnadu.

WINTER

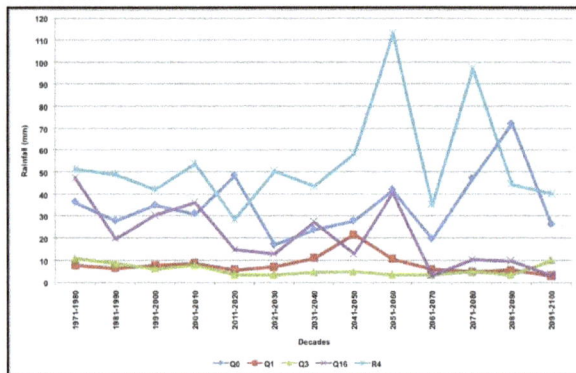

SUMMER

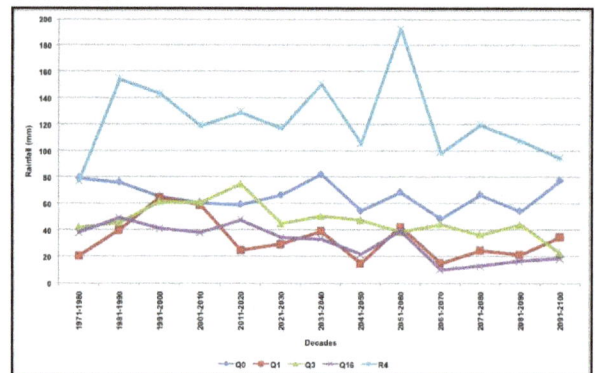

SWM

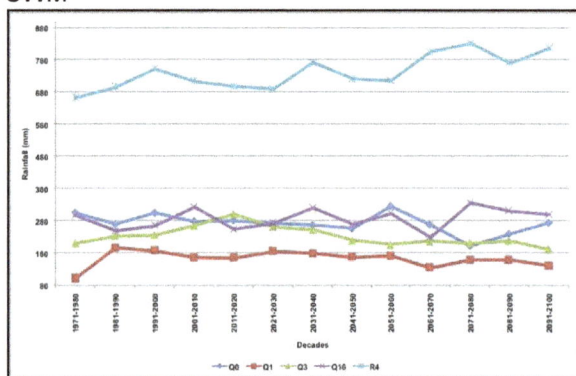

NEM

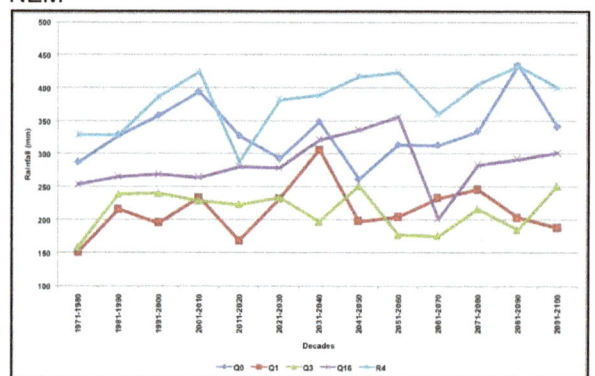

Figure 6. Rainfall projections at different time scales over Tamilnadu.

Rainfall (Figure 6) was projected to increase in few members of the ensemble, whereas it was not seen in other members of the ensemble for the same season. There was no consistency for providing information in respect of summer, winter and SWM rainfall projected by these ensembles. Interestingly NEM was projected to increase (1.2 to 51.85 mm) by all the ensembles studied. For precipitation the uncertainty is much larger, with significant natural variability. These findings are consistent with those published elsewhere (Akhtar et al., 2010). Improved rainfall projections represent a key bottleneck to reduce uncertainties in projections for impact studies. In crop growing season, rainfall is able to explain two-thirds of the variation in crop production. With a change in growing season precipitation, as much as a 10% change in production was reported by Lobell and Burke (2008).

Solar radiation (Figure 7) ranged between -0.02 to 0.38 MJm^{-2} for winter, -0.49 to 0.41 MJm^{-2} for summer, -1.38 to 0.30 MJm^{-2} for SWM and -1.22 to 0.90 MJm^{-2} for NEM. Wind speed (Figure 8) ranged between -0.57 to 0.21 Kmph for winter, -0.58 to 0.34 Kmph for summer, -0.26 to 0.19 Kmph for SWM and 0.18 to 0.19 Kmph for NEM. Relative humidity (Figure 9) ranged between -2.02 to 0.70% for winter, -4.59 to -0.87% for summer, -2.99

to 1.11% in SWM and -2.57 to 1.91% in NEM. Hence, the result of seasonal climate projections for maximum temperature, minimum temperature, rainfall, solar radiation, wind speed and relative humidity can be further used for impact studies on agriculture, for planning adaptation and mitigation strategies to sustain the agriculture production over the study area.

Conclusion

Solar radiation, wind speed and relative humidity had no consistent increase or decrease in the PRECIS ensembles and RegCM4 projections studied. The projection indicated that maximum temperature and minimum temperature did show definite increase adding confidence to the range predicted. The information about rainfall is consistent only for NEM. It can be concluded from the study that uncertainty in climate projections can be sorted out to some extent by using PRECIS ensembles and RegCM4 models. These ranges of uncertainty have to be taken into account while framing adaption strategies, since seasonal climate plays vital role in most of the food crops production.

WINTER

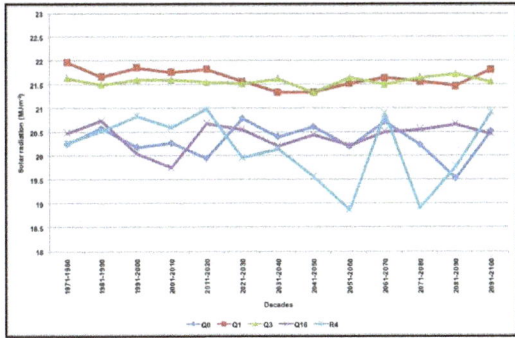

SUMMER

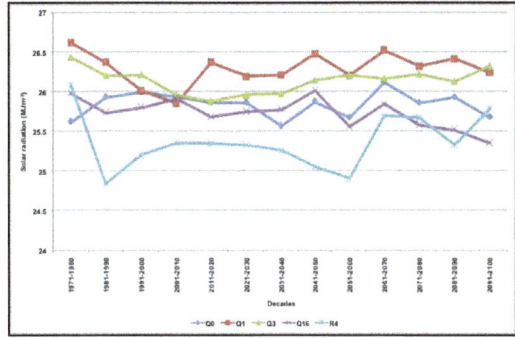

SWM

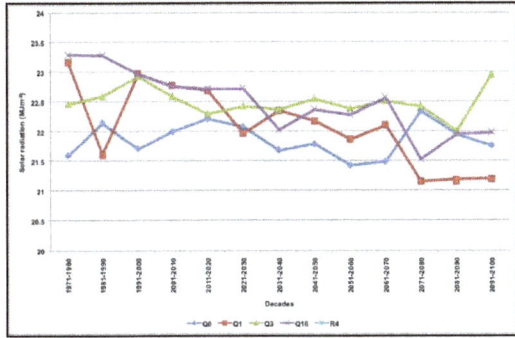

NEM

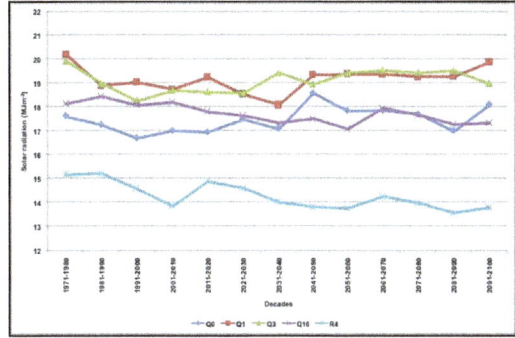

Figure 7. Solar radiation projections at different time scales over Tamilnadu.

WINTER

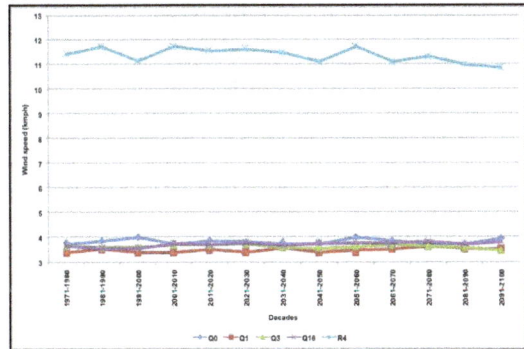

SUMMER

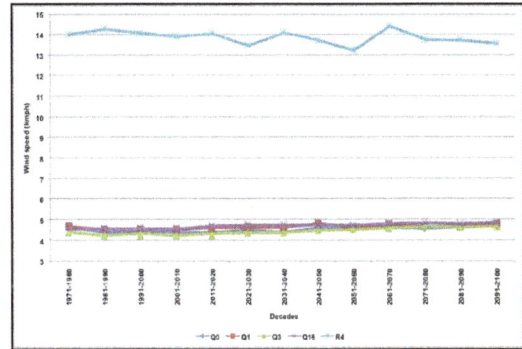

SWM

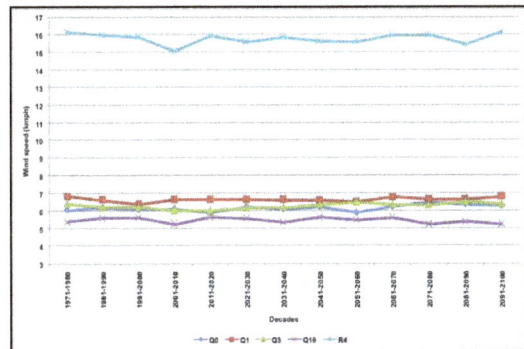

NEM

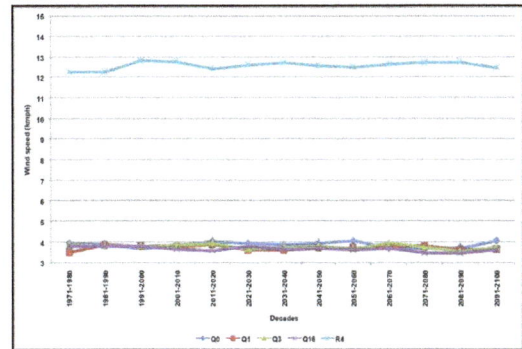

Figure 8. Wind speed projections at different time scales over Tamilnadu.

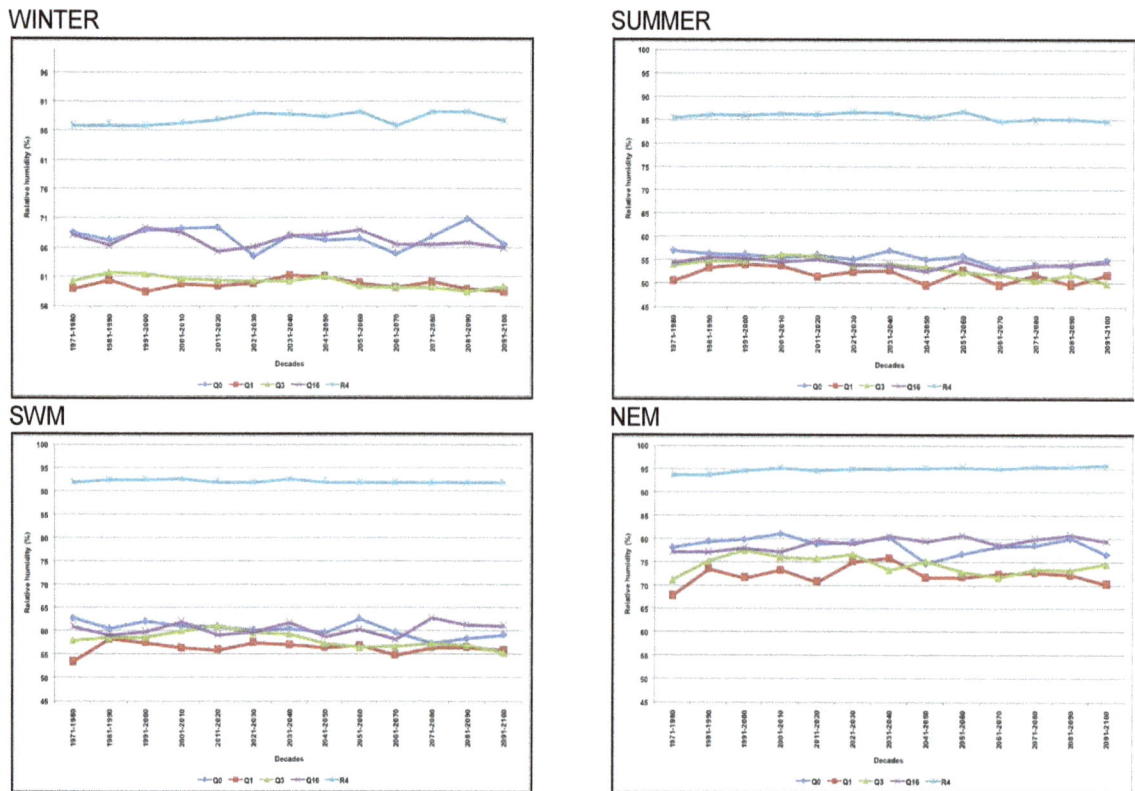

Figure 9. Relative humidity projections at different time scales over Tamilnadu.

ACKNOWLEDGEMENTS

Authors are thankful to Hadley Centre, UK Met office for providing boundary conditions for the study and the Ministry of Foreign Affairs, Norway and the Royal Norwegian Embassy, New Delhi for financial support to undertake the study through ClimaRice project.

REFERENCES

Aggarwal PK, Mall RK (2002). Climate change and rice yields in diverse agro-environments of India. II. Effect of uncertainties in scenarios and crop models on impact assessment. Clim. Change 52(3):331-343.

Akhtar M, Ahmad N, Booij MJ (2010). Use of regional climate model simulations as input for hydrological models for the Hindukush-Karakorum-Himalaya region, Hydrol. Earth Syst. Sci. 13:1075-1089.

Broccoli AJ, Dixon KW, Delworth TL, Knutson TR, Stouffer RJ (2003). Twentieth-century temperature and precipitation trends in ensemble climate simulations including natural and anthropogenic forcing, J. Geophys. Res. 108(D24):4798-4811.

Elguindi N, Bi X, Giorgi F, Nagarajan B, Pal J, Solmon F, Rauscher S, Zakey A (2010). RegCM Version 4.0, User's Guide, ICTP, Trieste, Italy.

Gahukar RT (2009). Food security: The challenges of climate change and bioenergy. Cur. Sci. 96(1):26-28.

Geethalakshmi V, Lakshmanan A, Rajalakshmi D, Jagannathan R, Gummidi Sridhar, Ramaraj AP, Bhuvaneswari K, Gurusamy L, Anbhazhagan R (2011). Climate change impact assessment and adaptation strategies to sustain rice production in Cauvery basin of Tamil Nadu. Cur. Sci. 101(3):1-6.

Giorgi F, Mearns LO (1991). Approaches to the simulation of regional climate change: a review. Rev. Geophys. 29:191-216.

IPCC (2005). Guidance Notes for Lead Authors of the IPCC Fourth Assessment Report on Addressing Uncertainties. www.ipcc.ch.

IPCC (2007). Climate Change 2007: Impacts, Adaptation and Vulnerability. Contribution of Working Group II to the Fourth Assessment Report of the Intergovernmental Panel on Climate Change, M.L. Parry, O.F. Canziani, J.P. Palutikof, P.J. Van der Linden and C.E. Hanson, Eds., Cambridge University Press, Cambridge, UK, pp. 976.

Lakshmanan A, Geethalakshmi V, Rajalakshmi D, Bhuvaneswari K, Srinivasan R, Sridhar G, Sekhar NU, Annamalai H (2011). Climate change adaptation strategies in the Bhavani basin using the SWAT model. Appl. Eng. Agric. 27(6):887-893.

Lin E, Xiong W, Ju H, Xu Y, Li Y, Bai L, Xie L (2005). Climate change impacts on crop yield and quality with CO_2 fertilization in China. Phil. Trans. R. Soc. B. 360:2149-2154.

Lobell BD, Burke BM (2008). Why are agricultural impacts of climatechange so uncertain? The importance oftemperature relative to precipitation. Environ. Res. Lett. 3:1-8.

May W (2002). Simulated changes of the Indian summer monsoon under enhanced greenhouse gas conditions in a global time-slice experiment. Geophys. Res. Lett. 29(7):22.1-22.4.

McSweeney C, Jones R (2010). Selecting members of the 'QUMP' perturbed-physics ensemble for use with PRECIS. Met Office Hadley Centre, UK. P. 9.

Murphy JM, Sexton DM, Barnett DN, Jones GS, Webb MJ, Collins M, Stainforth DA (2004). Quantification of modelling uncertainties in a large ensemble of climate change simulations. Nature 430(7001):768-772.

Oden T, Robert M, Omar G (2010). Computer Predictions with Quantified Uncertainty. ICES REPORT10-39. pp. 2-3.

Rajalakshmi D, Jagannathan R, Geethalakshmi V, Ramaraj AP (2012).

Rice yield response to the projected climate change over Cauvery Delta Zone.Journal of Agrometeorology (Special issue), 14:443-448.

Ramaraj AP, Jagannathan R, Dheebakaran GA (2009). Assessing Predictability of PRECIS Regional Climate Model for Downscaling of Climate Change Scenarios, ISPRS Archives XXXVIII-8/W3 Workshop Proceedings: Impact of Climate Change on Agriculture held during 17-18 December 2009, Space Application Centre, Ahmedabad, India. pp. 80-85.

Rupa KK, Sahai AK, Krishna Kumar K, Patwardhan SK, Mishra PK, Revadekar JV, Kamala K, Pant GB (2006). High-resolution climate change scenarios for India for the 21st century. Cur. Sci. 90:334-345.

Schenk N, Lensink S (2007). Communicating uncertainty inthe IPCC's greenhouse gas emissions scenarios. Clim. Change 82:293-308.

Singh J, Bimal KB, Manoj K (2009). Long term trend analysis of surface insolation and evaporation over selected climate types in India.ISPRS Archives XXXVIII-8/W3 Workshop Proceedings: Impact of Climate Change on Agriculture held during 17-18 December 2009, Space Application Centre, Ahmedabad, India. pp. 366-370.

Stainforth DA, Aina T, Christensen C, Collins M, Faull N, Frame DJ, Kettleborough JA, Knight S, Martin A, Murphy JM, Piani C, Sexton D, Smith LA, Spicer RA, Thorpe AJ, Allen MR (2005). Uncertainty in Predictions of the Climate. Responses to Rising Levels of Greenhouse Gases. Nature 433:403-406.

Webster DM, Anrei PS (2000). A methodology for quantifying uncertainty in climate projections. Clim. Change 46:417-446.

Wilson S, Hassell D, Hein D, Jones R, Taylor R (2008). Installing and using the Hadley Centre regional climate modelling system, PRECIS. Version 1(7):11-154.

Wiltshire A, Camilla M, Jeff R, Claire W, Carol M, Pankaj K, Daniela J (2010). Technical report on analysis of uncertainty. High Noon Project No 227087.

Yoshimori M, Stocker TF, Raible C, Renold M (2005). Externally Forced and Internal Variability in Ensemble Climate Simulations of the Maunder Minimum. J. Clim. 18:4253-4270.

8

Reengineering agricultural technology education based on constructivism, engineering design and systems thinking in farming systems of Khuzestan province, Iran

Ahmad Reza Ommani and Azadeh N. Noorivandi

Department of Agricultural Management, Shoushtar Branch, Islamic Azad University, Shoushtar, Iran.

The main objective of research was reengineering agricultural technology education based on constructivism, engineering design and systems thinking in farming systems of Khouzestan province, Iran. The method of research was correlative descriptive and causal relation. A random sample of agricultural extension educators of Khouzestan province, Iran (n=105) were selected for participation in the study. Based on the constructivism the top six ranked items in favorable conditions were: (1) Emphasizing discourse and collaboration; (2) Ability to communicate ideas; (3) Developing self thought and reliance; (4) Discovering knowledge through group work; (5) Programs place a high value on field work, and (6) Using problem solving methods. Based on engineering design the top six ranked items in favorable conditions were: (1) Communication and analytical skills; (2) Interpersonal skills: teamwork, group skills, attitude, and work ethic; (3) Problem-solving and creative thinking; (4) Formative evaluation; (5) Contributions from both social and natural science in the educational process, and (6) The ability to negotiate and influence, and self-management. Also based on systems thinking, the top six ranked items in favorable conditions were: (1) Focus on the needs of farmers; (2) Negotiating assessment processes; (3) Considering farm as system; (4) Need for a more holistic understanding; (5) Helping learners confront personal beliefs and create their own theories of learning, and (6) Providing learners with opportunities to examine, analyze, and reflect on their own thinking.

Key words: Agricultural technology education, constructivism, engineering design, systems thinking.

INTRODUCTION

A national employer survey identified desired job skills needed in today's workforce. Today's jobs require a portfolio of skills in addition to academic and technical skills. These include communication skills, analytical skills, problem-solving and creative thinking, inter-personal skills, the ability to negotiate and influence, and self-management (Kelley and Kellam, 2009). Dearing and Daugherty (2004) conducted a study to identify the core engineering-related concepts by surveying 123 professionals in technology education, and engineering education. The top five ranked concepts were:

1. Interpersonal skills: teamwork, group skills, attitude, and work ethic,
2. Ability to communicate ideas: verbally, physically, and visually,
3. Ability to work within constraints/ parameters,
4. Experience in brainstorming and generating ideas,

5. Product design assessment: Does a design perform its intended function?

Groves (2008) pointed out, using a constructivist approach in the education where the educator encourages learner to discover knowledge through group work, inquiry, and experimentation. Educator does not simply answer the questions though, he or she encourages the learner to think beyond the question and find the answer through self thought and dialogue with others. In the constructivist approach, the instructor's objective is also to prepare the information for the level of current understanding of the learner. In fact the learner is of utmost importance. In constructivism, the learner guides the lesson and the educator analyses and realizes the individuality of each learner and so tailors their lesson to best reach them.

Crawford (2001) wrote, there are five key elements to actively engaging learner in a constructivist approach to teaching. These five basic elements are:

(1) Relating: Learning in the context of one's life experiences or preexisting knowledge;
(2) Experiencing: Learning by doing, or through exploration, discovery, and invention;
(3) Applying: Learning by putting the concepts to use;
(4) Cooperating: Learning in the context of sharing, responding, and communicating with others
(5) Transferring: Using knowledge in a new context or novel situation; one that has not been covered in class.

Kelley and Kellam (2009) have attempted to provide a philosophical framework for technology education that embraces new philosophies of learning and thinking (constructivism, engineering design, and systems thinking). If technology educators determine that their purpose is to help prepare farmers to live and work in this global society, then these educators should consider carefully defining a philosophical framework upon which to build a new curriculum. The authors wish for technology educators to consider the proposed framework as a foundation for technology education as it has much promise in preparing farmers to function in today's technological society.

Bawden (1991) wrote the systemic paradigm calls for us to rethink our views of our world. If this rethinking is to lead to the sort of innovative and regenerative processes leading to large-scale improvements in the quality of relationships between people and their environments, it must come from a belief that new ways are crucial to produce new knowledge. As agricultural scientists, we must be prepared to question critically our beliefs about what we really think constitutes improvements to agriculture. We must also be prepared to enter into debates about what should be as well as creating visions about what could be. Our focus must extend beyond what is effective and efficient to embrace the ethical. We must be prepared to state what we think is good and what we

think is bad, and we certainly must be ready to discuss what is aesthetically acceptable and what is not.

The key elements of systems thinking in farming systems include a holistic approach, orientation towards the needs of defined target groups, high levels of farmer participation and hence co-learning by farmers and specialists. It is now widely acknowledged that the farming systems research approach has made significant contributions to the improvement of agricultural research and education systems throughout the World (Collinson, 2000).

Based on different researchers (Kelley and Kellam, 2009; Dearing and Daugherty, 2004; Crawford, 2001; Bawden, 1991), the main factors that affected on technology education, is expressed in Figure 1.

MATERIALS AND METHODS

The method of research was correlative descriptive and causal relation. A random sample of agricultural extension educators of Khouzestan province, Iran (n=105) were selected for participation in the study. A questionnaire was developed to gather information regarding reengineering in agricultural technology education.

Questions were generated from the literature review. The survey was divided into two sections to gather data on personal characteristics of extension educators and the degree of current and favorable regarding agricultural technology education based on constructivism, engineering design and systems thinking from extension educators' perspective. Responses for 2nd section were categorized using a Likert-type scale from point 1 to 5 representing very low important to very high important respectively. Content and face validity were established by a panel of experts from faculty members. Questionnaire reliability was estimated by calculating Cronbach's alpha. Reliability was Cronbach's alpha=0.85. Data collected were analyzed using the Statistical Package for the Social Sciences (SPSS). Appropriate statistical procedures for description (frequencies, percent, means, and standard deviations) were used.

RESULTS AND DISCUSSION

Reengineering agricultural technology education based on constructivism

Based on agricultural extension educators' perception, current and favorable conditions regarding agricultural technology education based on constructivism were analyzed. Based on the results that were explained in Table 1, current condition regarding all items of technology education based on constructivism is not favorable. Prioritization is based on the coefficient of variation (CV). The coefficient of variation is less about any subject; this subject is a higher priority. The top six ranked items in favorable conditions were: (1) Emphasizing discourse and collaboration (M=4.67, Sd=0.89), (2) Ability to communicate ideas (M=3.98, Sd=0.78); (3) Developing self thought and reliance (M=4.42, Sd=0.89); (4) Discover knowledge through group work (M=4.56, Sd=0.96); (5) Programs place a high value on field work (M=4.71, Sd=1.06), and (6)

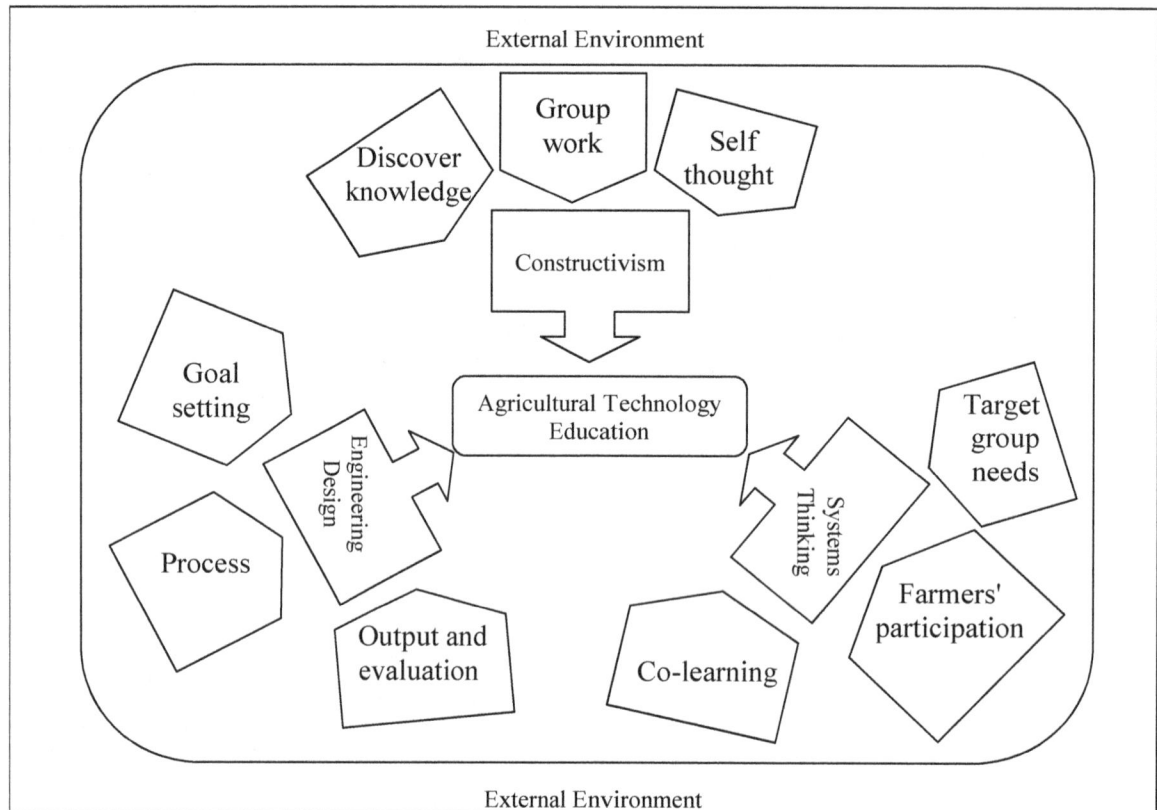

Figure 1. Conceptual framework of research.

Table 1. Reengineering agricultural technology education based on constructivism.

Items	Current condition		Favorable condition*		CV[a] of FC[b]	Priority
	Mean	Sd	Mean	Sd		
Emphasizes discourse and collaboration	2.65	1.05	4.67	0.89	0.19	1
Ability to communicate ideas	2.46	1.03	3.98	0.78	0.20	2
Developing self thought and reliance	2.65	0.91	4.42	0.89	0.20	3
Discover knowledge through group work	2.34	0.94	4.56	0.96	0.21	4
Programs place a high value on field work	2.88	0.99	4.71	1.06	0.23	5
Using problem solving methods	2.54	1.09	4.67	1.09	0.23	6
Learner Is of utmost importance	2.65	0.90	3.98	0.90	0.23	7
Role of teacher is a guide, facilitator and co-explorer	2.81	0.98	4.23	0.99	0.23	8
Developing teamwork practices	2.23	0.98	4.09	0.98	0.24	9
Self-Direction, self-monitoring, self-assessment to engage learners on a personal level	2.55	0.93	4.44	1.07	0.24	10
Learners can learn how to understand other's opinions	2.76	0.91	4.06	0.98	0.24	11
Information seeking by farmers	2.21	0.92	4.02	0.98	0.24	12
Promote learner-centered instruction	2.34	1.09	4.06	0.99	0.24	13
Programs include opportunities for reflection about the various discussions, and experiences	3.08	0.96	4.10	1.03	0.25	14
Goal of education Is HRD	2.76	1.06	3.89	1.01	0.26	15
Using visual techniques for education	2.45	1.08	4.09	1.08	0.26	16
Using action research in technology education	2.12	0.93	4.09	1.09	0.27	17

5, Very high important; 1, very low important; *must be considered; [a]*Coefficient* of variation; [b]Favorable condition.

Table 2. Causal comparative between current and favorable conditions in agricultural technology education based on constructivism by Wilcoxon signed ranks.

Items	Z	Sig
Developing teamwork practices	10.100	0.000**
Ability to communicate ideas	10.247	0.000**
Using problem solving methods	11.314	0.000**
Developing self thought and reliance	9.450	0.000**
Information seeking by farmers	10.154	0.000**
Discover knowledge through group work	11.091	0.000**
Learner is of utmost importance	9.892	0.000**
Using visual techniques for education	10.128	0.000**
Goal of education is HRD	9.098	0.000**
Promote learner-centered instruction	10.540	0.000**
Emphasizes Discourse and Collaboration	10.671	0.000**
Programs place a high value on field work	11.092	0.000**
Using action research in technology education	11.231	0.000**
Learners can learn how to understand other's opinions	10.923	0.000**
Role of teacher is a guide, facilitator, and co-explorer	10.491	0.000**
Self-direction, self-monitoring, self-assessment to engage learners on a personal level	10.991	0.000**
Programs include opportunities for reflection about the various discussions, and experiences.	3.890	0.020*

*P≤0.05, **P≤0.01.

Table 3. Causal comparative between overall items of constructivism education in current and favorable conditions by Wilcoxon signed ranks.

Item	Z	Sig
Overall items of constructivism education	14.100	0.000**

*P≤0.05, **P≤0.01.

Using problem solving methods (M=4.67, Sd=1.09).

In inferential analysis, Wilcoxon signed ranks were used for analyzing causal comparative of educational constructivism elements in agriculture between current and favorable conditions. Based on the results of Tables 2 and 3 in each and overall items (Z=14.100, P=0.000) there were significant differences between current and favorable conditions. Thus, there must be providing condition to development constructivism education. It is a learning or meaning-making theory which offers an explanation of the nature of knowledge and how human beings learn.

Knowledge is obtained by participating in content instead of imitation or repetition. Learning activities in constructivist settings are characterized by active engagement, inquiry, problem solving, and collaboration with others. The role of educator is a guide, facilitator, and co-explorer who encourages learners to question, challenge, and formulate their own ideas, opinions, and conclusions rather than a distributor of knowledge (Flynn,2006; Dangel and Guyton, 2004).

Reengineering agricultural technology education based on systems thinking

Based on agricultural extension educators' perception, current and favorable conditions regarding agricultural technology education based on systems thinking were analyzed. Based on the results that were explained in Table 4, also current condition regarding all items of technology education based on systems thinking is not favorable. The top six ranked items in favorable conditions were: (1) Focus on the needs of farmers (M=4.92, Sd=0.84); (2) Negotiating assessment processes (M=4.89, Sd=0.98); (3) Considering farm as system (M=4.69, Sd=0.90); (4) Need for a more holistic understanding (M=4.50, Sd=0.93); (5) Helping learners confront personal beliefs and create their own theories of learning (M=4.94, Sd=1.09), and (6) Providing learners with opportunities to examine, analyze, and reflect on their own thinking (M=3.94, Sd=0.96).

In inferential analysis, Wilcoxon signed ranks was used for analyzing causal comparative of systems thinking

Table 4. Reengineering Agricultural Technology Education Based on Systems Thinking.

Items	Current condition		Favorable condition*		CV of FC	Priority
	Mean	Sd	Mean	Sd		
Focus on the needs of farmers	2.89	1.10	4.92	0.84	0.17	1
Negotiating assessment processes	2.03	1.09	4.89	0.93	0.19	2
Considering farm as system	3.89	0.77	4.69	0.90	0.19	3
Need for a more holistic understanding	3.08	0.84	4.50	0.93	0.21	4
Helping learners confront personal beliefs and create their own theories of learning	3.10	0.91	4.94	1.09	0.22	5
\Providing learners with opportunities to examine, analyze, and reflect on their own thinking	2.94	0.83	3.94	0.96	0.24	6
Encouraging learners to self-assess, learning from their successes and mistakes	3.43	0.67	4.08	1.01	0.25	7
Emphasis on continuous evaluation	2.95	0.77	3.89	1.03	0.26	8
Participatory learning between farmers	2.09	0.97	4.02	1.09	0.27	9
Co-learning by farmers and specialists	3.08	1.11	3.90	1.30	0.33	10
Orientation towards the needs of defined target groups	2.79	1.02	4.09	1.83	0.45	11
High levels of farmer participation	2.99	0.98	4.37	1.97	0.45	12

5, Very high important; 1, very low important; *, must be considered.

Table 5. Causal comparative between current and favorable conditions in agricultural technology education based on systems thinking by Wilcoxon signed ranks.

Item	Z	Sig
Orientation towards the needs of defined target groups	11.902	0.000**
High levels of farmer participation	10.095	0.000**
Co-learning by farmers and specialists	4.901	0.018*
Emphasis on continuous evaluation	8.092	0.000**
Participatory learning between farmers	9.893	0.000**
Negotiating assessment processes	10.430	0.000**
Focus on the needs of farmers	9.682	0.000**
Considering farm as system	8.092	0.000**
Need for a more holistic understanding	5.904	0.021*
Providing learners with opportunities to examine, analyze, and reflect on their own thinking	10.093	0.000**
Helping learners confront personal beliefs and create their own theories of learning	11.009	0.000**
Encouraging learners to self-assess, learning from their successes and mistakes	4.676	0.014*

*$P \leq 0.05$, **$P \leq 0.01$.

elements in agriculture between current and favorable conditions. Based on the results in each (Table 5) and overall (Table 6) items (Z=12.159, P=0.000) there were significant differences between current and favorable conditions. The key elements of farming systems thinking include a holistic approach, orientation towards the needs of defined target groups, high levels of farmer participation and hence co-learning by farmers and specialists (Petheram and Clark, 1998). Different researches such as Collinson (2000) stated that the farming systems approach has made significant contributions to the improvement of agricultural research and education systems throughout the World.

Reengineering agricultural technology education based on engineering design

Also current and favorable conditions regarding agricultural technology education based on engineering design were analyzed. Based on the results that were explained in Table 7 current condition regarding all items

Table 6. Causal comparative between overall items of systems thinking in current and favorable conditions by wilcoxon signed ranks.

Item	Z	Sig
Overall items of constructivism education	12.159	0.000**

*P≤0.05, **P≤0.01.

Table 7. Reengineering agricultural technology education based on engineering design.

Items	Current condition		Favorable condition*		CV of FC	Priority
	Mean	Sd	Mean	Sd		
Communication and analytical skills	2.98	0.84	4.49	0.68	0.15	1
Interpersonal skills: teamwork, group skills, attitude, and work ethic	3.01	0.97	3.99	0.76	0.19	2
Problem-solving and creative thinking	3.08	1.09	4.91	0.96	0.20	3
Formative evaluation	3.34	0.82	4.69	1.03	0.22	4
Contributions from both social and natural science in the educational process	2.49	0.91	4.20	0.93	0.22	5
The ability to negotiate and influence, and self-management	3.11	1.11	4.95	1.14	0.23	6
Ability to communicate ideas: verbally, physically, and visually	2.92	1.09	4.03	0.94	0.23	7
Product design assessment	2.09	0.93	3.92	0.92	0.23	8
Summative evaluation	3.01	0.93	4.29	1.12	0.26	9
Participation of farmers in goal setting	3.02	1.04	3.98	1.09	0.27	10
Using engineering participatory methods in need assessment	3.04	1.01	4.01	1.12	0.28	11
Experience in brainstorming and generating ideas	2.95	0.99	3.49	1.03	0.30	12

5, Very high important; 1, very low important; *, must be considered.

of technology education based on engineering design is not favorable. The top six ranked items in favorable conditions were: (1) Communication and analytical skills (M=4.49, Sd=0.68); (2) Interpersonal skills: Teamwork, group skills, attitude, and work ethic (M=3.99, Sd=0.76); (3) Problem-solving and creative thinking (M=4.91, Sd=0.96); (4) Formative evaluation (M=4.69, Sd=1.03); (5) Contributions from both social and natural science in the educational process (M=4.20, Sd=0.93), and (6) The ability to negotiate and influence, and self-management (M=4.95, Sd=1.14).

In inferential analysis, Wilcoxon signed ranks were used for analyzing causal comparative of engineering design elements in agriculture between current and favorable conditions. Based on the results in each (Table 8) and overall (Table 9) items (Z=12.159, P=0.000), there were significant differences between current and favorable conditions. Multiple researchers such as Kelley and Kellam (2009) and Daugherty (2005) pointed out, engineering design for addressing the standards for technological literacy, creating a new model that attracts and motivates learner from all literacy levels. Today's job forces require a set of skills. These include communication skills, analytical skills, problem-solving and creative thinking, interpersonal skills, the ability to

negotiate and influence, and self-management (Kelley and Kellam, 2009).

CONCLUSIONS AND RECOMMENDATION

The key elements which need to be incorporated into a new approach to technology education in farming systems are: Constructivism (constructivist design is a professional model that responds to the call for "new philosophy model of agricultural education" that emphasize learner centered strategies), systemic thinking (the need for a more holistic understanding of the context of farming and rural livelihoods), engineering design (the active participation and partnership of farmers and other key stakeholders in the process of design, planning, implementing, monitoring and evaluating research and communication skills, analytical skills, problem-solving and creative thinking, interpersonal skills, the ability to negotiate and influence, and self-management). By incorporating these principles, we can provide appropriate condition for technology education and diffusion. In this article, the author has attempted to reengineer technology education based on constructivism, engineering design, and systems thinking in farming

Table 8. Causal comparative between current and favorable conditions in agricultural technology education based on engineering design by Wilcoxon signed ranks.

Item	Z	Sig
Using engineering participatory methods in need assessment	10.443	0.000**
Participation of farmers in goal setting	10.434	0.000**
Experience in brainstorming and generating ideas	3.455	0.035*
Product design assessment	12.453	0.000**
communication and analytical skills	10.554	0.000**
problem-solving and creative thinking	8.094	0.000**
The ability to negotiate and influence, and self-management	9.763	0.000**
Formative evaluation	10.881	0.000**
Summative evaluation	10.554	0.021*
Contributions from both social and natural science in the educational process	8.986	0.000**
Interpersonal skills: teamwork, group skills, attitude, and work ethic	9.776	0.000**
Ability to communicate ideas: verbally, physically, and visually	10.566	0.000**

*$P \leq 0.05$, **$P \leq 0.01$.

Table 9. Causal comparative between overall items of engineering design in current and favorable conditions by Wilcoxon signed ranks.

Item	Z	Sig
Overall items of constructivism education	13.901	0.000**

*$P \leq 0.05$, **$P \leq 0.01$.

systems and provide a philosophical framework for technology education that holds true to adult education approaches that are at the heart of the success of technology education in farming systems.

Based on the results, current condition regarding all items of technology education based on constructivism is not favorable. The top six ranked items in favorable conditions were: (1) Emphasizing discourse and collaboration; (2) Ability to communicate ideas; (3) Developing self thought and reliance; (4) Discovering knowledge through group work; (5) Programs place a high value on field work, and 6) Using problem solving methods.

Also current condition regarding all items of technology education based on systems thinking is not favorable. The top six ranked items in favorable conditions were: (1) Focusing on the needs of farmers; (2) Negotiating assessment processes; (3) Considering farm as system; (4) Need for a more holistic understanding; (5) Helping learners confront personal beliefs and create their own theories of learning, and (6) Providing learners with opportunities to examine, analyze, and reflect on their own thinking.

In addition, current condition regarding all items of technology education based on engineering design is not favorable. The top six ranked items in favorable conditions were: (1) Communication and analytical skills;

(2) Interpersonal skills: Teamwork, group skills, attitude, and work ethic; (3) Problem-solving and creative thinking; (4) Formative evaluation; (5) Contributions from both social and natural science in the educational process, and (6) The ability to negotiate and influence, and self-management.

REFERENCES

Bawden RJ (1991). Systems Thinking and Practice in Agriculture. J. Dairy Sci. 74:2362-2373.

Collinson M (2000). A History of Farming Systems Research. FAO/IFSA/CABI.

Crawford ML (2001). Teaching contextually: Research, rationale, and techniques for improving student motivation and achievement in Mathematics and Science. Waco, TX: CCI Publishing, Inc.

Dangel JR, Guyton E (2004). An Emerging Picture of Constructivist Teacher Education. Constructivist 15(1):1-35.

Daugherty MK (2005). A changing role for technology teacher education. Journal of Industrial Technology Education, 42(1):41-58.

Dearing BM, Daugherty MK (2004). Delivering engineering content in technology education. Technol. Teach. 64(3):8-11.

Flynn P (2006). The Constructivist Design Model for Professional Development. Institute for Learning Centered Education.

Groves M (2008). The Constructivist Approach in Adult Education. California State University.

Kelley T, Kellam N (2009). A Theoretical Framework to Guide the Re-Engineering of Technology Education. J. Technol. Educ. 20(2):37-49.

Petheram RJ, Clark RA (1998). Farming Systems Research: Relevance to Australia. Aust. J. Exp. Agric. 38:101-115.

9

Can design improve the agricultural work conditions in the developing countries? A study based in the development of an animal traction seeder

Eduardo Romeiro Filho and Aline Capanema Barros

Industrial Engineering Departament, Universidade Federal de Minas Gerais, Integrated Laboratory of Design and Product Engineering, Escola de Engenharia, bloco 1, sala 3104, Av. Presidente Antônio Carlos, 6627 - Campus Pampulha, 31270-901 - Belo Horizonte – MG, Brazil.

This paper presents a study conducted as a part of development of an animal traction seeder machine, using principles related to disciplines such as usability, Ergonomics Work Assessment (EWA) and Antropotechnology, in a perspective of Design for Sustainability (DfS) and product development suitable for the Base of the Pyramid (BoP). The study provided information for a series of design changes to be introduced in the equipment, in order to improvement of working conditions in small family farming properties. From this approach analysis on an existing implement, are suggested ways to adaquate the product and presented a proposal for a new animal traction seeder machine.

Key words: Design for sustainability, base of the pyramid, ergonomics work assessment.

INTRODUCTION

Human labour is still the main source of energy used in agricultural work in developing countries (Jafry and O'Neill, 2000). It is also responsible for approximately half of the cultivated area in the world (Ramaswamy, 1994). In developing countries like Brazil and others in Latin America, Africa and Asia, there is a need to create sustainable ways of development and income generation for the "Bottom of the Pyramid" (Prahalad, 2009) people. Employ improvement in rural areas, through the incentive to family farming on small farms, is a much used alternative and represents a potential increase in food production.

In Brazil, family farming produces 87% of manioc, 70% of beans, 46% of corn, 38% of coffe, 34% of rice and 21% of wheat (IBGE, 2006). However, it is important to highlight the specific features of this kind of farming, such as the restriction to investment on equipment and limited

access to formal education that often undermines the use of more sophisticated technologies. In small farms like these, agricultural machinery with low cost and technological adequacy that makes it easy to be operated by the farmer is essential. In these cases, machinery moved by animal traction is a good alternative to the one using machanical traction, since it uses renewable resources, the implementation costs are usually low and it does not need a sophisticated technical system. In the mid-1990's draft animals saved the equivalent to US$ 6 billion in fossil fuel with more than 300 million animals used (Wilson, 2003).

In Mexico alone, over 3,765,000 animals are used in agriculture (Ortiz-Laurel and Rössel, 2007). The tools and implements for animal traction available in the Brazilian Market, however, are characterized mainly by outdated technological solutions and design (Araújo et al., 1999), as is the case of planters (Figure 1).

The Brazilian industry of agricultural machinary is of great importance to the country and to its export potential. Nevertheless, it still lacks a system of systematized product development (Romano et al., 2005) and provides an opportunity for the application of new design tools and approaches that take into account the need for new solutions that fit the characteristics of small producers, ensuring standards of efficiency and financial return to the investment made in an appropriate period of time. Furthermore, the focus should be on the user of the equipment. Therefore, aspects of safety, comfort and efficiency should always be considered, using principles of ergonomics and industrial design.

From the foregoing, this paper aims to present a case of development of a fertilizer seeder powered by animal traction from an approach that considers the principles of usability, Ergonomics and Anthropotechnology, with views to suitability for agriculture sustainable and its application in small properties, characteristics of the Base of the Pyramid (Prahalad and Hart, 2002).

This study was developed based on design of a "chassis to keep implements", built in Brazil by IAC - Agronomic Institute of Campinas, and described by Peche Filho et al. (1987) with the objective of providing to the small farmer a model of low cost animal traction equipment (Figure 2). The target proposal was a modular product: starting from a common base (the chassis), a variety of tools could be coupled to meet diverse demands that emerge from different methods and phases of farming. In this way, solutions have been developed for plows, cultivators, fertilizer and planter (which were the basis for this study). The technical system used as a basis for the design of the new drill seeder was also developed by the IAC (Figure 3).

The choice for the planter comes from the fact that it presents a series of problems related to use, such as postural needs resulting from the weight of the equipment and the need for balance during movement (the deposits modify the center of gravity of the machine). Add to that the constant flow of information related to the conducting of the animal, the ground conditions, the speed and direction of the trajectory as well as the operation of the input-output system of seeds and fertilizer. Finally, the fact that they are no technically updated, low cost and suitable for small farmers' equipment available in Brazil was crucial for the choice.

Demand for better product-user interfaces

Although the term UCD has greater application in the field of software engineering, its principles can be applied to any device or product, in studies of human-computer interface (Nielsen, 1993). This is done by recognizing the importance of users, their needs, capabilities and limitations, and the contexts in which they will relate to the product. It is also important to bear in mind that UCD represents not only techniques, methods, processes and procedures to design products and "usable" systems, but mainly the philosophy that puts the user at the center of the design process (Rubin, 1994). In this aspect, Ergonomics and Usability can be considered important concepts in a vision of UCD. Adler and Winograd (1992) define usability as the ability of a product or device to take advantage of the skills of its users, working effectively in a given range of real work situations, going accordingly to the principles adopted in ergonomics. Although they can not be regarded as similar disciplines both can be incorporated into a design perspective that brings the user as a central concern in a real situation activity with the product.

In this study, factors related to anthropotechnology are also considered relevant (Wisner, 1985, 1997; Geslin, 2004), once they evaluate the impacts resulting from the transfer of technology between different regions, either by its geographical features, economic, social or cultural. In the case of agricultural labour this aspect becomes even more important in view of the differences in education, tradition, conditions of use and technical knowledge among farmers. When developing a technical solution (or set of solutions) one should take these factors into account, together with an "Anthropotechnological" approach.

Nowadays, among the various fields dedicated to improving the product-user interface, usability is perhaps one of the most widely used that can provide results in this sense. Nielsen (1993) states that this approach is possible for any object, product, system, or service used by humans which have potential problems in their use and that they should be subjected to some form of "Usability Engineering". Despite the fact that in the literature review there were more references about studies related to the product-user interface for software development, studies devoted to equipment, durable and capital capital, such as equipment for power transmission (Costa, 2006), CNC milling machines (Shinno, 2002), electric screwdrivers (Freund et al., 2000) and medical equipment (Rose et al., 2005; Carrol et al., 2002; Garmer et al., 2004; Liu and Osvalder, 2004), were also collected. Data from literature suggest that despite these equipment operators receive training, the application of methods to improve the usability of these products is appropriate and beneficial to its design and to its users.

In this case, the application of principles of Ergonomics in developing technical solutions is especially recommended, either in a human-centered approach (Dreyfuss, 2001), to the physiological aspects of labour (Kroemer and Gradjean, 2005; Iida, 2005) or to human factors (Nemeth, 2004). Methodologies of participatory nature are also mentioned as useful in developing appropriate solutions to the rural environment (Kogi, 2006), especially in industrially underdeveloped countries (Jafry and O'Neill, 2000). In addition, Ergonomic Work

Figure 1. Example planter produced by John Deere in the 1920s (above), very similar to those currently available on the market (below). Source: author's file and http://www.marchesan.com.br

Assessement (EWA) (Guérin et al., 2001; Wisner, 1987) can provide interesting opportunities and contributions to the development of agricultural equipment (Cerf and Sagory, 2004). While the farmers are (in the situation assessed) the owners of the means of production, the forms of work organization (in particular those related to the "optimal time" for planting) are influenced by factors beyond their control in a much more evident way from that of the work performed in factories, for example.

Moreover, it is natural for technology (as well as organizational structures) from certain countries not to be easily adaptable to others, as stated by Wisner (1985, 1987) in his studies related to anthropotechnology. Shahnavaz (1991) includes in this assessment an approach that is more closely linked to human factors and to impacts of technology transfer between countries of different levels of development.

Several experiments demonstrate that when a technology designed for a certain reality is transferred to another, in a different context, it must undergo significant changes in order to adapt itself to the conditions peculiar to the region it is taken to and to its people. Each population has its own culture and traditions, different levels of formal education, technological expertise and production methods. Therefore, each one requires unique technical solutions, developed for its own reality. This problem is even more severe in agriculture, where climatic, geographic and cultivars influence directly the adequacy (or not) of imported technologies. Within a single country there may be large regional diferences, including different ethnic groups, which is the case of Brazil that should be considered in the design, planning and implementation of technologies. The issue, in these cases, lies in investigating the real situation of rural

Figure 2. Prototype of the "chassis to keep implements" in field studies. (Source: author's file).

Figure 3. Prototype of a mechanical traction seeder. Both developed by IAC - Agronomic Institute of Campinas. (Source: author's file).

Figure 4. Prototype of the planter built by IAC, coupled to the chassis to keep implements. (Source: author's file).

labour including variables not normally foreseen by the designers, such as using situations in bad conditions or maintenance restrictions, common situation in developing countries. Thus, a user-oriented approach aims to contribute to the research of real situations of equipment use, providing new elements to the project team.

Animal traction

Animal traction was adopted as a parameter in this study because it is of great importance to the development of agriculture, especially in small and medium farms in pioneer or with unfavorable topography regions (Pereira et al., 2010). This indicates that there is great potential for using this form of energy, which is a segment that lacks catering from implemente factories, more focused on the development of equipment suitable for use in large properties.

Despite the economical and technical constraints to the use of animal power, it presents significant advantages which deserve to be addressed: it is an abundant source of renewable, decentralized and mobile energy which

does not depend on inputs (such as fuel) or imported equipment that entail external dependency. The investment cost is low when compared to alternative technologies, such as mechanical tractor which is not accessible to most smallholder farmers in different regions of the world. Furthermore, it lends itself to be used in sloping areas where mechanization is not appropriate. Also the employ generation is much larger in comparison to the moto-mechanization. It is an important factor to be considered where there is large availability of skilled workforce in need to generate income.

Finally, the use of animal traction as an energy source in establishments served only by human strength is undoubtedly a substantial technological progress and a large gain in productivity. Moreover, there is an undeniable improvement in the working conditions of the farmer from the use of this type of traction, since much of the physical effort is transferred to the animal.

RESEARCH METHODOLOGY

The first phase of the study consisted of free (non-systematic) observation, which is a step of the EWA method (Guérin et al., 2001; Béguin and Daniellou, 2004). These observations showed that the variables relevant for understanding the activity were offset, posture, information taking and gaze direction, which was verified on detailed observations. In fact, subsequent observations, as well as consulted references (Santos, 1986) demonstrated that monitoring the flow of seed and fertilizer is essential to the satisfactory completion of the proposed task. Therefore, it was decided that the most appropriate location for the driver would be behind the mechanism of the planter so that the continuous monitoring of machine operation was made possible.

In order to design the new seeder using a user centered approach, several observations in field were performed (Figure 2) and the chassis was used in other applications (such as spring cultivators) over about a month, once the seeder prototype had not been built. In addition, the main features (technical and dimensional) of some of the animal traction seeders available in the Brazilian market were identified and their use in planting was also accompanied by the research group. Given the fact that the conditions of use of planters have very different characteristics from those observed in other stages of the cultivation process, a literature review on the topic was conducted, especially on postural needs and information taking in agricultural equipment. This was especially important in view of the fact that the project was conducted from the adaptation of a technical system originally developed for mechanical traction (Figure 3) to animal traction. A prototype of the seeder was built, but only to evaluate how the proposed technical system would work (Figure 4).

The ergonomics approach (particularly EWA) requires a detailed analysis of the activity in the real situation held from field observations. However, there was no previous situation of seeder use in such chassis (wider than other models on the market). The project included the use of the seeder on two rows, which made the machine more stable than the existing ones that work only on one row. The research for similars in the market did not indentify any seeder with such characteristics. The intervention configures itself in this way, in an approach lying among those labeled by Iida (2005) as "ergonomic correction" (since it was an adequacy of the existing chassis) and "ergonomic design" (since the seeder presented several unprecedented aspects in relation to others on the market).

In this case, the ergonomic approach was complemented by some tools adapted from those described in ISO standards 18529 and 13407 (Human-Centred Design Process for Interactive Systems, 1999): Watching users; Questionnaires, Interviews and Evaluation Expert. The way of applying questionnaires differed from the interviews and verbalizations provided in EWA. The questions were directed to researchers (in this case considered "experts") and sought to address technical issues related to the use of equipment such as the compatibility of different planting systems on the use of a system of soil beading, which would eventually interfere with the effort required and the "optimal" period for planting. Thus, it was possible to build an information base that served to the construction of the situation addressed in the project in various ways.

As an essential element to complement the approach, a literature review on the implications of technology transfer between different realities was used so that anthropotechnology would be incorporated as "design philosophy" to the technical solution development process. This aspect proved to be essential to product acceptance by target users, emphasizing the importance of the adequacy of the final solution to the different realities found in the Brazilian countryside. In this case, once again, using questionnaires sent to researchers proved to be relevant to the development of the new machine. Finally, for the dimensioning of machine controls and determination of viewing angles needed to take the necessary information to work activity, anthropometric tables available, such as Dreyfuss (2001) and Ferreira (1988), were used.

RESULTS AND DISCUSSION

Seeder machine

For the development of the animal traction seeder it is essential to consider the real needs of the users and topographical, climatic, demographic, cultural and sociological data from each region. Failing to do that, one would not be able to adequately transfer and spread the new solution, especially when it comes to technologies related to rural areas, traditionally more conservative. When considering the implementation of agricultural animal traction equipment in several regions of Brazil, it should be considered that:

(i) The existence of different levels of competence between populations (some workers more familiar to mechanical aspects than others, some illiterate, some more qualified to training different animal species) can lead to the development of different education processes, in order to satisfactorily meet the needs of the various groups;
(ii) Geographic differences, different soil types, terrain characteristics, can cause some equipment to be much more accepted and have a better performance in one region than in another;
(iii) Climate differences can result in a specific labour organization (number of breaks, resting periods) for each situation. In areas where the worker is exposed to higher temperatures, heavier equipment that may demand a greater physical strain from the farmer can present major drawbacks in relation to its acceptance among users;
(iv) Demographic differences (availability of workforce),

existing means of communication, access to training programs may require a couple of operators to drive the equipment and animals rather than a single person. Moreover, the animal (or animals) used may require a second person to direct it.

In addition, other factors influence in the increase in labour costs with animal traction (Santos, 1986):

(i) The weather and its variations;
(ii) The time constraints (deadlines to perform work due to the optimal condition of the physical environment, harvest periods);
(iii) The physical effort required by the equipment, and the postures necessary to get information;
(iv) The animal used (and therefore the speed and traction force) and operating conditions that require more physical exertion by the operator (eg arched ground, stone, etc.);
(v) The difficult maneuvers imposed to the end of the lines by certain equipment and the use of joints of animals;
(vi) The experience and competence of the operators (during the learning phase of operation of the new machine he may have a greater energy expenditure).

Issue of fatigue

Fatigue can be considered as a reversible decrease in functional capacity of an organ or a body as a result of an activity (Iida, 2005). It can usually be retrieved after a period of rest or a pause. The concept of fatigue includes both objective reduction of the capacity of the neuromuscular mechanisms, regarding subjective feelings of discomfort and tiredness. In the rural workers situation, when using the animal traction planter, the fatigue comes primarily of certain factors as mentioned below:

(i) Efforts to stabilize the machine, since the conventional seeder of only one row, are quite high, which makes them very unstable. Even when they are still, it is often difficult to keep them standing. This is an item that requires constant attention and effort by the operator, at the risk of tipping the machine;
(ii) Performing maneuvers at the end of each crop row. The machine should have its back raised and rotate 180 degrees on the front wheel. When there is use of a pair of bulls this work is even more difficult, given the specific characteristics of these animals. If one takes into consideration the weight of these machines, which reaches seventy pounds being empty, one will have an idea of the effort involved;
(iii) Lifting of machine to remove the straw from the planting device when there is straw accumulation;
(iv) Corrections of the trajectory of the machine with stabilization. The operator must also be aware of the

trajectory taken by the machine, having to constantly make corrections to maintain the quality of work and the parallelism of the rows;

(v) Supplying of deposits of seeds and fertilizer, performed several times per hectare, according to seed size and / or density of planting. The farmer takes to the field several bags of seeds, leaving them at strategic points to refuel the machine at the right time.

Postural aspect

In terms of evaluating the activity, posture assumes a central role once it is an easily observable variable (Guérin et al., 2001) and is directly related to the needs of the activity, such as applying forces and taking information that is relevant to the development of the work. Harris (1982) and Iida (2005) analyze the changes after the application of mechanical traction systems in agriculture.

Reversing the position of the farmer (now placed in front of the implement) leads to the need for continuous twisting of the torso in order to follow the evolution of cultivation by the implement pulled by tractor. This is inadequate, since the action of the implement in the ground is an essential source of information, needed so that the task is conducted appropriately. Even in old advertisements, dating from the early twentieth century, the animal traction planters in two or more rows predicted that the farmer should be sitting on the equipment.

Based on the EWA with the seeder, one can see that the positions taken during the work of seeding denote the importance of control of the output seed tank by the operator. As consequence of the visual exploration required performing the seeding, the operator of the seeder (positioned behind the implement) assumes a forward leaning posture for more than 50% of the working time. It was found that the operator controls continuously the output of seeds and every ten minutes, on average, checks the level of seeds in storage.

Another important finding regarding the posture is the position of the handlebars (instruments of handle of the operator in animal traction implements), not always allowing adjustments in height and / or width appropriate. Its adequate design could greatly reduce postural costs to the operator.

Recommendations for improving the equipment

It is necessery to improve the equipment so as to increase their performance and provide users with better work conditions. The user must, always when possible, work in the adequate posture without making excessive effort and with good visibility to control the animal traction system. The new designs of agricultural animal traction equipment should also:

(i) Improve the design of the implement so that it can

"pivot" and facilitate the maneuvers, great sources of energy expenditure;

(ii) Simplify and improve the regulation systems of the machine, so the maintenance is performed more quickly and efficiently;

(iii) Prioritize and facilitate access to forms of information necessary for developing their strategies (e.g. seed output area). In these cases, the evaluation of the operator on the results achieved by the equipment is essential to the development of the activity;

(iv) Develop implements elements incorporated from the farmer's "know-how", so that they can be better adapted and more accepted by the farmers;

(v) Develop alternative technical solutions in order to meet different demands made by different users. In this case, the design of a common base (chassis) allows a form of "modular design," reducing the costs of production and broadcast of equipment. Thus making it widely adaptable to regional differences or even the characteristics of each farmer.

In addition to this, other improvements can be introduced in the agricultural activities in terms of education, extension services and work organization:

(i) Enable suppliers of equipment, whether private or public companies (as extension services), to diagnose technical (and dimensional) solutions most appropriate to each farmer, based on the concept of modularity of the implements attached to the chassis;

(ii) Disseminate new agricultural techniques to farmers, so that they have a greater amount of options for developing their work, using cultivars or varieties that allow, for example, a longer growing season;

(iii) Disseminate operation manuals of the machines to operators (these should be adequate so as to used satisfactorily even by the illiterate).

(iv) Avoid work during the hottest hours of the day, whether through forms of work organization such as the adoption of cultivars (or varieties) that allow a longer period of planting;

(v) Inform farmers not only the logic of machine usage, but also the logic of its mechanical operation, enabling them to diagnose faults and perform minor repairs on the equipment in the field.

Starting of the proposed design from a seeder originally concepted for use with mechanical traction (Figure 3), a more compact and lightweight version was built to be coupled in the original chassis (Figure 4). However, a prototype with the proposed features based on the study for field tests in real situation was not built. In its place, a model with the proposed changes was developed (Figure 5). Although this is a limitation to the evaluation of the study results, interesting changes in the design can be observed in order to make the equipment more suited to real work situations in the field, considering the different components of the workload (e.g. the physical effort in the seedering activities, cognitive and psychic aspects

Figure 5. Model of the equipment after the proposed changes. (Source: author's file).

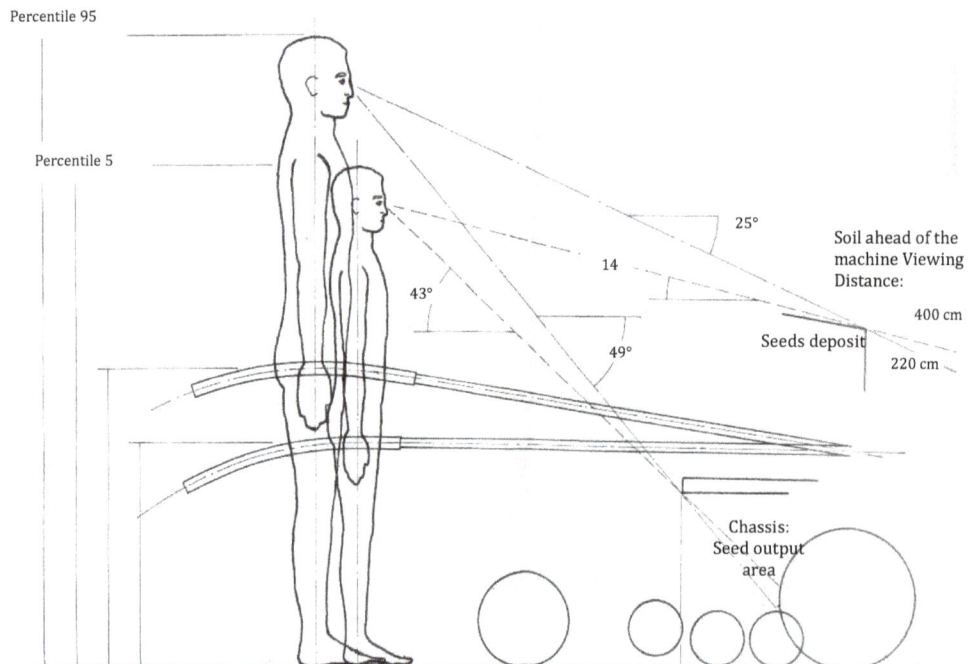

Figure 6. Schematic drawing of the result of applying some principles of dimensioning to percentiles ranging from 5 to 95%. Dimensions defined based on Dreyfuss (2001) and Ferreira (1988).

related to work, as observation of the seed output area and seeding preparation).

As an example for reducing the physical load component the size and the curvature of the handlebars, aiming at a comfortable grip in extreme situations (Figure 6), deserve attention. Handlebars that are longer than traditional ones were proposed. This allows maneuvering and shifts in equipment direction to be more easily made.

The use of two or three planting rows also gives more stability to the product, further decreasing the effort of the user. The longer handlebars also allow a better and continuous monitoring of work performed, without further postural efforts. The diameter adopted for handle was 25 mm (defined based on Dreyfuss, 2001), considering the use of rubber coverage.

An important aspect of the cognitive component of the work is that the farmer has continuous access to information related to the equipment operation and to planting conditions. Bearing in mind the total height of the seeds deposit, are considered the taking of information from the seeds outlets and soil viewing distance in front of the machine. In this particular case an important issue was raised, because the view in front of the machine is reduced as the height of the conductor decreases. Moreover, one should consider the existence of a traction animal, which would make it impossible for the operator to see the ground ahead of the machine. Although it may lead to a greater physical expenditure, it was decided to keep the user in his traditional work posture, standing behind the planter. This decision can be justified by the small area of cultivation planned for the use of the machine.

Considering some specific characteristics of the product development, some data could not be collected in order to bring statistical conclusive results. However, is possible to compare qualitatively the design solution with some of the traditional seeders found in Brazil. Although it has not been built the final prototype of the machine, it is expected that their weight is between 70 and 80 kg, which is compatible with other existing models as described in Almeida et al. (2002) and Almeida and Silva (1999). Aspects such as travel speed and number of seeds should vary according to the culture, topography and animal adopted, but also a function of operator experience and their physical and cognitive conditions. Regarding the developed model, the first observations demonstrate that simultaneous use of two planting lines becomes seeder more stable, which reduces the effort of driving. On the other hand, the ends of line maneuvers become more complex, requiring (as observed in the field) more space to revolution than the traditional equipment, and a finer degree of attention by the operator. Moreover, it is expected that learning about the new technology may facilitate such operation.

Finally, the mechanical aspects of the implement should be taken into account for a greater adequacy to the user. The use of less vulnerable to bushing cropping systems (accumulation of straw and crop residues in the planting system) facilitates the planting operation and reduces operator effort, once it reduces the need for cleaning the machine, shortens the time and facilitates the intire planting process. This will ultimately bring benefits to the workload. In addition, the adoption of simple mechanical systems allows for easier diffusion of the implement, since a larger number of farmers will be able to understand it and use it easily.

Conclusions

Based on the literature review and observations made during the field research as well as interviews with technicians and farmers, a number of suggestions for improving a model of animal traction seeder were presented. The objective was to adress the largest possible number recommendations for meeting the needs of the users and minimizing their substantial workload. For a long time, because of greater financial return and government policies of incentive, manufacturers concentrated their efforts on the development of implements for large farms, making the technology intended for small production remains stagnated. It is therefore the opportunity for improvement of these devices. However, we tried to emphasize here that the cultural, social, geographic and economic differences between regions and the diversity of cultivars and climate lead to the need for developing technical solutions to very peculiar characteristics, so as to satisfactorily cater for the existing conditions.

It should be emphasized that this work was centered on the presentation of opportunities for applying UCD principles in the design of agricultural machinery, and not the improvement of mechanical planting. This is a challenge that also depends on other product design tools such as ways of design for modularity. Starting from a basic chassis numerous technical and usability solutions that may be suited to the different conditions observed, as well as to the different features of the users, can be developed. Using a modular product system the costs of production and acquisition would be reduced, while the rural extension services could be responsible for guiding farmers on the best options available in every situation. It would thus be possible to build appropriate solutions to regional differences in terms of climate, terrain, culture and needs even with respect to the different characteristics of users.

In the case presented the choice of an animal traction machine is mainly due to a larger number of variables involved, as professionals involved with mechanical traction machines usually have (or should have) specific training for the task. Therefore, it is believed that in the case of animal traction there is greater relevance of traditional knowledge, not formal, connected to the farmer's experience. This eventually made the research more robust and rich in its results. Design principles can and should be applied to the development of any technical solution linked to rural activity, whether in small properties with strong technical and / or financial constraints or in large enterprises.

It is noteworthy that the main contribution of the study presented focuses on a theoretical approach to the problem and on the proposal of recommendations based on (1) literature review and (2) observations of the use of other implements in field. These contributed to the preparation of a technical solution that would incorporate the features of the proposed adaptation to users and would meet the conditions observed in the field. However, it was not possible to build a complete prototype of the seeder based on the suggestions for improvements. This makes it impossible to compare the results obtained in terms of, for example, reduction of time or physical effort required for the various operations in planting, or more qualitative assessments, such as the perception of comfort by users in a real situation of machine usage compared to other implements of similar function.

It is important to note that the approach used and the results suggested show that the development of agricultural implements intended for small farmers is suitable. In a user centered approach there are, therefore, several aspects that must be analyzed before a project linked to the agriculture. This should consider specific aspects of rural life and work, which differ widely in relation to urban work environment, which is usually studied by ergonomics. The number of variables applied to rural work, related to climatic, geographic and cultivars factors, for example, makes the assessment become more complex, requiring specific strategies and methodological tools in the search for technical solutions that meet the needs of the sector. Applying this approach in the design of agricultural machinery can contribute to the solution of various demands of the field and the improvement of working and life conditions of the rural population.

ACKNOWLEDGEMENT

The authors would like to thank the staff of the Section of Harvesting Machine and Processing of Agricultural Products Division of Agricultural Engineering of the Agronomic Institute of Campinas, through Dr. Claudio Alves Moreira, without whom it would have been impossible to accomplish this work.

REFERENCES

Adler PS, Winograd TA (1992). Usability: Turning Technologies into Tools. Oxford: Oxford University Press, Inc.

Almeida RA, Leão PGF, Barcellos LC, Silva JG (2002). Development and evaluation of an animal-traction planter (Desenvolvimento e Avaliação de uma Semeadora Adubadora à Tração Animal). Pesquisa Agropecuária Trop. 32 (2):81-87.

Almeida RA, Silva JG (1999) Operating performance of a seeder fertilizer for animal traction, with different planter systems, adjustments of cutting and mulching disk, no-till bean (Phaseolus vulgaris L.). (Desempenho operacional de uma semeadora adubadora à tração animal, com diferentes sistemas de sulcadores, regulagens de disco de corte e coberturas mortas, no plantio direto do feijoeiro - Phaseolus vulgaris L.). Pesquisa Agropecuária Trop. 29(2):73-80.

Araújo AG, De Figueiredo PRA, Júnior RC (1999). Field Evaluation of

Animal Traction Equipment for Soil Tillage in Brazil. AMA-AGR MECH ASIA AF 30(3):23-27.

Béguin P, Daniellou F (2004). Metodologia da Ação Ergonômica: Abordagens do Trabalho Real (Ergonomis Action Methodology: Real Work Approach), in Falzon (ed.) Ergonomia (Ergonomics). São Paulo: Editora Blucher, pp. 281-301.

Carrol C, Marsden P, Soden P, Naylor E, New J, Dornan T (2002) "Involving users in the design and usability evaluation of a clinical decision support system". Comput. Meth. Prog. Biol. 69(2):123-135.

Cerf M, Sagory P (2004). Agricultura e Desenvolvimento Agrícola (Agriculture and Agricultural Development), in Falzon (ed.) Ergonomia (Ergonomics). São Paulo: Editora Blucher, pp. 535 – 544.

Costa SF (2006). Usabilidade no Projeto de Produto: um estudo de caso na área de transmissão de energia. (Usability in the Product Design: A Case Study on Energy Transmission Area) Masters Dissertation – Universidade Federal de Minas Gerais.

Dreyfuss H (2001). The Measure of Man and Woman: Human Factors in Design. New York: Dreyfuss Associates.

Ferreira DMP (1988). Pesquisa Antropométrica e Biodinâmica dos Operários da Indústria de Transformação – RJ (Anthropometric and Biodynamic Research on Transformation Industry Workers – RJ). Rio de Janeiro: Instituto Nacional de Tecnologia.

Freund J, Takala EP, Toivonen R (2000). Effects of two ergonomic aids on the usability of an in-line screwdriver. Appl. Ergon. 31(4):371-376.

Garmer K, Ylvén JIC, Karlsson M (2004). User participation in requirements elicitation comparing focus group interviews and usability tests for eliciting usability for medical equipment: a case study. Int. J. Ind. Ergonom. 33(2):85-98.

Geslin P (2004). Anthropotechnology. In: Neville S. Alan H, Karel B, Salas E, Hendrick (eds). Handbook of Human Factors and Ergonomics Methods H. London: CRC Press, pp. 87-1– 87-7

Guérin F, Laville A, Daniellou F, Duraffourg J, Kerguelen A (2001) Compreender o Trabalho para Transformá-lo: A Prática da Ergonomia (Understand the work to change it: the Ergonomics Practise). São Paulo: Editora Edgard Blücher.

IBGE - Instituto Brasileiro de Geografia e Estatística (2009). Censo Agropecuário 2006: Brasil, Grandes Regiões e Unidades da Federação (2006 Agricultural Census: Brazil, Major Regions and Federation Units). Rio de Janeiro, IBGE.

Harriss J (ed.) (1982). Rural Development: Theories of Peasant Economy and Agrarian Change. London: Hutchinson University Library.

Iida I (2005). Ergonomia – Projeto e Produção (Ergonomics: Design and Production). São Paulo: Editora Edgard Blücher.

ISO/TR 13407 (1999) Human-centred design process for interactive systems. Genève: International Organization for Standartzation.

Jafry T, O'Neill DH (2000). "The application of ergonomics in rural development: a review". Appl. Ergon. 31(3):263-268.

Kogi K (2006). Participatory methods effective for ergonomic workplace improvement. Appl. Ergon. 37(4):547-554.

Kroemer KHE, Grandjean E (2005). Manual de Ergonomia: Adaptando o Trabalho ao Homem (Ergonomics: Fitting the Task to the Man). Porto Alegre: Bookman.

Liu Y, Osvalder AL (2004). Usability evaluation of a GUI prototype for a ventilator machine. J. Clin. Monitor. Comp. 3(18):5-6.

Nemeth CP (2004). Human Factors Methods for Design: Making Systems Human-Centered. New York: CRC Press.

Nielsen J (1993). Usability engineering. London: Academic Press.

Ortiz-Laurel H, Rössel D (2007). "Current status of animal traction in Mexico". Ama-Agr. Mech. Asia. AF 2:(38)83-88.

Peche Filho A, Moreira, CA, Bernardi JÀ, Giomo V (1987) Chassi Porta Implementos Triangular: Desenvolvimento e Desempenho Operacional (triangle chassis to keep implements: development and operational performance). Anais XVI Congresso Brasileiro de Engenharia Agrícola (Proceedings of XVI Brazilian Congress of Agricultural Engineering), Jundiaí: CONBEA, pp. 266-271.

Pereira RGA, Carvalho JOM, Mendes AM, Leonidas FC (2010). Incorporação da tração animal em sistemas agroecológicos em Rondônia (Incorporation of animal traction agroecological systems in Rondônia) Cadernos de Agroecologia 1(5):1-4.

Prahalad CK (2009). The Fortune at the Bottom of the Pyramid: Eradicating Poverty through P rofits, rev. and Updated 5th

Anniversary Edition: Eradicating Poverty through Profits. New Jersey: Pearson Prentice Hall.

Ramaswamy NS (1994). Draught animals and welfare. Rev. Sci. Tech. 13 (1):195–216, http://www.ncbi.nlm.nih.gov/pubmed/8173096 (14 February 2013).

Romano LN, Back N, Ogliari A, Marini VK (2005). An Introduction to the Reference Model for the Agricultural Machinery Development Process. Product: Management and Development 3(2):109-132.

Rose AF, Schnipper JL, Park ER, Poon EG, Li Q (2005). Using qualitative studies to improve the usability of an EMR. J. Biomed. Inform. 38(1): 51-60.

Rubin J (1994). Handbook of usability testing: how to plan, design and conduct effective tests. New York: John Wiley & Sons, Inc.

Santos V (1986). A Abordagem Ergonômica da Utilização da Tração Animal em Pequenas Explorações Agrícolas Brasileiras. (The Ergonomic Approach of Animal Traction Use in Brazilian Small Farms) Paris: Conservatoire National des Arts e Métiers.

Shahnavaz H (1991). Transfer of Technology to Industrial Developing Countries and Human Factors Considerations. Lulea: Center for Ergonomics of Developing Countries.

Shinno H (2002) Product development methodology for Machine Tools. Tokyo: Japanese Society of Mechanical Engineering.

Wilson RT (2003). The environmental ecology of oxen used for draught power. Agric. Ecosyst. Environ. 97(1-3):21-37.

Wisner A (1997). Anthropotechnologie: vers un monde industriel pluricentrique. Collection Travail. Paris: Octares.

Wisner A (1985). Consecuencias de la transferência de tecnicas sobre as condiciones de trabajo. (Consequences of Technology Transference on Work Conditions) Paris: CNAN.

Wisner A (1987). Por Dentro do Trabalho: Ergonomia: Método & Técnica. (Inside the Work: Ergonomics, method and techniques) São Paulo: FTD - Oboré.

10

Molecular cloning and tissue expression analyses of a *UF3GT* gene from *Capsicum annuum* L.

Hui ZHOU[1], Ming-hua DENG[1,3], Jin-fen WEN[2], Jin-long HUO[4], Xue-xiao ZOU[3] and Hai-shan ZHU[1]

[1]College of Horticulture and Landscape, Yunnan Agricultural University, Kunming 650201, China.
[2]Faculty of Modern Agricultural Engineering, Kunming University of Science and Technology, Kunming 650224, China.
[3]Institute of Vegetable Crops, Hunan Academy of Agricultural Science, Changsha 410125, China.
[4]Faculty of Animal Science and Technology, Yunnan Agricultural University, Kunming 650201, China.

In the present study, a novel gene designated as *UF3GT* was isolated from *Capsicum annuum* L. Sequence and structural analysis determined that the *UF3GT* protein, which contained 1401 base pairs encodes 447 amino acids and belongs to the Glycosyltransferase-GTB-type superfamily. The isoelectric point and the molecular weight of this gene are 5.97 and 49589, respectively. The deduced amino acid sequence shows highly identity with known anthocyanin synthases in other species. The phylogenetic tree analysis revealed that the *C. annuum* L. *UF3GT* has a closer genetic relationship with the *UF3GT* of petunia (*Petunia x hybrida*). The RT-PCR gene expression analysis indicated that the *UF3GT* gene was mostly expressed in pericarp, moderately expressed in stem, leaf, flower, placenta and seed, whereas no expression signal was detected in root. The study established the primary foundation for further research on this *C. annuum* L. gene.

Key words: *Capsicum annuum* L., semi-quantitative RT-PCR, glycosiltransferase gene, tissue expression profile analysis.

INTRODUCTION

Capsicum annuum L. (commonly known as hot pepper, chili, chili pepper, and bell pepper) is a dicotyledonous flowering plant. It belongs to family Solanaceae, which includes potato, tomato, eggplant, African eggplants etc. (Knapp, 2002; Hunziker, 2001). In China, people living in Yunnan province have planted this crop for thousands of years. The whole plant is violet black, especially the fruit. Shades ranging from violet to black pigmentation in *C. annuum* L. are attributed to anthocyanin accumulation.

Anthocyanins are plant pigments widely distributed in colored fruits and flowers. They also exhibit antioxidant activities and therefore may contribute to the prevention of heart disease, cancer, and inflammatory disease (Hou,

2003; Bagchi et al., 2004; Katsube et al., 2003). They are usually localized in the vacuoles of petal epidermal cells (Goto et al., 1982; Tanaka et al., 2004) and are responsible for diverse pigmentation from orange to red, purple and blue in flowers, fruits and vegetables. Often, these compounds also occur in leaves, stems, seeds, and other tissues. Contributing to the colorful appearance of flowers, fruits and vegetables, anthocyanins help them to attract animals, leading to seed dispersal and pollination (Harborne and Williams, 2000). Mazza and Miniati (1993) reported that anthocyanins might be important in protecting plants against ultraviolet-induced damage, as well. In addition, they play roles as antioxidants and in protecting DNA and the photosynthetic apparatus from high radiation fluxes. Recently, using the known genetic sequence to synthesize primers, several genes, such as *CHS*, *CHI*, *F3H*, *DFR*, *ANS*, etc have been cloned in grapes, apples, strawberries, and other plants (Jaakola et al., 2002; Li et al., 2001).

UF3GT (UDP glucose flavonoid 3-glucosyltransferase) has been shown to be one of the key enzymes in anthocyanins biosynthesis derived from the phenylpropanoid pathway. The UF3GTs play important roles not only in modifying flower color but also in increasing the solubility and stability of hydrophobic flavonoids (Hondo et al., 1992; Yoshida et al., 2000).

UF3GT have been isolated from flowers of many ornamental plants, including *Gentiana triflora* (Tanaka et al., 1996) and *Petunia x hybrida* (Yamazaki et al., 2002). They have also been isolated from *Zea mays* (Goto et al., 1982), *Antirrhinum majus* (Martin et al., 1991), *Vitis vinifera* (Ford et al., 1998), *Hordeum vulgare* (Wise et al., 1990), *Perilla frutescens* (Gong et al., 1997) and *Fragaria ananassa* (Almeida et al., 2007). Many *UF3GT* genes have been cloned and heterologously expressed, which makes in-depth study of the molecular mechanism controlling anthocyanin biosynthesis possible. Semi-quantitative RT-PCR has been conducted to analyze the relative expression levels in various tissues. This study provides primary information for further understanding the biochemical functions of *UF3GT* in *C. annuum* L.

MATERIALS AND METHODS

Sample collection

All plants used in the study were derived from College of Horticulture, Yunnan Agricultural University. Yunnan Purple Pepper No.1 (*C. annuum* L.) tissues (root, stem, leaf, blossom, pericarp, placenta, and seed) were instantly frozen in liquid nitrogen and stored at −80°C before use.

Total RNA extraction and first-strand cDNA synthesis

Total RNA from Yunnan Purple Pepper No.1, was extracted by the Trizol procedure (TaKaRa) and cDNA was synthesized using High Fidelity PrimeScript RT-PCR Kit (TaKaRa) according to the manufacturer's protocol.

PCR amplification

The PCR was performed to isolate the Yunnan Purple Pepper No.1 gene using the pooled cDNAs from different tissues. The 20 μl reaction mixture contained 1.5 μl (25 ng/μl) DNA, 1 μl 2.5 mM mixed dNTPs, 2 μl 10 x Taq DNA polymerase buffer (MgCl₂ plus), 0.4 μl 10 μM forward and reverse primer, 0.3 μl 5 U/μl Taq DNA polymerase and 13.4 μl sterile water. The PCR program initially started with a 94°C/4 min, followed by 35 cycles of 94°C/1 min, 57°C/45 s, 72°C/90 s, and then 72°C extension for 10 min, finally 4°C to terminate the reaction. The mRNA and amino acid sequences for UF3GT genes from various plant species available in the databank of National Centre for Biotechnology Information

(http://www.ncbi.nlm.nih.gov) were used to locate conserved boxes by multiple sequence alignment through CLUSTALW 1.8 and primers with the following sequence were designed. UF3GT-F: ATGACTACTTCTCAACTTCATATT, UF3GT-R: TYAAGTARGCTTGTGACATTTAA (C/T= Y, A/G= R).

RT-PCR for expression profile

The primers of Yunnan Purple Pepper No.1 *UF3GT* gene which were used to perform the RT-PCR for tissue expression profile analysis was same discussed above. The 20 μl mixture contained was: 4 μl 5×PrimeSTAR PCR Buffer, 0.4 μl dNTP Mixture (10 mM each), 0.2 μl of forward and reverse primer (10 μM), 0.2 μl of PrimeSTAR HS DNA Polymerase (2.5 U/μl) (TaKaRa), 2 μl aliquot of cDNA and 13 μl RNase Free dH₂O water. The products of amplification were checked on a 1.5% agarose gel and visualized with ethidium bromide.

Bioinformatics analysis

PCR amplification was repeated 5 times independently. The products were cloned into pMD18-T vector (TaKaRa) and sequenced bidirectionally. At least 10 independent clones were sequenced for each PCR product. Sequencing data were edited and aligned using DNASTAR software (DNAStar Inc., Madison, Wisc.). The cDNA sequence was predicted using the GenScan software (http://genes.mit.edu/GENSCAN.html). Putative protein theoretical molecular weight (Mw) and isoelectric point (pI) prediction, signal peptide prediction, subcellular localization prediction and transmembrane topology prediction were performed using the Compute pI/Mw Tool (http://us.expasy.org/tools/pi_tool.html), SignalP 3.0 server (http://www.cbs.dtu.dk/services/SignalP/), PSort II (http://psort.hgc.jp/), TMpred (http://www.ch.embnet.org/software/TMPRED_form.html), respectively. The Blastp program and Conserved Domain Architecture Retrieval Tool were used to search for similar proteins and conserved domain, respectively (http://www.ncbi.nlm.nih.gov/Blast). The alignment of the nucleotide sequences and deduced amino acid sequences were computed using ClusterX, and the phylogenetic trees were computed using the ClustalX and Mega 4.0 softwares with standard parameters. Secondary structures of deduced amino acid sequences were predicted with SOPMA (http://npsa-pbil.ibcp.fr/). The 3D structures were predicted based on the existed 3D structures by the amino acids homology modeling on swiss server (http://swissmodel.expasy.org/).

RESULTS

RT-PCR results for *C. annuum* L. *UF3GT* gene

Through RT-PCR with pooled tissue cDNAs, for *C. annuum* L. *UF3GT* gene, the resulting PCR products were about 1430 bp long (Figure 1).

cDNA nucleotide sequence analysis using the BLAST software at NCBI server (http://www.ncbi.nlm.nih.gov/BLAST) revealed that this gene was not homologous to any of the known *C. annuum* L. genes .The sequence prediction was carried out using the GenScan software and the results showed that the cDNA sequences encoded 447 amino acids.

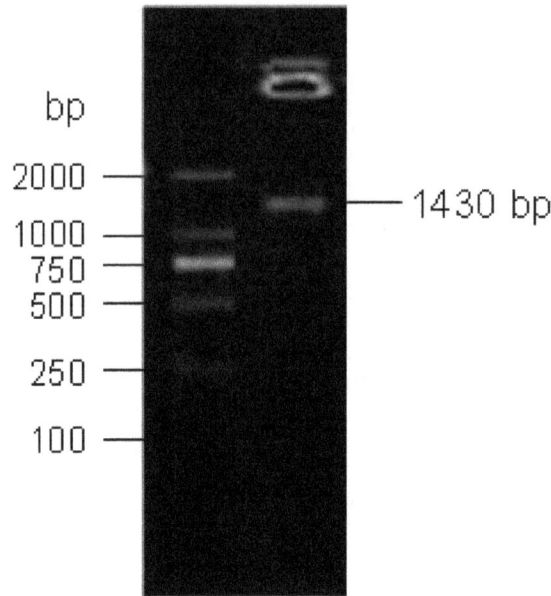

Figure 1. Analysis of PCR products on 1.5% agarose gels.

Figure 2. The complete coding sequence of *Capsicum annuum* L. *UF3GT* gene and its encoding amino acids * indicates the stop codon.

The theoretical pI and Mw of this deduced protein of this *C. annuum* L. gene was computed using the Compute pI/Mw Tool. The pI of this gene was 5.97 while molecular weight of the putative proteins was 49,589.

Prediction of protein properties

The complete coding sequence (CDS) of this gene and the encoded amino acids are presented in Figure 2.

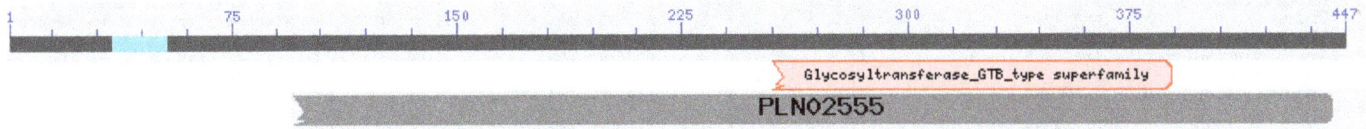

Figure 3. The conservative domains of the protein encoded *Capsicum annuum* L. *UF3GT* gene.

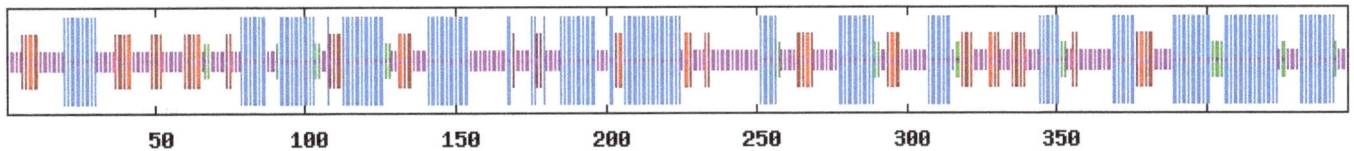

Figure 4. Secondary structure of UF3GT protein in Capsicum annuum L. Helices, strands and coils are indicated, respectively, with long, middle and short vertical lines.

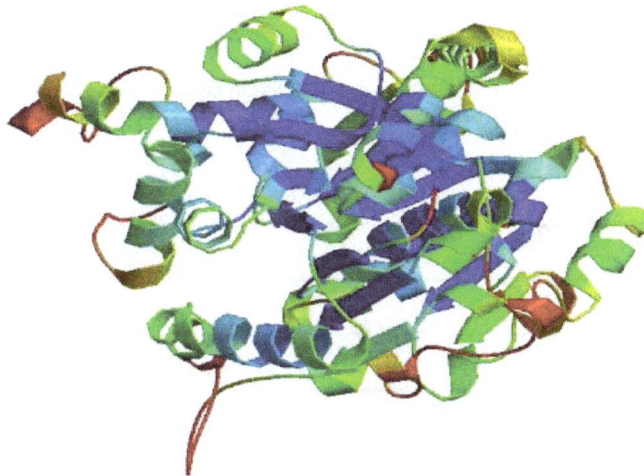

Figure 5. The tertiary structure of *UF3GT*.

The putative protein was also analyzed using the Conserved Domain Architecture Retrieval Tool of Blast at the NCBI server (http://www.ncbi.nlm.nih.gov/BLAST) and the conserved domain was identified as PLNO2555 domain (Figure 3). According to Figure 3, we can see it belongs to Glycosyltransferase-GTB-type superfamily.

The secondary structure analysis of the deduced amino acid sequence by GOR algorithm indicated that the protein consisted of 40.94% α-helix, 37.36% random coils and 17.00% β-sheets (Figure 4). The signal peptide prediction performed by SignalP 3.0 on the basis of a combination of several artificial neural networks and hidden Markov models revealed that *C. annuum* L. *UF3GT* contained potential signal peptide with 99.9% probability (Bendtsen et al., 2004). Transmembrane topology prediction made by TMpred (http://www.ch.embnet.org/software/TMPRED_form.html) , showed that *C. annuum* L. *UF3GT* had N-terminus outside with 4 strong transmembrane helices. For subcellular localization analysis, the amino acid sequence was submitted to the PSORT program, and Reinhardt's method showed *UF3GT* was probably located in the cytoplasm with up to 76.7% probability (Nakai and Horton, 1999).

Homology modeling

In order to better understand the detailed structures of *UF3GT*, the homology modeling was performed to estimate its 3D structure. The 3D structure of *UF3GT* (2-443AA) by homology modeling was based on template-3hbjA (chain A) with 43.1% sequence identity (Guex and Peitsch, 1997; Schwede et al., 2003; Arnold et al., 2006). The 3D structure analysis may provide a basis for further studying the relationship between structure and function of *UF3GT* (Figure 5).

Capsicum annuum · Prunus avium · Vitis labrusca · Arabidopsis thaliana · Petunia x hybrida · Dianthus caryophyllus

(protein sequence alignment data)

Figure 6. The alignment of the protein encoded by *C. annuum* L. *UF3GT* with *UF3GT* from other species.

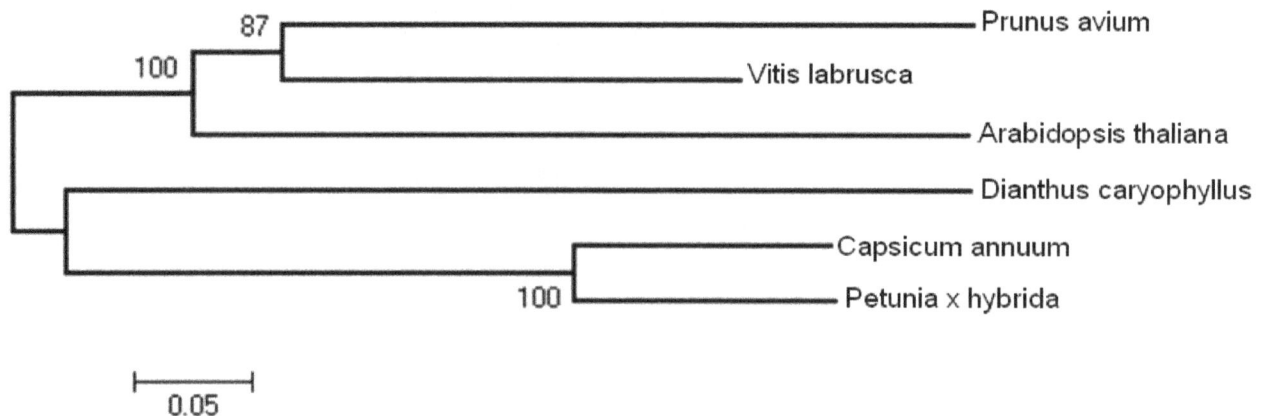

(phylogenetic tree diagram)

Prunus avium
Vitis labrusca
Arabidopsis thaliana
Dianthus caryophyllus
Capsicum annuum
Petunia x hybrida

87
100
100

0.05

Figure 7. Phylogenetic tree analysis for selected *UF3GT* protein.

Sequence analysis and evolutionary relationships

Further BLAST analysis of this protein revealed that *C. annuum* L. *UF3GT* has high homology with the anthocyanin synthases protein *UF3GT* of 5 species - *Prunus avium*, *V. labrusca*, *Arabidopsis thaliana*, *Dianthus caryophyllus*, *P. × hybrida* (Figure 6).

Based on the result of the alignment of *UF3GT*, the phylogenetic tree was constructed using the ClustalW software (http://www.ebi.ac.uk/clustalw), as shown in Figure 7. The phylogenetic tree analysis revealed that the *C. annuum* L. *UF3GT* has a closer genetic relationship with the *UF3GT* of petunia.

Gene expression profile analysis was carried out and results revealed that the *C. annuum* L. *UF3GT* gene was highly expressed in pericarp, moderately expressed in stem, leaf, flower, pericarp, placenta and seed. There was no expression of *C. annuum* L. *UF3GT* in root (Figure 8).From the tissue expression profile analysis in our experiment it can be seen that this gene was obviously differentially expressed in some tissues and there were no expression in some tissues. As we did not study functions at protein levels yet, there might be many possible reasons for differential expression of this gene. The suitable explanation for this under current conditions is that at the same time those biological activities related

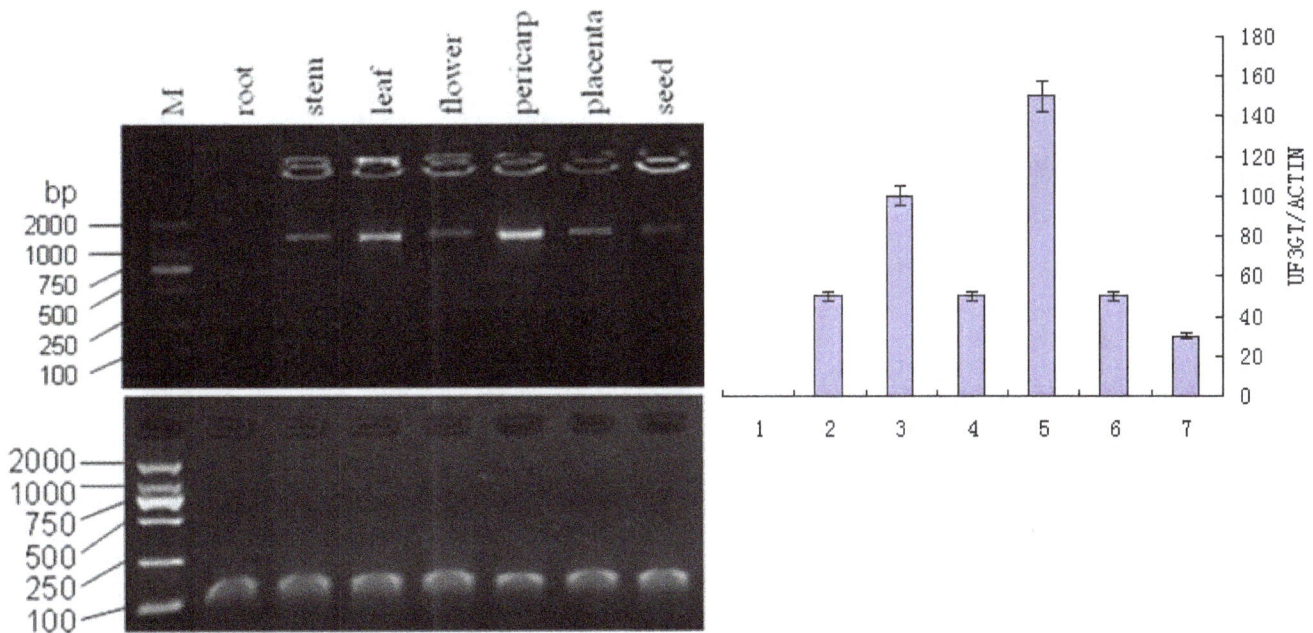

Figure 8. Tissue expression distribution of *C. annuum* L. *UF3GT* gene. M, DL2000 markers.

to anthocyanosides synthesis of this gene were presented diversely in different tissues.

DISCUSSION

UF3GT gene have been cloned and characterized from horticulture crops, such as Petunia (Weiss et al., 1995; Moalem-Beno et al., 1997), Periwinkle (Ohlsson and Berglund, 2001), Rose (Zieslin et al., 1974), and a few other species. In this study, a 1430-bp full-length cDNA gene *UF3GT* was isolated from *C. annuum* L.. The deduced amino acid sequence of *UF3GT* showed extensive similarity to their counterparts in other species. Through tissue expression analyses, we detected *UF3GT* expression in stem, leaf, flower, pericarp, placenta and seed where anthocyanin was concentrated, the expression of *UF3GT* was higher in pericarp. This phenomenon is in accordance with the anthocyanin content in these tissues (data not given). These data suggest that *UF3GT* maybe a specific tissue expression gene. And it was consistent with previous research (Hughes and Hughes, 1994). Anthocyanin content might have close correlation with *UF3GT* expression levels in response to tissue specificity and genetic background.

In this report, the complete coding sequences of the *C. annuum* L. *UF3GT* gene was isolated based on the conserved sequence information of the *P. ×hybrida* and some referenced *C. annuum* L. ESTs. Sequence identification further validated that comparative genomics method is one useful tool to isolate the unknown genes especially the conserve coding region of genes for *C. annuum* L. or other plants.

In conclusion, we have isolated the *C. annuum* L. *UF3GT* gene and performed necessary functional analysis and tissue expression profile analysis. This established the primary foundation for further research on *C. annuum* L. *UF3GT* gene.

ACKNOWLEDGEMENTS

This study was supported by the Earmarked Fund for Modern Agro-industry Technology Research System, National Department Public Benefit Research Foundation (200903025), the National High Technology Research and Development Program of China (863 Program, 2006AA100108-4-10, 2006AA10Z1A6-3), National Natural Science Foundation of China (Grant No. 31160394), the Foundation of Natural Science of Yunnan Province, China (2009CD055).

REFERENCES

Almeida JR, D'Amico E, Preuss A, Carbone F, de Vos CH, Deiml B, Mourgues F, Perrotta G, Fischer TC, Bovy AG, Martens S, Rosati C (2007). Characterization of major enzymes and genes involved in flavonoid and proanthocyanidin biosynthesis during fruit development in strawberry (Fragaria × ananassa). Arch Biochem. Biophys. 465:61-71.

Arnold K, Bordoli L, Kopp J, Schwede T (2006). The SWISS-MODEL Workspace: A web-based environment for protein structure homology modeling. Bioinformatics 22:195-201.

Bagchi D, Sen CK, Bagchi M, Atalay M (2004). Antiangiogenic, antioxidant, and anti-carcinogenic properties of a novel anthocyanin-rich berry extract formula. Biochemistry (Mosc). 69(1):75-80.

Bendtsen JD, Nielsen H, von Heijne G, Brunak S (2004). Improved prediction of signal peptides: SignalP 3.0. J. Mol. Biol. 340:783-795.

Ford CM, Boss PK, Hoj PB (1998). Cloning and characterization of *Vitis vinifera* UDP-glucose: flavonoid 3-O-glucosyltransferase, a homologue of the enzyme encoded by the maize Bronze-1 locus that may primarily serve to glucosylate anthocyanidins *in vivo*. J. Biol. Chem. 273:9224-9233.

Gong Z, Yamazaki M, Sugiyama M, Tanaka Y, Saito K (1997). Cloning and molecular analysis of structural genes involved in anthocyanin biosynthesis and expressed in a forma-specific manner in *Perilla frutescens*. Plant Mol. Biol. 35:915-927.

Goto T, Kondo T, Tamura H, Imagawa H, Lino A, Takeda T (1982). Structure of gentiodelphin, an acylated anthocyanin isolated from *Gentiana Makinoi* that is stable in dilute aqueous solution. Tetrahedron Lett. 23:3695-3698.

Guex N, Peitsch MC (1997). SWISS-MODEL and the Swiss-PdbViewer: An environment for comparative protein modeling. Electrophoresis 18:2714-2723.

Harborne JB, Williams CA (2000). Advances in flavonoid research since 1992. Phytochemistry 55:481-504.

Hondo T, Yoshida K, Nakagawa A, Kawai T, Tamura H, Goto T (1992). Structural basis of blue-color development in flower petals from *Commelina communis*. Nature 358:515-518.

Hou DX (2003). Potential mechanisms of cancer chemoprevention by anthocyanins. Curr. Mol. Med. 3(2):149-159.

Hughes J, Hughes MA (1994). Multiple secondary plant product UDP-glucose glucosyltransferase genes expressed in cassava (*Manihot esculenta* Crantz) cotyledons. DNA Seq. 5:41-49.

Hunziker AT (2001). Genera solanacearum: The genera of Solanaceae illustrated, arranged according to a new system. Gantner Verlag, Ruggell, Liechtenstein, P. 516.

Jaakola L, Määttä K, Pirttilä AM, Törrönen R, Kärenlampi S, Hohtola Al (2002). Expression of genes involved in anthocyanin biosynthesis in relation to anthocyanin, proanthocyanidin, an d flavonol levels during bilberry fruit development. J. Plant Physiol. 130(2):729-739.

Katsube N, Iwashita K, Tsushida T, Yamaki K, Kobori M (2003). Induction of apoptosis in cancer cells by bilberry (*Vaccinium myrtillus*) and the anthocyanins. J. Agric. Food Chem. 51(1):68-75.

Knapp S (2002). Tobacco to tomatoes: A phylogenetic perspective on fruit diversity in the Solanaceae. J. Exp. Bot. 53:2001-2022.

Li YJ, Sakiyama R, Maruyama H, Kawabata S (2001). Regulation of anthocyanin biosynthesis during fruit development in Nyoho strawberry. J. Jan. Soc. For Hort. Sci. 70(1):28-32.

Martin C, Prescott A, Mackay S, Bartlett J, Vrijlandt E (1991). Control of anthocyanin biosynthesis in flowers of *Antirrhinum majus*. Plant J. 1:37-49.

Mazza G, Miniati E (1993). Anthocyanins in fruits, vegetables, and grains. CRC press, Boca Raton, Florida, USA.

Moalem-Beno D, Tamari G, Leitner-Dagan Y, Borochov A, Weiss D (1997). Sugar-dependent gibberellin-induced chalcone synthase gene expression in petunia corollas. Plant Physiol. 113:419-424.

Nakai K, Horton P (1999). PSORT: a program for detecting the sorting signals of proteins and predicting their subcellular localization. Trends Biochem. Sci. 24:34-35.

Ohlsson AB, Berglund T (2001). Gibberellic acid-induced changes in glutathione metabolism and anthocyanin content in plant tissue. Plant Cell Tissue Org. 64:77-80.

Schwede T, Kopp J, Guex N, Peitsch MC (2003). SWISS-MODEL: an automated protein homology-modeling server. Nucleic Acids Res. 31:3381-3385.

Tanaka Y, Katsumoto Y, Brugliera F, Mason J (2004). Genetic engineering in floriculture. Plant Cell Tiss. Org. 80:1-24.

Tanaka Y, Yonekura K, Fukuchi-Mizutani M, Fukui Y, Fujiwara H, Ashikari T, Kusumi T (1996). Molecular and biochemical characterization of three anthocyanin synthetic enzymes from *Gentiana triflora*. Plant Cell Physiol. 37:711-716.

Weiss D, van der Luit A, Knegt E, Vermeer E, Mol JNM, Kooter JM (1995). Identification of endogenous gibberellins in petunia flower, induction of anthocyanin biosynthetic gene expression and the antagonistic effect of abscisic acid. Plant Physiol. 107:695-702.

Wise RP, Rohde W, Salamini F (1990). Nucleotide sequence of the Bronze-1 homologous gene from *Hordeum vulgare*. Plant Mol. Biol. 14:277-279.

Yamazaki M, Yamagishi E, Gong Z, Fukuchi-Mizutani M, Fukui Y, Tanaka Y, Kusumi T, Yamaguchi M, Saito K (2002). Two flavonoid glucosyltransferases from *Petunia x hybrida*: molecular cloning, biochemical properties and developmentally regulated expression. Plant Mol. Biol. 48:401-411.

Yoshida K, Toyama Y, Kameda K, Kondo T (2000). Contribution of each caffeoyl residue of the pigment molecule of gentiodelphin to blue color development. Phytochemistry 54:85-92.

Zieslin N, Biran I, Halevy AH (1974). The effect of growth regulation on the growth and pigmentation of 'Baccara' rose flowers. Plant Cell Physiol. 15:341-349.

11

Prevalence and distribution in different agro-ecologies and identification of resistance source for wheat stripe rust

Vishal Gupta[1], R. A. Ahanger[1], V. K. Razdan[1], B. C. Sharma[2], Ichpal Singh[1], Kavaljeet Kaur[1] and M. K. Pandey[3]

[1]Division of Plant Pathology, Faculty of Agriculture, Sher-e-Kashmir University of Agricultural Sciences and Technology of Jammu, Chatha-180 009, Jammu, India.
[2]Division of Agronomy, Faculty of Agriculture, Sher-e-Kashmir University of Agricultural Sciences and Technology of Jammu, Chatha-180 009, Jammu, India.
[3]Division of Genetics and Plant Breeding, Faculty of Agriculture, Sher-e-Kashmir University of Agricultural Sciences and Technology of Jammu, Chatha-180 009, Jammu, India.

To assess the prevalence and distribution of wheat stripe rust (*Puccinia striiformis*), a survey of the randomly selected wheat fields located in sub-tropical to temperate agro-ecologies of Jammu province of Jammu and Kashmir was conducted consecutively for three years during *rabi* seasons of 2009-2010, 2010-2011 and 2011-2012. The results revealed that, by and large, all the wheat growing areas under study were found infested by this disease with maximum and minimum Area Under Rust Progress Curve (AURPC) values of 2422.21 and 351.70 reported from Jammu sub-tropics and inter-mediate Doda districts, respectively. The cropping season of *rabi* 2010-2011 recorded highest disease prevalence, probably due to conducive environmental conditions coupled with virulent pathotypes out-break and monoculture of wheat varieties. Among twenty one germplasms of wheat screened for source of disease resistance against stripe rust under epiphytotic conditions, Agra local and PBW-343 were found susceptible, RSP-561 showed moderate resistance, whereas, all the other cultivars were moderately susceptible to *Puccinia striiformis*.

Key words:*Puccinia striiformis*, stripe rust, wheat, distribution, resistance.

INTRODUCTION

In India, wheat (*Triticum aestivum* L.) is the second most important cereal crop next to rice. It is cultivated over an area of 29.9 million hectare with present production of 93.90 million tones and projected demand of 100 millions tones by 2030 AD (Sharma, 2011). Out of various biotic factors, stripe rust (yellow rust) caused by *Puccinia striiformis* West. f. sp. *tritici* is one of the main constraints in realizing the genetic potential of most of the cultivated varieties both in respect of yield and grain quality. Mono-culture of wheat cultivars coupled with broad adaptation of the pathogen to diverse climatic conditions and emergence of new pathotypes, make the disease a matter of great concern (Kolmer, 2005). In early 1970 and mid 1980's, the stripe rust epidemics occurred in North Africa, Indian subcontinent, Middle east, East Africa highland and China due to the breakdown of resistance genes (*Yr2* and *Yr9*) which were present in most of the cultivated varieties (Saari and Prescott, 1985; McIntosh, 2009; Chen et al., 2009). Presently, stripe rust is spreading rapidly in vast tracts stretching from Turkey, Syria and Northern Iraq to Southern Uzbekistan and the potential crop loss is in billions of dollars. In most wheat producing areas yield losses caused by stripe rust have ranged from 10 to 70 per cent, depending on the susceptibility of cultivar, earliness of the initial infection, rate of disease development and duration of disease (Chen, 2005). The distribution of the disease depends much upon climatic factors such as rainfall, humidity, temperature, etc. (Emge and Johnson, 1972). Lower temperature of 6 to 18°C, wet conditions along with intermittent rains favors stripe rust epidemics (Cortazar, 1985).

In India, yellow rust was observed in Punjab, Haryana, Western Uttar Pradesh and Jammu and Kashmir and was adjudged to be responsible for considerable yield losses (Nagarajan et al., 1984). During cropping season of 2010-11, the stripe rust appeared in severe form in almost all the wheat growing areas of Jammu region of Jammu and Kashmir state of India, wherein, wheat was grown over an area of 239 thousand hectares with a production and productivity of 465.33 thousand tones and 19.47q ha^{-1}, respectively (Anonymous, 2011). Further, variation in climatic conditions had led to a breakdown of disease resistance in mega wheat variety PBW-343 as well as out-breaks of new pathotypes, which caused a significant loss to the production of wheat. Timely application of newly identified fungicide molecules restricted the spread of stripe rust (Chen, 2005) but its use was found to be uneconomical and ecologically unsafe. Hence, growing of wheat cultivars having resistance genes against the *P. striiformis* remains the most inexpensive, efficient and sustainable method to manage the disease (Ittu, 2000). Generally, race-specific resistance genes and non-specific resistance genes have been detected and incorporated into commercial cultivars for providing effective protection against the disease. As the resistance is transitory because of outbreak of new virulent races (Stubbs, 1985), there is a need to investigate new source of resistance against the *P. striiformis*. Keeping in view, the role of this staple crop in food security and its economic value for the region, the present study was undertaken to know the distribution and prevalence of stripe rust of wheat and determination of its source of resistance.

MATERIALS AND METHODS

Occurrence of the stripe rust of wheat

The major wheat growing areas of Jammu province, that is, sub-tropical: Kathua, Samba, and Jammu districts with an altitude of 300 to 800 m above msl and rainfall of 1069 mm; intermediate: Rajouri, Udhampur and Doda districts having altitudinal range of 800 to 1500 m above msl and rainfall of 1478 mm; and temperate

Kishtwar district with altitude of more than15000 m above msl and 1052 mm rainfall were surveyed to ascertain the distribution of stripe rust, during *rabi* seasons of 2009-2010, 2010-2011 and 2011-2012. Disease incidence and disease severity (percent plant part) infected were recorded at fifteen days interval from January onwards till harvesting of the crop by simple random sampling using Modified Cobb's Scale (Peterson et al., 1948), and the value for area under rust progress curve (AURPC) was calculated using the formula given by Pandey et al (1989). The infected leaves of different cultivars from different locations were collected, pressed and dried in the laboratory and were submitted to Directorate of Wheat Research (ICAR), Regional Station Flowewrdale, Shimla for the pathotype identification.

Screening of germplasm

Twenty one germplasms of wheat *viz.* PBW-343, PBW-550, Raj-3765, Raj-3077, PBW-502, PBW-373, PBW-396, VL-804, TL-2966, HPW-289, VL-916, HS-508, TL-2942, TL-2963, HS-240, VL-907, HS-507, Agra Local, TBW-17, V-21 and RSP-561 were screened under artificial epiphytotic conditions for stripe rust resistances at Research Farm, SKUAST-J, Chatha, of Jammu and Kashmir state (India) during cropping season of *rabi* 2009 to 2010. The varieties were grown in lines of 1.5 m length with a row to row spacing of 22.5 cm in last week of November. Whole of the experimental area was surrounded by two rows of susceptible cultivar (A 9-30-1). The spore dust (mixed inoculums of various pathotypes) of urediospores of stripe rust was obtained from Regional Rust Research Centre (ICAR), Flowerdale, Shimla. The suspension was prepared by suspending the inoculum in distilled sterile water (95 ml) with small amount of Tween-20 (5 ml). 10 to 12 seedlings were maintained in each of the earthen pots filled with sterilized soil. The spore suspension so prepared was sprayed on two weeks old seedlings of susceptible wheat variety (A 9-30-1) thoroughly using an atomizer. Some plants were also inoculated directly by using lancet needle. The inoculated plants were misted with the help of hand sprayer and the polythene bags containing plants were kept in glass house for 48 h. These inoculated plants were then transferred in green house and irrigated regularly to create humid conditions. Regular monitoring was done for symptom development. On the development of symptoms, the inoculated pots were immediately transferred to the field to carry out the screening studies. The data on disease severity and host reaction was recorded according to modified Cobb's Scale (Peterson et al.,1948) and was combined to calculate the coefficient of infection (CI) by multiplying the severity value by 0.10, 0.4, 0.8 or 1.00 for host response ratings of resistant (R), moderately resistant (MR), moderately susceptible (MS) or susceptible (S), respectively (Pathan and Park, 2006) to classify the wheat varieties into different groups. The relationships of final rust severity (FRS) and AURPC with coefficient of infection were determined by correlation.

RESULTS AND DISCUSSION

The data presented in Table 1 revealed that the stripe rust of wheat was found- prevalent in all the major wheat growing areas of Jammu province in all the three cropping seasons of 2009-2010, 2010-2011 and 2011-2012. During 2009-2010, Jammu district recorded the maximum disease incidence, disease severity, final rust severity (FRS) and area under rust progressive curve (AURPC) with their corresponding values of 23.35, 22.42, 42.83 and 1312.55%, respectively, whereas, minimum disease incidence, disease severity, final rust severity

Table 1. Prevalence of stripe rust of wheat in Jammu province during 2009 to 2011.

District	Disease severity (%)			Disease severity (%)			Final rust severity (%)			AURPC Value		
	2009-2010	2010-2011	2011-2012	2009-2010	2010-2011	2011-2012	2009-2010	2010-2011	2011-2012	2009-2010	2010-2011	2011-2012
Kathua	15.07	21.88	21.77	14.37	39.43	37.43	34.67	80.33	75.33	817.4	1733.62	1333.12
Samba	21.3	25.97	21.97	17.15	44.5	41.11	33.92	82.95	80.95	987.25	1788.37	1754.17
Jammu	23.35	41.57	39.57	22.52	59.9	56.78	42.83	88.83	85.83	1312.55	2998.05	2956.05
Rajouri	11.72	13.72	12.92	11.05	17.55	16.05	36.75	40.75	38.57	397.43	477.32	462.22
Udhampur	10.91	12.91	12.67	9.57	12.97	12.57	35.33	48.77	45.77	405.72	428.32	418.32
Doda	10.8	12.8	11.66	7.14	11.84	11.09	28.57	38.67	35.67	321.28	378.12	355.72
Kishtwar	13.39	16.39	15.67	12.87	22.24	21.74	38.75	41.75	40.75	781.37	853.22	837.37
Mean± S.E	15.22± 1.67	20.75± 3.94	19.46± 3.71	13.52± 1.93	29.77± 6.94	28.11± 6.53	35.83± 1.66	60.29± 8.52	57.55± 8.34	717.57±119.33	1236.72±370.58	1159.56±358.46
Range	10.8-23.35	12.80-41.57	11.66-39.57	7.14-22.52	11.84-59.90	11.14-59.78	28.57-42.83	38.67-88.83	35.67-85.83	321.28-1312.55	378.12-2998.05	355.72-2956.05

(FRS) and area under rust progressive curve (AURPC) values of 10.80, 7.14, 28.57 and 321.28%, respectively were recorded in district Doda. In the next cropping season of 2010 to 2011, the stripe rust showed increased incidence and severity in all the areas as compared to the preceding year of 2009 to 2010 with maximum incidence and severity of 41.57 and 59.90%, respectively, in Jammu district and minimum incidence and severity of 12.80 and 11.84%, respectively, in Doda. However, the disease incidence and severity showed decreasing trend in 2011 to 2012 but it was found to be higher than first year (2009-10) and the values for disease incidence and disease severity were 39.57 and 56.78% in Jammu district and 11.66 and 11.09%, respectively, in Doda district. It was found that all the major cultivars succumbed to the disease, which may be due to the advent of new race(s) or changes in genetic makeup of the existing races as the rust pathogen itself is highly mutable and equipped with means to recombine this variability into a new lineage group. The capacity to develop durable and efficient control methods is largely based on the knowledge of the pathogen population structure and its potential for adapta-

tion to new cultivars (Brown, 1996). Out of different pathotypes, two races of the stripe rust, 78S84 and 46S119, were observed to be prevalent in the area, however, 78S84($Yr27$) was predominant and more aggressive in the Jammu province due to the widespread cultivation of susceptible wheat variety (PBW-343). New pathotypes with specific adaptation facilitated by acquisition of specific single gene virulence, and virulence combinations, were major factors underlying regional epidemics. The acquisition of virulence for $Yr2$ in the 1970s, $Yr9$ in 1990s, and $Yr27$ in recent years contributed significantly to regional and continental epidemics, resulting in heavy yield losses (Wellings et al., 2009). The information regarding the surveys of pathogen populations and the genetic characterization of virulence genes provide valuable information to design breeding strategies and prioritize which physiological races are to be targeted for durable resistance in wheat. Hailuar and Fininsab (2006) have also reported that stripe rust epidemics vary significantly between locations, seasons and also among cultivars. The weather data of Jammu sub-tropics observed during the three cropping season and given in Figures 1 to 4 revealed that the

average maximum and minimum temperatures of 23.1 and 10.1°C; maximum and minimum relative humidity of 86.5 and 52.00%, respectively and rainfall of 32.8 mm recorded during 2010 to 2011 were found to be quite favorable for the development of stripe rust epidemics as compared to the years 2009-2010 and 2011-2012. Further, the cultivation of susceptible varieties on large scale in the irrigated areas of sub-tropics along with intermittent rains during February, and March played a major role in the build-up of inoculums for development of the stripe rust. Although the replacement of susceptible varieties was through on mass scale but the presence of inoculum increased the disease severity and incidence in the successive year, whereas in the intermediate region, the maximum area under cultivation of wheat was rainfed and the average rainfall (84.30 mm) was also high and was in consonance with the critical water requirement stages of the crop , but due to low relative humidity, the rate of disease development was slow and the hosts showed tolerance response against the disease. In temperate zone, all the abiotic factors remained favorable for the development of the disease and the favorable

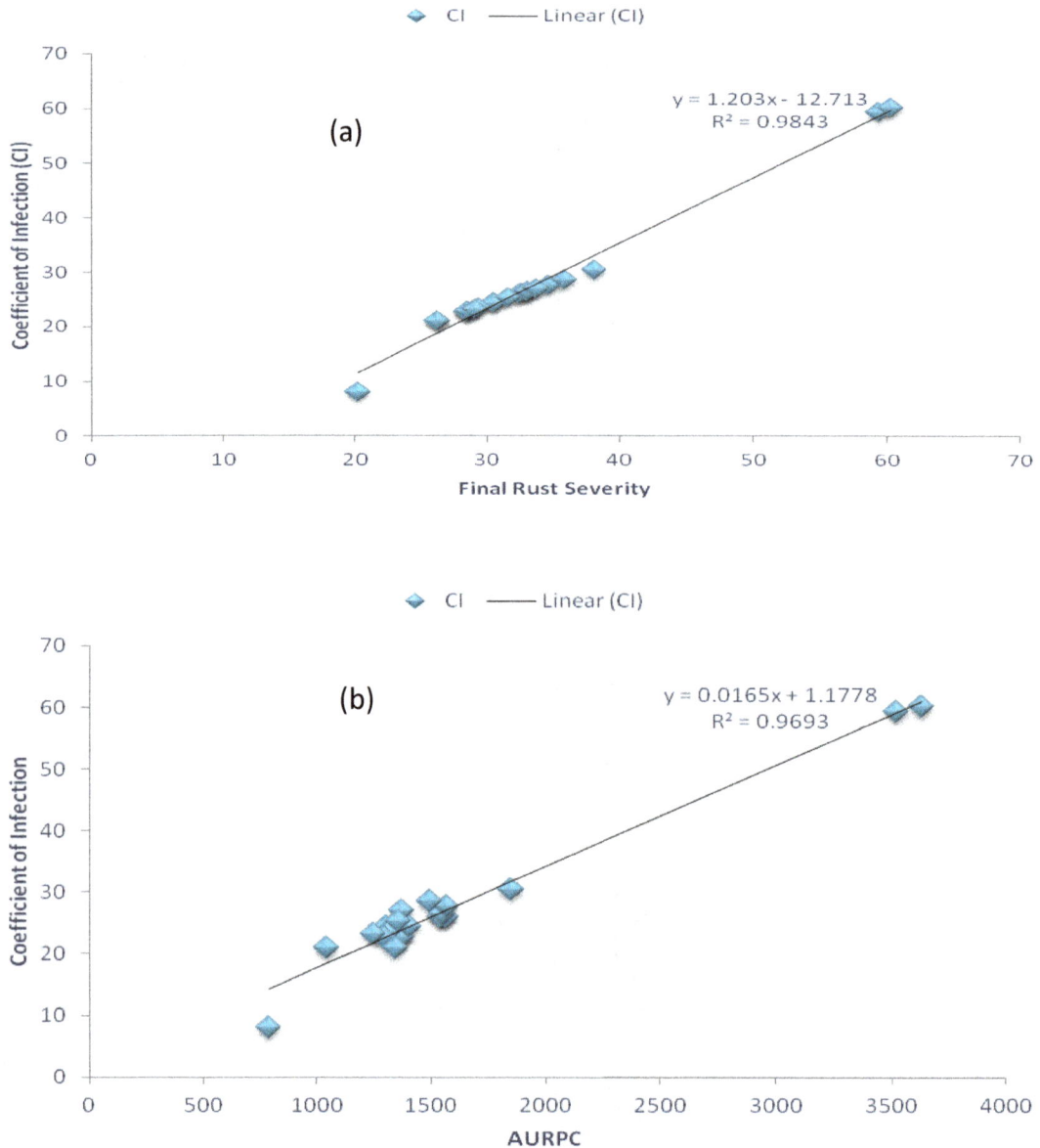

Figure 1. (a) Association between final rust severity (FRS) and coefficient of infection (CI). (b) Association between area under rust progress curve (AURPC) and coefficient of infection (CI).

weather conditions along with cultivation of susceptible and primitive varieties, usual delayed sowing, and presence of green bridges all might have contributed to the development of disease in all the three cropping seasons. Ali et al. (2009) recorded considerable variation in the expression of partial resistance to stripe rust across the locations having varied climatic conditions. Qamar et al. (2012) found that the temperature in January and February is positively correlated with yellow rust severity.

Among the twenty one germplasms of wheat screened against stripe rust (Table 2), the disease severity ranged from 0.57 to 19.26% in 1st week of February but by the end of crop season the disease severity on different cultivars varied from 20.21 to 60.23% with highest in Agra local (60.23%) followed by PBW-343 (59.32%) and lowest on RSP-561 (20.61%). From these results, it is apparent that stripe rust infection was well established in all the wheat germplasms screened for the disease. The germplasm viz., RSP-561 was ranked as moderately resistant with Area Under Rust Progress Curve (AURPC) of 787.13, whereas, PBW-550, PBW-373, HS-508, VL-804 TL-2963, HS-240, VL-907, Raj-3765, Raj-3077, PBW-502, PBW-396, TL-2966, HPW-289, VL-916, HS-508, TL-507, VL-21 and TBW-17 were ranked as moderately susceptible to yellow rust with AURPC ranging between 1043.93 and 1850.78, while Agra local

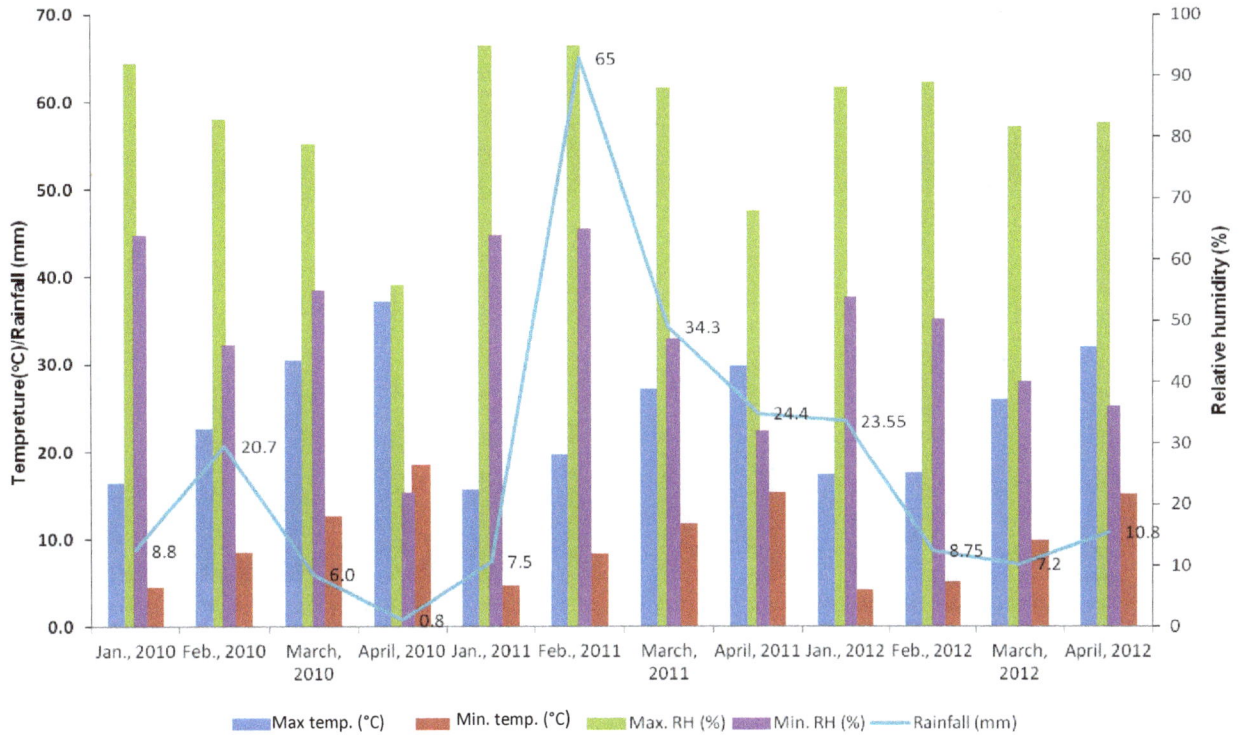

Figure 2. Meteorological data of Sub-tropics region during 2010-2011 and 2012.

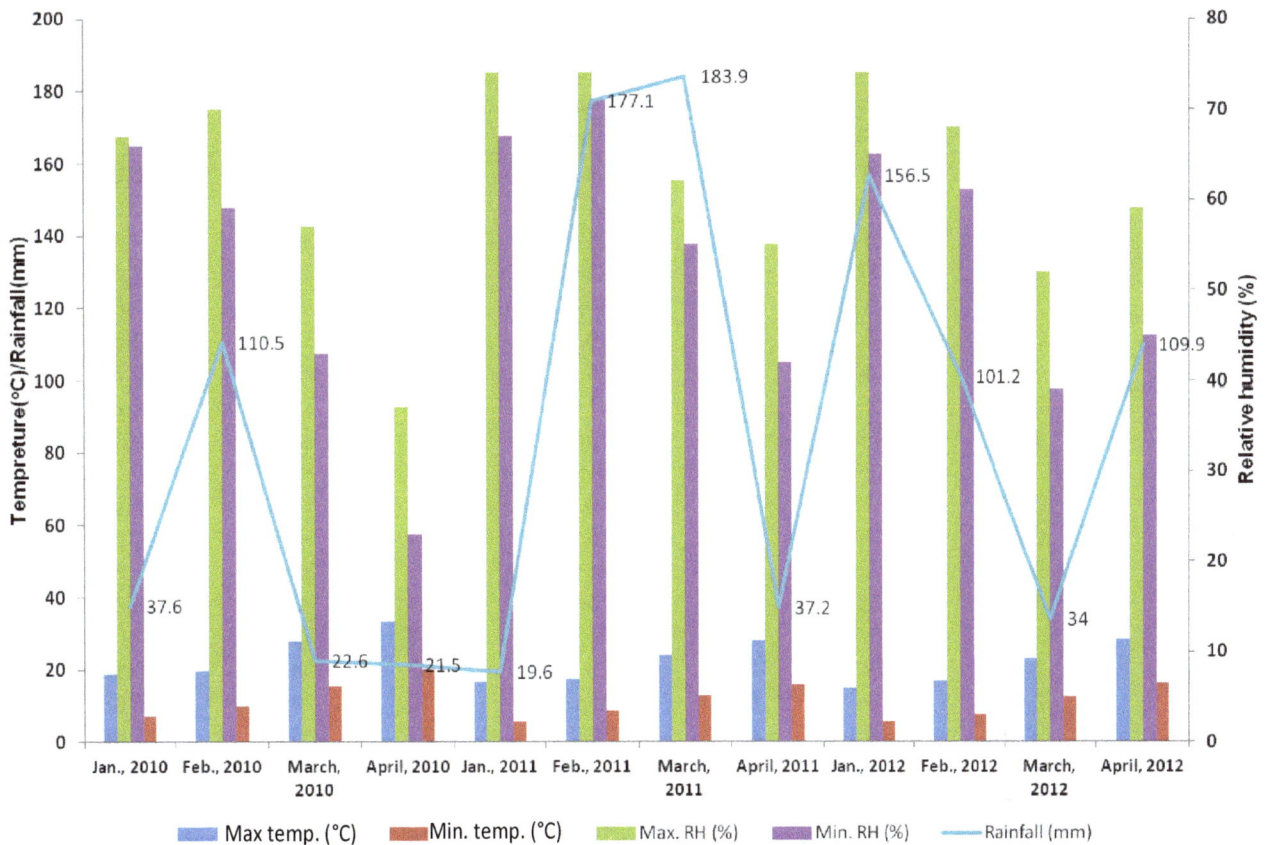

Figure 3. Meteorological data of Intermediate region during 2010-2011 and 2012.

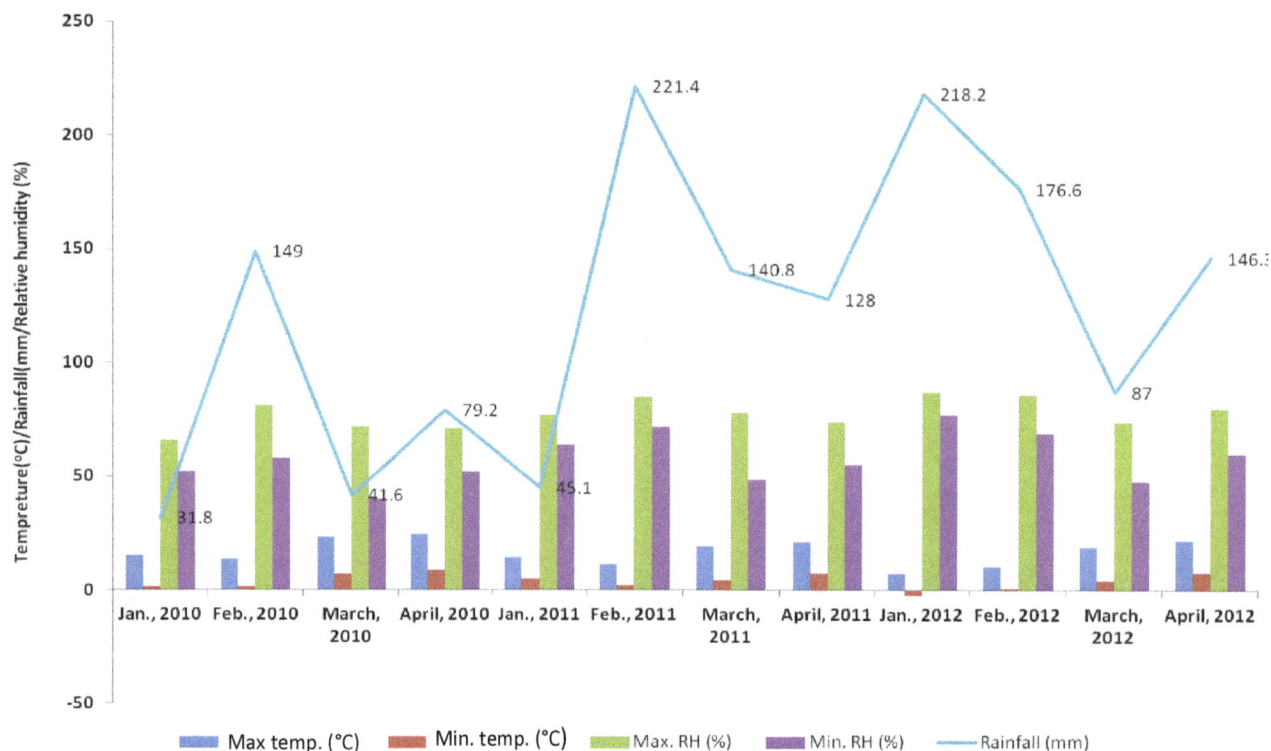

Figure 4. Meteorological data of temperate region during 2010-2011 and 2012.

Table 2. Screening of wheat germplasm against stripe rust under epiphytotic conditions, during 2009-10.

Germplasm	Percent disease severity at different intervals						AURPC	C.I.
	1st Feb	15th Feb	2nd March	17th March	1st April	16th April		
PBW-343	18.39	36.69	45.62	54.97	58.32	59.32	3516.83	59.32
PBW-550	5.03	15.83	17.83	21.65	23.54	26.21	1043.93	20.97
Raj-3765	3.13	10.17	12.33	25.33	27.31	28.46	1364.03	22.77
Raj-3077	3.03	9.17	15.67	28.67	31.22	32.58	1538.03	26.06
PBW-502	2.34	10.17	15.33	30.00	31.65	32.41	1567.88	25.93
PBW-373	3.87	10.83	12.67	30.00	31.56	33.01	1552.50	26.41
PBW-396	3.24	10.50	12.33	25.33	27.37	28.41	1370.33	22.73
VL-804	2.15	8.50	12.00	27.33	29.45	30.52	1404.23	24.42
TL-2966	2.43	7.89	14.33	30.33	32.66	33.06	1544.33	26.45
HPW-289	3.46	10.30	20.33	35.33	36.67	38.05	1850.78	30.44
VL-916	3.98	11.33	12.00	30.00	31.78	34.59	1565.93	27.67
HS-508	0.57	5.89	12.33	25.33	27.64	30.49	1300.80	24.39
TL-2942	0.99	6.33	14.67	28.67	31.45	35.78	1492.58	28.62
TL-2963	1.76	8.36	12.00	25.00	26.34	29.29	1308.38	23.43
HS-240	1.21	6.67	12.33	26.33	28.65	33.72	1371.68	26.98
VL-907	2.85	9.36	12.00	23.33	25.67	28.47	1290.30	22.78
HS-507	2.65	8.54	12.00	22.00	24.72	29.04	1246.58	23.23
Agra local	19.26	38.60	47.43	56.74	59.21	60.23	3625.88	60.23
TBW-17	1.43	6.33	12.67	25.33	29.46	31.62	1354.73	25.30
V-21	5.26	8.60	17.43	22.74	25.21	26.23	1345.88	20.98
RSP-561	3.43	9.83	12.33	18.83	19.54	20.21	787.13	8.08

AURPC: Area under rust progress curve; C.I: Coefficient of infection.

Table 3. Response of different wheat germplasm against stripe rust, during 2009-2010.

Disease score	Disease reaction	Germplasm
1	Immune (0)	Nil
2	Nearly immune (1 to 5%)	Nil
3	Resistant (6-10%)	Nil
4	Moderately resistant (11-25%)	RSP-561
5	Moderately susceptible (26-40%)	TL-2963, HS 240, VL-907, Raj-3765, Raj-3077, PBW-502, PBW-396,VL-804, TL-2966, HPW-289, VL-916, HS-508, HS 507, TL-507, TBW-17, PBW-373, PBW-550, V-21.
6	Susceptible (41-65%)	Agra local, PBW-343
7	Highly susceptible (>65%)	Nil

and PBW-343 having AURPC of 3625.88 and 3516.83, respectively, were ranked as susceptible to yellow rust. None of the germplasm showed resistant reaction against the disease (Table 3). The positive correlation coefficient of infection (Figure 1a and b) was found with FRS and AURPC with R^2 value of 0.984 and 0.969, respectively, which was found to be in conformity with findings of Ali et al. (2008) who had also found strong association between coefficient of infection (CI), with FRS and AURPC. Sinha et al. (2006) have found that the popular variety PBW 343 grown in the state of Punjab had fallen susceptible to the newly identified race 78S84 of yellow rust. Dadrezaei and Torabi (2001) on the other hand reported that the disease appeared late on the resistant cultivars as compared to the susceptible cultivars. The delay in the appearance of yellow rust and its slow development on resistant cultivars resulted in late secondary infections until unfavorable climatic conditions occurred. Ahmad et al. (2006) reported that there was a considerable amount of genetic variation among various entries ranging from immune to susceptible response. They observed that among the entries screened, 19 entries (17.60%) were found to be totally immune (average coefficient of infection value = 0.00), 13 entries (12.03%) as resistant (C.I $\leq$ 3), 29 entries (26.85%) susceptibility (CI < 10), and the rest were rated as highly susceptible (CI > 10).

REFERENCES

Ahmad M, Alam SS, Alam S, Khan IA, Ahmad N (2006). Evaluation of wheat germplasm against yellow rust (*Puccinia striiformis* f. sp. *tritici*) under natural conditions. Sar. J. Agric. 22(3):662-65.

Ali S, Shah SJA, Maqbool K (2008). Field based assessment of partial resistance to yellow rust in wheat germplasm. J. Agric. Rural Dev. 6(1-2):99-106.

Ali S, Shah SJA, Rahman H (2009). Multilocation variability in Pakistan for partial resistance in wheat to Puccinia striiformis f. sp. tritici. Phyto. Medit. 48:269-279.

Anonymous (2011). *Database Digest of Statistics*. Directorate of Economics and Statistics, Govt. of Jammu and Kashmir. P. 121.

Brown JKM (1996). The choice of molecular marker methods for population genetics studies of plant pathology. New Phytol. 133:181–195.

Chen WQ, Wu LR, Liu TG, Xu SC. (2009). Race dynamics, diversity and virulence evolution in *Puccinia striiformis* f.sp. *tritici*, the causal agent of wheat stripe rust in China from 2003 to 2007. Plant Dis. 93:1093-1101

Chen XM (2005). Epidemiology and control of stripe rust (*Puccinia striiformis* f. sp. *tritici*) on wheat. Can. J. Plant Pathol. 27:314–337.

Cortazar SR (1985). Relation between rainfall and temperature and incidence of three wheat rusts, at La Plantian Experimental Station. Agric. Tecnia. 45:273-277.

Dadrezaei ST, Torabi M (2001). Study of yellow rust epidemiology in Khuzestan Province of Iran. Meeting the challenge of yellow rust in cereal crops. *Proceedings* of the First Regional Conference on Yellow Rust in the Central and West Asia and North Africa Region. pp. 223-226. Karaj, Iran.

Emge RG, Johnson DR (1972). Epiphytology of stripe rust of wheat, caused by *Puccinia striiformis*, in north east Oregon during 1971. Plant Dis. Rep. 56:1071-1073.

Hailuar D, Fininsab C (2006) . Epidemics of stripe rust (*Puccinia striiformis*) on common wheat (*Triticum aestivum*) in the highlands of Bale, Southeastern Ethiopia. Sci. Dir. 26(8):1209-1218.

Ittu M (2000). Components of partial resistance to leaf rust in wheat. Acta. Phytopathologica et Entomologica Hungarica 35:161-168.

Kolmer JA (2005). Tracking wheat rust on a continental scale. Curr. Opin. Plant Biol. 8:441-449.

McIntosh RA (2009) History and status of the wheat rusts. In RA McIntosh (ed) Proc Borlaug Global Rust Initiative 2009 Technical Workshop BGRI Cd Obregon, Mexico, pp.11-23.

Nagarajan S, Bahadur P, Nayar SK (1984). Occurrence of a new virulence, 47S102 of *Puccinia stridormis* West., in India during the crop year 1982. Cereal Rusts 12:28-31.

Pandey HN, Menon TCM, Rao MV (1989). A single formula for calculating Area Under Rust Progress Curve. *Rachis* 8:36-39.

Pathan AK, Park RF (2006). Evaluation of seedling and adult plant resistance to leaf rust in European wheat cultivars. Euphytica 149:327-342.

Peterson RF, Campbell AB, Hannah AE (1948). A diagrammatic scale for estimating rust intensity of leaves and stem of cereals. Can. J. Res. 26:496-500.

Saari EE, Prescott JM (1985). World distribution in relation to economic losses. In: Roelfs AP, Bushnell WR (eds) The cereal rusts Vol II. Academic Press Inc, Orlando, pp. 259-298.

Sharma I (2011). Vision 2030. Directorate of Wheat Research (ICAR), Karnal-Haryana. P. 1.

Sinha VC, Parashar M, Sharma RK (2006). Profile of yellow rust pathotypes prevalent in the Northern Plains of India and the resistance genes to counter them. SABRAO J. Bree.Gen. 38(1):59-62.

Stubbs RW (1985). Stripe rust: In Roelfs, AP, Bushnell and WR (eds.) The Cereal Rusts, Volume II, Diseases, Distribution, Epidemiology and Control. Academic Press, Inc., Orlando, Florida. pp. 61-101.

Qamar M, Ahmad SD, Asif M (2012). Determination of levels of resistance in Pakistani bread wheat cultivars against stripe rust (*Puccinia striiformis*) under field conditions. Afr. J. Agric. Res. 7(44):5887-5897.

Wellings CR, Singh RP, Yahyaoui A, Nazari K, McIntosh RA (2009). The development and application of near-isogenic lines for monitoring cereal rust pathogens. In: McIntosh RA (ed) Proc Borlaug Global Rust Initiative Technical Workshop, BGRI Cd Obregon, Mexico, pp. 77-87.

12

Studies on agricultural waste management through preparation and utilization of organic manures for maintaining soil quality

Sruti Karmakar[1] , Koushik Brahmachari[2] and Aniruddha Gangopadhyay[1]

[1]Department of Earth and Environmental Studies, National Institute of Technology, Durgapur – 713209, Burdwan, West Bengal, India.
[2]Department of Agronomy, Bidhan Chandra Krishi Viswavidyalaya, Mohanpur – 741252, Nadia, West Bengal, India.

Solid waste management has become one of the vital issues to protect health and public safety. Preparation of organic manures like vermicompost, Farm Yard Manure (FYM) etc. from various organic wastes (agricultural wastes) will save our environment from pollution as well as application of these manures in agricultural land prevent those lands from the harmful effect of chemical fertilizers. With these views keeping in background for saving our environment from ill effects of indiscriminate use of chemical fertilizers by substituting them partially or entirely through applying organic manures after converting agricultural wastes into wealth (organic manures), an experiment was carried out in the farmer's field at village Shikarpur (P.O. Bhagirathi Shilpashram, Dist. Nadia, Pin. 741248, W.B., India) during the year 2008 to 2010 with two crops (rice –rainy season and Lentil –winter season). The experiment was laid out in randomized block design with 5 treatments (T_0- without fertilizer or manure, T_1-100% organic through vermicompost, T_2- 100% organic through FYM, T_3-100% chemical through fertilizer and T_4-50% organic through mixed organic manure + 50% chemical through fertilizer) replicated 3 times. It has been found that the vermicompost treated soil showed better result in comparison to that demonstrated by the chemical fertilizers in terms of soil physical and chemical properties as well as productivity of soil.

Key words: Organic waste, vermicompost, farm yard manure, soil quality, chemical fertilizer.

INTRODUCTION

Wastes which arise virtually from all human activities can be classified conveniently with respect to their source. Major categories include household and consumer wastes (for example, municipal wastes), industrial wastes, agricultural wastes, extraction wastes, energy production wastes and sewage sludges. Waste produced by agricultural activities comprises crop residues, stubbles, straws, animal slurries, silage effluents, weeds etc. With the increase in global temperature, weeds which are mostly of C_4 types are sustaining in a better way due to increased rate of photosynthesis with decreasing photorespiration. Thus, these weeds are occupying fallow lands vis-à-vis are encroaching agricultural land in a vigorous way. Huge biomass of these weeds after destruction are being dumped in open site; creating nuisance to environment. India alone produces more than 400 million tones of agricultural wastes annually. It has got a very large percentage of total world production of rice husk, jute, stalk, baggase, groundnut shell and coconut fiber etc. (Raju et al., 2012).

Increase in cropping intensity results in generation of huge biomass throughout the year from agricultural fields. After harvesting, threshing and post harvest processing bulk amount of crop residues and stubbles remains dumped in the agricultural land or nearby areas. These wastes may cause tremendous environmental pollution which may affect health and wellbeing of living organisms including human. Accumulation and putrefaction of these wastes and consequent adverse effects on surroundings have become a serious issue. To get rid of such situation, proper management of these organic wastes is very essential. In one scenario, there is trouble associated with organic waste management. In the other scenario, there is a problem related to intensive use of chemical fertilizers which is creating toxicity in soil of agricultural fields. Sometimes these toxic chemicals accumulate in plants as a residue of fertilizer added.

Frequent application of chemical fertilizers is deteriorating bio- physico- chemical properties of soil. As a result, soil fertility is being diminished gradually. This in turn is leading to reduction in crop yield per unit area. So it is an urgent need to reduce the use of chemicals in agricultural fields by using organics more and more. Use of organic manures produced/prepared from various organic wastes will save our environment as a whole; simultaneously organic wastes can also be managed properly. Moreover, it enhance soil health which is the balance between soil function for productivity, environmental quality, and plant and animal health (Doran and Zeiss, 2000; Doran, 2002). In this context, use of organic manures such as vermicompost, FYM etc. may supply sufficient amount of micro nutrients in available form to crops and improve the quality of the agricultural produces (Maynard, 1993). Besides supplying various nutrients to the current crop, they often leave substantial residual effect to succeeding crops. Application of organic manures helps to improve health as well as quality of soil. According to Doran et al. (1998), soil quality is the capacity of a living soil to function within natural or managed ecosystem boundaries. Soil health and soil quality are functional concepts which indicate how fit the soil is to support plant and animal productivity, maintain water and air quality, and support plant and animal health. Thus, soil quality can be regarded as soil health (Doran et al., 1996).

In the era of globalization, time has come to think about organic agriculture or organic farming in India also to sustain in globalized market of quality agricultural products. Vermicomposting is the process of producing compost through the action of earthworm. It is an eco- biotechnological process that transforms energy-rich and complex organic substances into stabilized humus-like product vermicompost. Preparation of vermicompost is an efficient as well as easily adoptable technique of compost preparation. This composting system can not only decompose a huge amount of organic wastes but also help to maintain higher nutrient status in composted materials (Ceccanti and Masciandaro, 1999; Lazcano and

Domínguez, 2011; Hema and Rajkumar, 2012). Keeping all these thoughts in background with the broader objective of saving our environment as a whole through proper management of agricultural wastes, a field experiment was carried out at village Shikarpur, Bhagirathi Shilpashram, Nadia- 741248, West Bengal, India to prepare organic manure from low-cost locally available organic waste through vermiculture biotechnology with the intension of substituting chemical fertilizers partially or entirely, augmenting soil quality for sustainability in agricultural production and to study the physical and chemical properties of soil in agricultural fields and also to establish the efficacy of vermicompost in comparison to chemical fertilizers and FYM vis-à-vis other organic manures.

MATERIALS AND METHODS

The name and address of the owner of vermicompost and FYM unit

Mr Animesh Mondal, Shantinagar, Madanpur, Nadia -741245, West Bengal, India (latitude-23° 0' 15.69" N, longitude-88° 29' 24.32" E).

Composts preparation and analysis

Vermicompost and FYM were prepared by Heap (Basak et al., 2011) and Trench methods, respectively (Sahai, 2004). The chemical properties (organic carbon, total nitrogen, total phosphorus, and total potash) of these manures were analysed. The organic carbon was determined by Walkley and Black's rapid titration method (Jackson, 1973). Total nitrogen was estimated by modified macro Kjeldahl method (Jackson, 1973). Total phosphorus was determined by Olsen's method (Jackson, 1973) and total potash was determined by the flame photometer method (Jackson, 1973).

Location and soil type of the experimental site

Shikarpur, Bhagirathi Shilpashram, Nadia-741248, West Bengal, India (latitude-23° 1' 53.62" N, longitude-88° 30' 46.97" E). According to textural classes proposed by U. S. Bureau of soils (Sahai, 2004), soil of experimental site was sandy loam in texture because it consisted of 73.7% sand, 10% silt and 16.3% clay.

Analysis of soil

After collection (twice- before crop establishment and after harvesting of crops), the soil samples were prepared for analyses in the laboratory. For preparation of soil samples different procedures were involved such as: drying, grinding, mixing, partitioning, sieving etc. Different physical and chemical properties were analysed by using different methods. Bulk density was determined by the method of Blake and Hartge (1986). Total porosity was estimated from the bulk density and particle density. Mechanical analysis of soil samples was determined following the Boyoucos hydrometer method (Gee and Bauder, 1986). The water holding capacity (WHC) of the soil was measured with the help of Keen- Rackzowski box as described by Baruah and Barthakur (1997). Saturated hydraulic conductivity was calculated by Dracy's equation. Water stable aggregates and their distribution in each soil layer under

Table 1. Chemical composition of applied vermicompost and FYM.

Composts	Organic C (%)	Total N (%)	Total P_2O_5 (%)	Total K_2O (%)
Vermicompost	11.9	1.23	2.06	0.78
FYM	5.29	0.53	0.25	0.6

different treatments were determined by wet sieving method as described by Yoder (1936). The pH of the soil sample was measured with the help of Backman's pH meter. Organic carbon was determined by Walkley and Black's rapid titration method (Jackson, 1973). Available nitrogen was estimated by Kjeldahl method (Jackson, 1973). Available phosphorus was determined by Olsen's method (Jackson, 1973) and available potassium was estimated by the flame photometer method (Jackson, 1973).

About crops

Two crops namely rice / *Oryza sativa* L. (rainy season- July to November) and lentil / *Lens culinaris Medik.* (winter season-November to March) were selected and sown. Their varieties were IET-4094 (Khitish) and B-77 (Asha), respectively. The experiment was laid out in randomized block design with 5 treatments (T_0-without fertilizer or manure, T_1-100% organic through vermicompost, T_2- 100% organic through FYM, T_3-100% chemical through fertilizer and T_4-50% organic through mixed organic manure + 50% chemical through fertilizer) replicated 3 times. Yield was recorded and statistically analyzed during two successive cropping years (2008 to 2009 and 2009 to 2010).

RESULTS AND DISCUSSION

In present research, following chemical analysis of organic manures (Table 1), it was found that the applied vermicompost contained 11.9% organic carbon, 1.23% total nitrogen, 2.06% total phosphorus and 0.78% total potash. Similar results were observed by Purohit (2006) and Palaniappan and Annadurai (2008). They opined that depending upon the nature of substrate, on an average the vermicompost contained 10.12 to 11.98% organic carbon, 1.09 to 2.75% total nitrogen, 2 to 2.45% total phosphorus and 0.78 to 1.39% total potash. Chemical compositions of applied FYM in current study were 5.29% organic carbon, 0.53% total nitrogen, 0.25% total phosphorus and 0.6% total potash. This result is also in agreement with those observed by Sahai (2004) and Roychoudhury et al. (2010). In their study, FYM contained on an average of 5.1 to 5.4% organic carbon, 0.52 to 0.56% total nitrogen, 0.23 to 0.28% total phosphorus and 0.58 to 0.63% total potash. Table 2 exhibits the comparison of various physical properties of soil for different treatments (T_i- initial property of soil or property of soil before crop establishment,T_0- without fertilizer or manure, T_1- 100% organic through vermicompost, T_2- 100% organic through FYM, T_3-100% chemical through fertilizer and T_4-50% organic + 50% chemical). From this table, it is clear that the bulk density value was found to be insignificantly (p>0.05) increased while results of treatment T_0 was compared with initial state of the soil before crop establishment (T_i). For all the treatments except T_3 (100% chemical through fertilizer), bulk density values significantly (p≤0.05) reduced in comparison to control that is, T_0 (without fertilize or manure). This value significantly (p≤0.05) reduced in case of T_1 comparison to T_3. This reduction in the values of bulk density might be due to the presence of organic materials in all those treatments (T_1, T_2, and T_4).

According to Miller et al. (2002) and Shirani et al. (2002), application of organic materials (manure and/or crop residues) can increase soil organic matter concentration and decrease bulk density. The value of porosity was significantly (p≤0.05) decreased in case of T_0 when compared with T_i. There was no significant (p>0.05) increase in porosity value while result of treatment T_0 (without fertilize or manure) was judged against the treatment T_3 (100% chemical through fertilizer). Significant (p≤0.05) increase in porosity values were found in case of treatment T_1 (100% organic through vermicompost) in comparison to both T_3 (100% chemical through fertilizer) and T_0 (without fertilize or manure). This indicates that treatment with 100% vermicompost is very much beneficial for enhancing porosity of soil. According to Sahai (2004), organic manure increases percentage of pore space in soil. Jadhav et al. (1993) noticed that application of vermicompost increased soil porosity. Table 2 also depicts that percentage of maximum water holding capacity was found to be significantly (p≤0.05) decreased while result of treatment T_0 (without fertilize or manure) was evaluated against treatment Ti (initial state of the soil before crop establishment). This value was increased significantly (p≤0.05) when T_1 (100% organic through vermicompost), T_2 (100% organic through FYM), T_3 (100% chemical through fertilizer) and T_4 (50% organic + 50% chemical) were compared with T_0 (without fertilize or manure). Percentage of maximum water holding capacity significantly (p≤0.05) increased in case of T_1 (100% organic through vermicompost) in comparison to T_3 (100% chemical through fertilizer). According to Biswas and Khosla (1971), addition of organic manures significantly improved water holding capacity of soil, compared to sole inorganic fertilizer application. Change was found to be insignificant (p>0.05) while saturated hydraulic conductivity of initial state of the soil before crop establishment (Ti) was compared with treatment T_0 (without fertilize or manure). Value of saturated hydraulic conductivity was significantly (p≤0.05) reduced for treatments T_1 (100% organic through vermicompost), T_2 (100% organic

Table 2. Comparison of various physical properties of soil for different treatments.

Bulk density (g.cm^{-3})

Replication	T_i	T_0	T_1	T_2	T_3	T_4	T_i/T_0	T_0/T_1	T_0/T_2	T_0/T_3	T_0/T_4	T_1/T_3
R_1	1.61	1.65	1.42	1.46	1.64	1.5	0.19	0.00	0.00	0.57	0.01	0.00
R_2	1.65	1.69	1.46	1.42	1.66	1.48						
R_3	1.66	1.67	1.41	1.52	1.68	1.57						
mean	1.64	1.67	1.43	1.47	1.66	1.52						
sd	0.026	0.020	0.026	0.050	0.020	0.047						

Porosity (%)

Replication	T_i	T_0	T_1	T_2	T_3	T_4	T_i/T_0	T_0/T_1	T_0/T_2	T_0/T_3	T_0/T_4	T_1/T_3
R_1	39.01	36.9	46.1	40.12	37.01	39	0	0	0	0.11	0	0
R_2	39.09	37	46.4	40.1	37.03	39.02						
R_3	39.11	36.8	46.1	40.02	37.02	39.01						
mean	39.07	36.9	46.2	40.08	37.02	39.01						
sd	0.053	0.1	0.173	0.053	0.01	0.01						

Maximum water holding capacity (%)

Replication	T_i	T_0	T_1	T_2	T_3	T_4	T_i/T_0	T_0/T_1	T_0/T_2	T_0/T_3	T_0/T_4	T_1/T_3
R_1	38.94	36.4	46.05	44.13	38.39	43.07	0.01	0	0	0.01	0	0
R_2	38.89	37.6	46.07	44.2	38.41	43.15						
R_3	38.93	37	46.09	44.12	38.46	43.14						
mean	38.92	37	46.07	44.15	38.42	43.12						
sd	0.026	0.6	0.02	0.044	0.036	0.044						

Saturated hydraulic conductivity (cm.h^{-1})

Replication	T_i	T_0	T_1	T_2	T_3	T_4	T_i/T_0	T_0/T_1	T_0/T_2	T_0/T_3	T_0/T_4	T_1/T_3
R_1	1.3	1.5	0.02	0.02	0.47	0.35	0.85	0.00	0.00	0.01	0.00	0
R_2	1.1	1.1	0.04	0.03	0.5	0.4						
R_3	1.2	1.09	0.03	0.07	0.5	0.39						
mean	1.2	1.23	0.03	0.04	0.49	0.38						
sd	0.1	0.234	0.01	0.026	0.017	0.026						

Aggregate ratio

Replication	T_i	T_0	T_1	T_2	T_3	T_4	T_i/T_0	T_0/T_1	T_0/T_2	T_0/T_3	T_0/T_4	T_1/T_3
R_1	0.6	0.61	0.75	0.3	0.24	0.53	0.39	0.00	0.00	0	0.01	0
R_2	0.62	0.67	0.79	0.35	0.27	0.55						
R_3	0.64	0.64	0.8	0.31	0.27	0.56						
mean	0.62	0.64	0.78	0.32	0.26	0.55						
sd	0.02	0.03	0.026	0.026	0.017	0.015						

Percentage aggregate stability (%)

Replication	T_i	T_0	T_1	T_2	T_3	T_4	T_i/T_0	T_0/T_1	T_0/T_2	T_0/T_3	T_0/T_4	T_1/T_3
R_1	37.15	37.03	38.24	27.6	20.07	32.1	0.10	0	0.00	0	0	0
R_2	37.2	37.12	38.25	27.4	20.14	32.8						
R_3	37.22	37.15	38.26	27.5	20.09	32.6						
mean	37.19	37.1	38.25	27.5	20.1	32.5						
sd	0.036	0.062	0.01	0.1	0.036	0.361						

Mean weight diameter (mm)

Replication	T_i	T_0	T_1	T_2	T_3	T_4	T_i/T_0	T_0/T_1	T_0/T_2	T_0/T_3	T_0/T_4	T_1/T_3
R_1	1.1	1.02	0.56	0.5	0.32	0.52	0.02	0	0	0	0	0.00
R_2	1.12	1.06	0.59	0.55	0.37	0.57						
R_3	1.17	1.04	0.59	0.51	0.39	0.56						
mean	1.13	1.04	0.58	0.52	0.36	0.55						
sd	0.036	0.02	0.017	0.026	0.036	0.026						

T_i- Initial property of soil or property of soil before crop establishment, T_0- without fertilizer or manure, T_1- 100% organic through vermicompost, T_2- 100% organic through FYM, T_3- 100% chemical through fertilizer and T_4- 50% organic + 50% chemical.

through FYM), T_3 (100% chemical through fertilizer) and T_4 (50% organic + 50% chemical) in comparison to treatment T_0 (without fertilize or manure). This value significantly ($p \leq 0.05$) reduced in case of T_1 comparison to T_3.

Change was found to be insignificant ($p > 0.05$) while soil aggregate ratios of initial state of the soil before crop establishment (Ti) was compared with treatment T_0 (without fertilize or manure). Apparent decrease in values of soil aggregate ratios (after two years) was statistically significant ($p \leq 0.05$) for all treatments except the treatment T_1 (100% organic through vermicompost) in comparison to treatment T_0 (without fertilize or manure). Significant ($p \leq 0.05$) increase was observed in case of treatment T_1 (100% organic through vermicompost) in comparison to the treatment T_3 (100% chemical through fertilizer). Significant ($p \leq 0.05$) changes in soil aggregate stability was found for any treatment under present study following two years in comparison to treatment T_0 (without fertilize or manure) except initial state of the soil before crop establishment (Ti). This value significantly ($p \leq 0.05$) reduced in case of T_3 comparison to T_1. The value of mean weight diameter was significantly ($p \leq 0.05$) decreased in case of T_0 (without fertilize or manure) comparison to treatment Ti (initial state of the soil before crop establishment). Significant ($p \leq 0.05$) decrease were observed in case of treatments T_1 (100% organic through vermicompost), T_2 (100% organic through FYM), T_3 (100% chemical through fertilizer) and T_4 (50% organic + 50% chemical) in comparison to T_0 (without fertilize or manure). In case of treatment T_1 (100% organic through vermicompost), this value was significantly ($p \leq 0.05$) increased compared with T_3 (100% chemical through fertilizer).

Application of vermicompost to soil gives a tremendous boost to soil physical health by improving water-holding capacity, structure formation and also by enhancing fertility (Jeyabal and Kuppuswamy, 2001; Edwards, 1998). Table 3 indicates the chemical properties of soil after harvesting of crops. This table enumerates the pH of soil was decreased insignificantly ($p > 0.05$) while pH value of soil before crop establishment (Ti) was compared with treatment T_0 (without fertilize or manure). Values of pH were decreased insignificantly ($p > 0.05$) for treatments T_1 and T_2 and increased insignificantly ($p > 0.05$) in case of T_3 contrasted with treatment T_0. In case of treatment T_1 (100% organic through vermicompost), this value was insignificantly ($p > 0.05$) decreased compared with T_3 (100% chemical through fertilizer). From this study, it was evident that organic manure alone can decrease the alkalinity of soil rapidly than chemical fertilizer. A study to know effect of FYM on soil pH showed that there was decrease in pH from 7.99 to 7.65 and each increment of FYM reduced the soil pH significantly due to organic acid production during its decomposition (Patil et al., 2003).

There was significant ($p \leq 0.05$) decrease in percentage of organic carbon while result of treatment T_0 (without fertilizer or manure) was judged against initial state of the soil

before crop establishment (Ti).

Significant ($p \leq 0.05$) increase in percentage of organic carbon was found in case of treatments T_1, T_2 T_3 and T_4 when it was evaluated against the treatment T_0. The percentage of organic carbon was significantly ($p \leq 0.05$) increased in case of treatment T_1 (100% organic through vermicompost) compared with T_3 (100% chemical through fertilizer). Reduction in percentage of total nitrogen was found to be insignificant ($p > 0.05$) in case of treatment T_0 (without fertilizer or manure) when it was compared with initial state of the soil before crop establishment (Ti). Percentages of total nitrogen was significantly ($p \leq 0.05$) increased for treatments T_1 (100% organic through vermicompost), T_3 (100% chemical through fertilizer) and T_4 (50% organic + 50% chemical) in comparison to treatment T_0 (without fertilizer or manure). These results indicate that in general organic manures as well as chemical fertilizers have positive impact on total nitrogen of soil. The percentage of total nitrogen was significantly ($p \leq 0.05$) higher in case of T_1 in comparison to treatment T_3 (100% chemical through fertilizer). From this table, it is clear that the significant ($p \leq 0.05$) reduction was found while amount of available phosphorus was compared between soil before crop establishment (Ti) and soil after two years following crop establishment with treatment T_0 (without fertilizer or manure). The amount of available phosphorus was significantly ($p \leq 0.05$) increased for treatments T_1 (100% organic through vermicompost), T_2 (100% organic through FYM), T_3 (100% chemical through fertilizer) and T_4 (50% organic + 50% chemical) in comparison to treatment T_0 (without fertilizer or manure). There was significant ($p \leq 0.05$) decrease in amount of available K while result of treatment T_0 (without fertilizer or manure) was compared with initial state of the soil before crop establishment (Ti).

In case of treatment T_1 (100% organic through vermicompost), this value was significantly ($p \leq 0.05$) increased compared with T_3 (100% chemical through fertilizer). The value of available K was found to be increased significantly ($p \leq 0.05$) for all the treatments (T_1, T_2, T_3 and T_4) compared to treatment T_0. Value of available K was significantly low for treatment T_3 in comparison to T_1. Thus, it is clear that both organic manures and chemical fertilizers can increase the amount of available K but efficacy is more in case of organics. Magdoff (1992) and Sahai (2004) reported that organic manure served as a reservoir of different types of nutrients which were essential for plant growth. According to Sudhakar et al. (2002), vermicompost contains micro sites rich in available carbon and nitrogen. Worm cast injected soils are also rich in water soluble phosphorous (Gratt, 1970) and contains two to three times more available potassium than surrounding soils (Sudhakar et al., 2002) which encourage better plant growth. Table 4 represents the pooled data of rice yield for consecutive two years of studies. Following statistical analysis, it was found that rice crop productions were significantly ($p \leq 0.05$) more in case of every treatment (T_1, T_2, T_3 and T_4) in

Table 3. Comparison of various chemical properties of soil for different treatments.

Replication	Treatments						Significance of differences (p-values)					
pH												
	T_i	T_0	T_1	T_2	T_3	T_4	T_i/T_0	T_0/T_1	T_0/T_2	T_0/T_3	T_0/T_4	T_1/T_3
R_1	7.1	6.9	6.9	6.9	7.2	7	0.13	0.21	0.06	0.49	1.00	0.07
R_2	7.2	7.1	6.8	6.8	7.1	7						
R_3	7.3	7.1	7	6.8	7	7.1						
mean	7.2	7.0	6.9	6.8	7.1	7.0						
sd	0.100	0.115	0.100	0.058	0.100	0.058						
Organic C (%)												
	T_i	T_0	T_1	T_2	T_3	T_4	T_i/T_0	T_0/T_1	T_0/T_2	T_0/T_3	T_0/T_4	T_1/T_3
R_1	0.47	0.4	0.95	0.98	0.56	0.73	0.01	0	0	0.00	0	0
R_2	0.48	0.42	0.99	0.96	0.58	0.75						
R_3	0.52	0.41	0.97	0.88	0.63	0.74						
mean	0.49	0.41	0.97	0.94	0.59	0.74						
sd	0.026	0.01	0.02	0.053	0.036	0.01						
Total N (%)												
	T_i	T_0	T_1	T_2	T_3	T_4	T_i/T_0	T_0/T_1	T_0/T_2	T_0/T_3	T_0/T_4	T_1/T_3
R_1	0.046	0.048	0.094	0.009	0.053	0.08	0.28	0.00	0.00	0.01	0.00	0
R_2	0.048	0.04	0.093	0.01	0.057	0.083						
R_3	0.045	0.041	0.098	0.011	0.058	0.083						
mean	0.05	0.043	0.095	0.01	0.056	0.082						
sd	0.002	0.004	0.003	0.001	0.026	0.003						
Available P (Kg.ha^{-1})												
	T_i	T_0	T_1	T_2	T_3	T_4	T_i/T_0	T_0/T_1	T_0/T_2	T_0/T_3	T_0/T_4	T_1/T_3
R_1	40.82	38.08	48.02	46.9	41.08	46.03	0	0	0	0	0	0
R_2	40.83	38.06	48.05	46.5	41.03	46.07						
R_3	40.87	38.05	48.04	46.8	41.09	46.05						
mean	40.84	38.06	48.04	46.73	41.07	46.05						
sd	0.026	0.015	0.015	0.208	0.032	0.020						
Available K (Kg.ha^{-1})												
	T_i	T_0	T_1	T_2	T_3	T_4	T_i/T_0	T_0/T_1	T_0/T_2	T_0/T_3	T_0/T_4	T_1/T_3
R_1	144.1	141.8	152.03	150.04	148.01	150.2	0	0	0	0	0	0
R_2	144.6	141.84	152	150.1	148.04	150.27						
R_3	144.8	141.79	152	150.13	148.04	150.25						
mean	144.5	141.81	152.01	150.09	148.03	150.24						
sd	0.361	0.026	0.017	0.046	0.017	0.036						

T_i-Initial property of soil or property of soil before crop establishment, T_0- without fertilizer or manure, T_1- 100% organic through vermicompost, T_2- 100% organic through FYM, T_3- 100% chemical through fertilizer and T_4- 50% organic + 50% chemical.

comparison to control (T_0). It was also found that production was not significantly higher for T_1 compared to T_3. The maximum rice yield was recorded under treatment T_1 where the lowest grain yield was observed in crop without fertilizer (T_0). It was found that the application of 100% vermicompost (T_1), 100% FYM (T_2), 100% chemical (T_3) and 50% organic + 50% chemical (T_4) increased the rice yield by 31.41, 30.56, 29.93 and 30.33%, respectively over control (the crop without fertilizer that is, T_0). Similarly, it was noticed that the application of 100% vermicompost (T_1), 100% FYM (T_2) and 50% organic + 50% chemical (T_4) increased rice yield by 2.12, 0.9 and 0.57%, respectively over 100% chemical through fertilizer (T_3). This may be due to the fact that organic manure, like vermicompost, is a nutritive plant food rich in NPK.

Comparing over all pool data of lentil crop productions (Table 5) under study, it was found that productions were significantly ($p \leq 0.05$) more in case of every treatment (T_1, T_2, T_3 and T_4) in comparison to control (T_0). It was also noted that production was significantly higher for T_1 compared to T_3. The highest seed yield was obtained

Table 4. Effect of nutritional management on grain yield of rice.

Replication	Treatments					Significance of differences (p-values)				
2008	T_0	T_1	T_2	T_3	T_4	T_0/T_1	T_0/T_2	T_0/T_3	T_0/T_4	T_1/T_3
R_1	2380	3400	3394	3356	3382	0	0	0	0	0.32
R_2	2398	3420	3410	3374	3385					
R_3	2389	3425	3411	3377	3403					
2009										
R_1	2460	3659	3589	3558	3575					
R_2	2462	3674	3600	3566	3586					
R_3	2485	3671	3584	3568	3588					
mean	2429	3541.5	3498	3466.5	3486.5					
sd	45.051	138.917	102.186	107.099	106.047					

T_0- Without fertilizer or manure, T_1- 100% organic through vermicompost, T_2- 100% organic through FYM, T_3- 100% chemical through fertilizer and T_4-50% organic + 50% chemical.

Table 5. Effect of nutritional management on seed yield of lentil.

Replication	Treatments					Significance of differences (p-values)				
2008	T_0	T_1	T_2	T_3	T_4	T_0/T_1	T_0/T_2	T_0/T_3	T_0/T_4	T_1/T_3
R_1	448	775	779	732	760	0	0	0	0	0.04
R_2	460	792	781	749	762					
R_3	460	797	795	733	785					
2009										
R_1	502	861	823	796	819					
R_2	522	877	847	819	817					
R_3	524	872	841	806	848					
mean	486	829	811	772.5	798.5					
sd	34.035	45.795	30.067	38.960	35.241					

T_0- Without fertilizer or manure, T_1- 100% organic through vermicompost, T_2- 100% organic through FYM, T_3- 100% chemical through fertilizer and T_4-50% organic + 50% chemical.

from the treatment T_1 and lowest seed yield was observed in crop without fertilizer (T_0). Application of 100% vermicompost (T_1), 100% FYM (T_2), 100% chemical (T_3) and 50% organic + 50% chemical (T_4) increased seed yield by 41.38, 40.07, 58.95 and 39.14%, respectively over control. Similarly, it was manifested that application of 100% vermicompost (T_1), 100% FYM (T_2) and 50% organic + 50% chemical (T_4) has increased the seed yield by 6.82, 4.75 and 3.26%, respectively over 100% chemical through fertilizer (T_3). These results are in accordance with those observed by Bwamiki et al. (1998) and Maynard (1993). They noticed that increase in productivity in the plots receiving organic manure/matter might be due to the fact that organic manure/matter not only provided additional nutrients other than N, P and K but also caused improvement in physical properties of soil. According to Suhane et al. (2008), vermicompost showed better results because exchangeable potassium (K) was over 95% higher in vermicompost compared to conventional compost. There were also over 60% higher amounts of calcium (Ca) and magnesium (Mg) which increased crop yield. In Guyana, an investigation into the

recycling of sugar cane bagasse and rice straw to produce compost, using vermitechnology and using the compost on *Phaseolus vulgaris*, was conducted by Ansari (2011). He concluded that physiochemical properties of rice straw and the combinations (bagasse with rice straw) were beneficial and enhanced growth and yield of *P. vulgaris*. His soil chemical analysis also indicated improvement in nutrient content.

Conclusion

Following results obtained from the period of two years of experimentation, it can be concluded that the application of vermicompost showed better result in comparison to chemical fertilizers in terms of soil physical and chemical properties as well as productivity of soil. Long-term use of chemical fertilizers which deteriorates the soil quality as well as diminishes the productivity of soil can be checked by using vermicompost. It may surely be concluded that recycling of organic wastes through vermicomposting is an effective and quick process for preparing organic manures. Application of vermicompost improves the soil quality as a whole which may be reflected through better crop production and use of vermicompost is better from all environmental aspects if compared with chemical fertilizer. It is envisaged that the problem of extensive use of chemical fertilizer can be solved to a great extent by increasing the use of organic manure produced from organic wastes by vermicomposting technology. Thus, one problem (generation and accumulation of organic wastes) can be used for solving another problem (toxicity of agricultural land and its reduction of crop production toward infertility) through proper management approach.

REFERENCES

Ansari AA (2011). Worm powered environmental biotechnology in organic waste management. Int. J. Soil Sci. 6(1):25-30.

Baruah TC, Barthakur HP (1997). A text book of soil analysis. New Delhi: Vikas Publishing House Pvt. ltd.

Basak RK, Pramanik M, Saha N (2011). Kenchosar tairi o tar byabahar. 2nd ed. Kalyani, T and G Printers.

Biswas TD, Khosla BK (1971). Building up of organic matter status the soil and its relation to the soil physical properties. In: Proceedings of International Symposium on Soil Fertility Evaluation, New Delhi, pp. 831-842.

Blake GR, Hartge KH (1986). Bulk density. In: Klute A (ed) Methods of soil analysis, Part I. Physical and Mineralogical Methods: Agronomy Monograph no. 9, 2nd ed. pp. 363-375.

Bwamiki DP, Zake JYK, Bekunda MA, Woomer PL, Bergstrom L, Kirchman H (1998). Use of coffee husks as an organic amendment to improve soil fertility in Ugandan banana production. Carbon and nitrogen dynamics in natural and agricultural tropical ecosystem. pp. 113-127.

Ceccanti B, Masciandaro G (1999). Vermicomposting of municipal and papermill sludges. Biocycle 6:71-72.

Doran JW, Sarrantonio M, Liebig MA (1996). Soil health and sustainability. Adv. Agron. 56:2-55.

Doran JW (2002). Soil health and global sustainability: translating science into practice. Agriculture, Ecosystems Environ. 88:119-127.

Doran JW, Liebig MA, Santana DP (1998). Soil health and global sustainability: in: proceedings of the 16th World Congress of Soil Science. Montepellier, France, 20-26 August 1998.

Doran JW, Zeiss MR (2000). Soil health and sustainability: managing the biotic component of soil qualify, Appl. Soil Ecol. 15:3-11.

Edwards CA (1998). Earthworm Ecology. U.S.A., CRC/Lewis Press; Boca Raton, FL.

Gee GW, Baudr JW (1986). Partical size analysis. In: klute A (ed). Methods of soil analysis. Part-I, ASA and SSSA, Madison (WI). Pp. 383-412.

Gratt JD (1970). Earthworm Ecology. Cultiv. Soils 10:107-123.

Hema S, Rajkumar N (2012). An assessment of Vermicomposting technology for disposal of vegetable waste along with industrial effluents. J. Environ. Sci. Comput. Sci. Eng. Technol. 1(1):5-8.

Jackson ML (1973). Soil chemical analysis. 2nd ed. New Delhi, Prentice Hall of India Ltd.

Jadhav SB, Jadhav MB, Joshi VA, Jagatap PB (1993). Organic farming in the light of reduction in use of chemical fertilizers. In: Proceedings of 43rd Annual Deccan Sugar Technology Association, Pune, Part I, SA 53-SA. P. 65.

Jeyabal A, Kuppuswamy G (2001). Recycling of organic wastes for the production of vermicompost and its response in rice–legume cropping system and soil fertility. Eur. J. Agron. 15(3):153-170.

Lazcano C, Domínguez J (2011). The use of vermicompost in sustainable agriculture: Impact on plant growth and soil fertility. In: Mohammad M (ed) Soil Nutrients. Nova Science Publishers, Inc.

Magdoff F (1992). Building Soils for Better Crops: Organic Matter Management. University of Nebraska Press, Lincoln, NE. P. 176.

Maynard AA (1993). Evaluating the suitability of MSW compost as a soil amendment in field-grown tomatoes. Compost Sci. Util. 1:34-36.

Miller JJ, Sweetland NJ, Chang C (2002). Hydrological properties of a clay loam soil after long-term cattle manure application. J. Environ. Qual. 31:989-996.

Palaniappan SP, Annadurai K (2008). Organic farming: Theory and Practice. Jodhpur, Scientific Publisers (India).

Patil PV, Chalwade PB, Solanke AS, Kulkarni VK (2003). Effect of fly ash and FYM on physico-chemical properties of Vertisols. J. Soils Crops 13(1):59-64.

Purohit SS (2006). Organic Farming in India. In: Purohit SS and Gehlot D (eds). Trends in Organic Farming in India. Jodhpur, Agrobios (India). pp 1-18.

Raju GU, Kumarappa S, Gaitonde VN (2012). Mechanical and physical characterization of agricultural waste reinforced polymer composites. J. Mater. Environ. Sci. 3(5):907-916.

Roychoudhury S, Karmakar S, Brahmachari K (2010). Nutrient Management through Organic and Inorganic Sources in sunflower under new alluvial soil. J. Interacademicia 14(3):316-320.

Sahai VN (2004). Fundamentals of soil. 3rd ed. New Delhi, Kalyani Publishers.

Shirani H, Hajabbasi MA, Afyuni M, Hemmat A (2002). Effects of farmyard manure and tillage systems on soil physical properties and corn yield in central Iran. Soil Tillage Res. 68:101-108.

Sudhakar G, Christopher Lourdura A, Rangasamy A, Subbian P, Velayuthan A (2002). Effect of vermicompost application on the soil properties, nutrient availability, uptake and yield of rice - A Review. Agric. Rev. 23(2):127-133.

Suhane RK, Sinha RK, Singh PK (2008). Vermicompost, Cattle-dung Compost and Chemical Fertilizers: Impacts on Yield of Wheat Crops. Communication of Rajendra Agriculture University, Pusa, Bihar, India.

Yoder RE (1936). A direct method of aggregate analysis of soils and the study of the physical nature of erosion losses. J. Am. Soc. Agron. 28:337-351.

13

Development of low cost technology for *in vitro* mass multiplication of potato (*Solanum tuberosum* L.)

E. P. Venkatasalam[1] , Richa Sood[1], K. K. Pandey[1], Vandana Thakur[1], Ashwani K. Sharma[2] and B. P. Singh[1]

[1]Central Potato Research Institute, Shimla-171 001 (H.P), India.
[2]Central Potato Research Station, Kufri, Shimla-171 012 (H.P), India.

Production of potato is constrained by lack of disease-free planting materials. This can be circumvented through tissue culture but the technology is costly limiting its adoption. As composition of culture medium used for shoot regeneration has a great influence on cost and there is a potential for use of locally available low cost resources as alternatives to the conventional costly laboratory resources. There is therefore, need to put in place interventions that will reduce the cost of production hence, making tissue culture products affordable. The present study describes a highly cost effective *in vitro* mass multiplication protocol for potato. The developed low cost medium can be used to boost the production of affordable disease-free potato seedlings, besides, this practice could be helpful in achieving more than 95% reduction in media cost.

Key words: Potato, micropropagation, low cost, medium, carbon source, tap water, plantlets.

INTRODUCTION

Potato is the third most important food crop in the world after rice and wheat which is consumed by more than a billion people worldwide. India is the second largest potato producer in the world after China; with the highest potato productivity among the top four potato producers in the world (CPRI vision, 2030). In potato cultivation, seed is the single most expensive input accounting for 40 to50% of production costs. On account of cumbersome methods involved in maintaining the nucleus stocks free from viruses and other diseases under conventional method of breeder's seed potato production, *in vitro* micropropagation techniques are gaining popularity. The recent advances in tissue culture techniques have facilitated the production, multiplication and maintenance of disease-free potato clones.

High production cost has been an impediment to tissue culture adoption which has further limited the technology to a few institutions and rich farmers while locking out the resource-challenged subsistence farmers. One factor contributing to the high cost of production is the cost of the culture nutrient medium which requires chemicals that are often very expensive (Savangikar, 2002). In order to increase application of tissue culture technology in potato farming, innovative approaches are needed to lower the cost of micro-propagule production. Various factors should be considered in developing plant tissue culture technique. One of the most important factors governing the *in vitro* shoots regeneration is largely determined by the composition of the culture medium (Rashid et al., 2000). Much number of researchers influenced to find alternatives materials to substitute alternatives to gelling agents, use of household sucrose, and some medium components objectively to reduce cost in media culture preparation. For example Raghu et al. (2007) have tried household sugar and tap water to substitution laboratory sucrose and double distilled water used in plant tissue culture. Besides successful in promoting the plantlet regeneration by using the substitution items provide cost reducing in media culture preparation. Many gelling agents are used for plant tissue culture media, such as agar technical Oxoid, agarose, phytagel and gelrite (Debergh, 1983).

Agar is the most commonly used as gelling agent for media preparation (Afrasiab and Jafar, 2011). From more than 100 years ago (Henderson and Kinnersely, 1988) until today, agar has been widely used as a gelling agent in plant tissue culture technique. This is because Henderson and Kinnersely (1988) reported its stability, high clarity and non toxic nature. As reported by Deb and Pongener (2010), agar substance often used in plant tissue culture as supporting agent, but because of the relatively high material costs caused them to do research for alternative materials with lower cost. Instead of high price of pure grade agar, there are some doubts about its nontoxic nature that influenced researcher to find alternatives material (Sharifi et al., 2010). Nagamori and Kobayashi (2001) used various types of starch and gums as cheaper alternatives for in commercial micro-propagation.

In addition to the gelling agents, the carbon sources such as grade sucrose that is often used in the micro-propagation of plants at laboratory contribute about 34% of the production cost (Demo et al., 2008). Sucrose has been reported as a source of both carbon and energy (Bridgen, 1994). Sugar was sucrose derived from sugar cane, in which the sucrose content in sugar is a combination of a glucose molecule with a molecule of fructose. Zapata (2001) cited in Kumara et al. (2010) reported success in reducing by 90% the cost of tissue culture banana trees are carried out due to the use of sucrose being replaced with table sugar. The quality and performance of plantlets grown on sucrose and on local commercial sugar resulted in all the sources of carbon fostered vigorous plantlet growth (Demo et al., 2008). Many research laboratories have used table sugar in the plant propagation medium (Kaur et al., 2005). Cost difference between household sugar and laboratory grade sucrose is quite big (97% difference). For the fact that laboratory grade sucrose is expensive, the use of 30 g/l household sugar can be proposed replacing the 30g/l sucrose as a way to reduce medium costs. Further studies should be conducted to identify the use of sugarcane juice directly to replace sugar. Buah et al. (2011) reported that costs of providing culture media using sugarcane juice are further cheaper than commercial sugar. In India, Ganapathi et al. (1995) reported that commercial grade sugar can replace analytical grade sucrose, with no significant change in the frequency of shoot formation in banana. The cost of commercial micropropagation has to be reduced drastically without compromising on the quality of

micropropagules especially in the developing countries (Kuria et al., 2008). However, no such low cost protocol has been developed for potato till date. Therefore, this study was undertaken to observe the interaction effects of the different gelling agents, type of water and carbon sources during *in vitro* mass multiplication of potato in order to evolve a low cost mass multiplication technology by adopting low cost substituent in the culture medium to make this system more economical and affordable.

MATERIALS AND METHODS

Three different experiments with different tetraploid (2n=4x=48) potato (*Solanum tuberosum* L. ssp. *tuberosum*) cultivars viz., Kufri Chandramukhi, Kufri Girdhari, Kufri Himalini, Kufri Bahar and Kufri Sindhuri belonging to different maturity groups were carried out at Central Potato Research Institute, Shimla during 2011 to 2012. Three double node cuttings dissected essentially from middle portion of the micro-plants were cultured per test tube (25 × 150 mm) containing 13 cm^3 MS medium (Murashige and Skoog, 1962). In the first experiment MS medium was prepared with nine types of water viz., rain, natural, tap, aquaguard, single distilled, double distilled, Type-I (Reverse osmosis), Type-II (Electronically de-ionized) and ultra-pure water, supplemented with sucrose at 30 gL^{-1} and solidified with agar (AR) at 7 gL^{-1}. In the second experiment MS medium was prepared with double distilled water, supplemented with seven types of carbon sources viz., commercial sugar, commercial sugar (sulphur less), sucrose, fructose, dextrose, sugar cubes and galactose at 30 gL^{-1} and solidified with agar (AR) at 7 gL^{-1}. In the third experiment MS medium was prepared with double distilled water, supplemented with sucrose at 30 gL^{-1} and solidified with four types of solidifying agent's viz., agar (PT), agar (bacteriological), agar (purified) and gelrite. The quantity of solidifying agent used was at 7 gL^{-1} except gelrite (2 gL^{-1}). All the three experiments were uniformly supplemented with 4.19 µM D-calcium pantothenate, NAA (0.05 µM) and GA₃ (0.29 µM). The experiment was carried out in a factorial completely randomized design (CRD) with four genotypes over a period of 28 days. The culture tubes were incubated under a 16 h photoperiod (irradiance of 60 µmol m^{-2} s^{-1}) at temperature of 22±1°C for 28 days in non-hermetic culture room.

After twenty-eight days of culturing, observations were recorded on morphological parameters such as micro-plant height (cm); number of green leaves, nodes and roots; inter-nodal and root length (cm); fresh as well as dry mass (mg) of microplants. As there were three micro-plants per culture tube, data was recorded for each micro-plant and averaged. In case of number of roots, only primary roots were counted, as there was secondary branching too. Root length was recorded for the longest root. Fresh and dry weight was taken for all the three plantlets. For dry weight, micro-plants were dried at 80°C for 48 h in the hot air oven and dry weight was recorded after bringing to room temperature. The experiment was repeated once again; data were pooled over individual experiments and analyzed statistically using the software AGRES for obtaining analysis of variance and means were separated according to the least significant differences at 0.05 level of probability.

RESULTS AND DISCUSSION

Carbon source

The analysis of variance showed that type of carbon

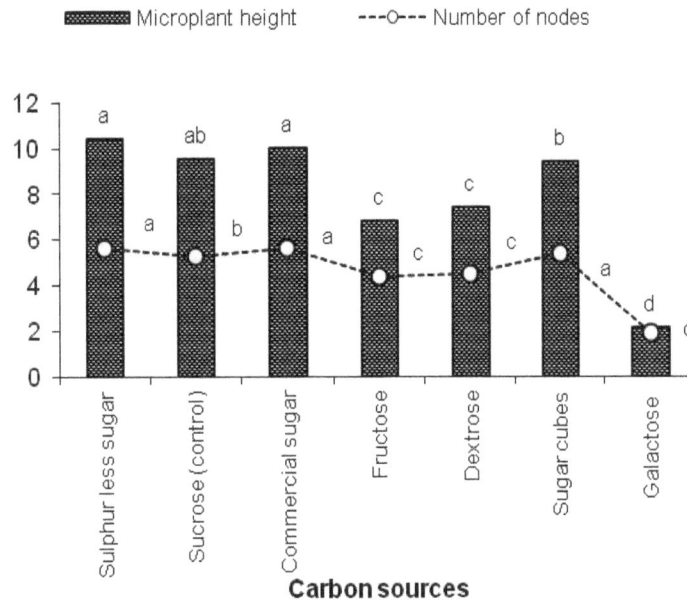

Figure 1. Effect of carbon sources on micro-plantlet height and number of nodes.

source had a major effect on all the morphological characters studied. Carbon source significantly ($P \leq 0.05$) influenced the microplant height, number of leaves, and number of nodes, inter nodal length, number of roots, root length, fresh as well as dry weight. Among the carbon sources sulphur less sugar (10.4 cm) significantly enhanced the microplant height which was also found at par with commercial sugar (10.0 cm) as well as standard control sucrose (9.5 cm), whereas, galactose recorded the minimum (2.1 cm). Sulphur less sugar as well as commercial sugar significantly increased the number of leaves as well as nodes as compared to other carbon sources. However, sugar cubes also exhibited at par results with respect to number of nodes (Figure 1). In general, galactose significantly reduced the number of leaves and nodes.

Almost all the carbon sources except galactose had statistically similar effect on inter-nodal length. This may be due to the increased availability of carbon in the form of purified sucrose that increases the intracellular sucrose concentration and it also has been reported to stimulate the in vitro growth of different crop species as a result of more negative water potential in the medium (Riek et al., 1997; Ebrahim et al., 1999). In addition to this, sucrose has been considered as one of the most common carbon source used in plant tissue culture due to its efficient uptake across the plasma membrane (Shimon et al., 2000; Sima and Desjardins, 2001; Yu et al., 2000). While, Mentha piperita cultured on media prepared with tap water + commercial sugar and double distilled water + tissue culture grade sucrose did not show any difference (Sunandakumari et al., 2004). Blanc et al (2002) reported

that, rapid hydrolysis of sucrose could increase the content of hexoses and storage compounds, directing the cells of embryogenic callus of Hevea brasiliensis to proliferate fast. Besides this, commercial sugar is impure sucrose and they may contain some other substances which may not suitable for tissue culture (Hossain et al., 2005). Our results are also in agreement with the earlier findings. Except dextrose and galactose, all other carbon sources exhibited same effect on number of roots as well as root length. This may be due to the decreased availability of carbon source in the medium that increased the number of roots and root length. Therefore, commercial sugar can be easily used for root proliferation and reduction of cost during micropropagation. Among different carbon sources sulphur less sugar (295 mg) recorded significantly maximum fresh weight in comparison to hexose and pentose carbon sources used in this study which was found to be at par with sucrose (292 mg), commercial sugar (280 mg) and sugar cubes (280 mg). Sucrose (22.4 mg) resulted in production of significantly maximum dry weight, however, it was found to be at par with sulphur less sugar (20.7 mg) (Table 1).

Solidifying agents

The analysis of variance showed that solidifying agents significantly ($P \leq 0.05$) influenced the microplant height, number of leaves and nodes, inter nodal length and fresh as well as dry weight. Gelrite (9.4 cm) significantly enhanced the microplant height in comparison to agar purified (standard) followed by bacteriological grade agar

Table 1. Effect of different carbon sources on morphological characters of potato microplants.

Characters/carbon sources	C_1	C_2	C_3	C_4	C_5	C_6	C_7	$LSD_{0.05}$
Number of leaves	5.5^a	5.2^b	5.4^{ab}	4.3^c	4.4^c	5.2^b	1.9^d	0.27
Inter-nodal length (cm)	1.9^a	1.8^{ab}	1.8^{ab}	1.5^c	1.7^{bc}	1.8^{ab}	1.2^d	0.16
Number of roots	5.4^a	5.5^a	5.2^a	5.3^a	4.1^b	5.5^a	0.0^c	0.39
Root length (cm)	7.6^a	7.5^a	7.4^a	5.9^c	6.7^b	7.3^a	0.0^d	0.45
Fresh weight (mg)	295.0^a	292.0^a	280.0^a	192.0^b	202.0^b	280.0^a	61.0^c	18.81
Dry weight (mg)	20.7^{ab}	22.4^a	19.2^b	14.9^c	14.2^c	19.3^b	7.5^d	2.69

C_1: Sulphur less sugar; C_2: Sucrose; C_3: Commercial sugar; C_4: Fructose; C_5: Dextrose; C_6: Sugar cubes; C_7: Galactose. Values are mean of four cultivars, three microplantlets and six replicates (test tubes). Values superscripted with the same letter in each column are not significantly different on the basis of least significant difference (P≤0.05).

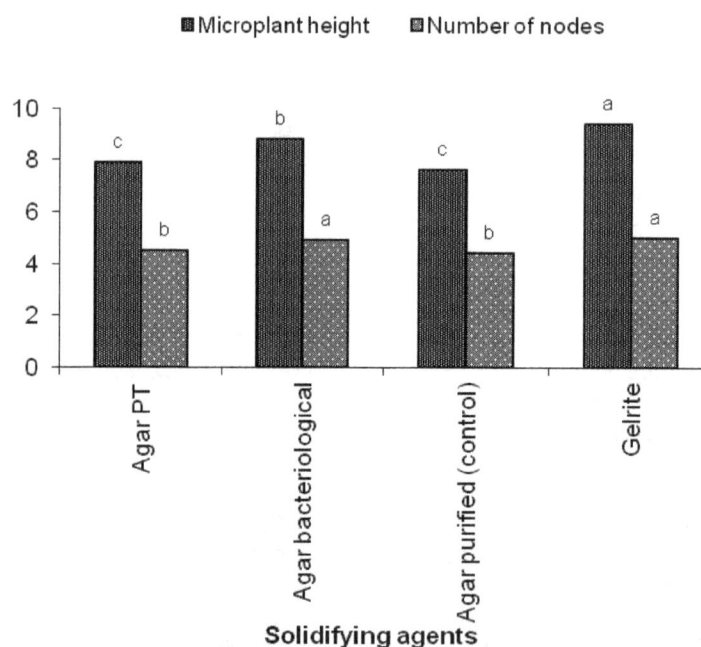

Figure 2. Effect of solidifying agents on micro-plantlet height and number of nodes.

(8.8 cm). Among the solidifying agents, gelrite significantly increased the number of leaves (4.6) and nodes (5.0) as compared to standard check however, it was found to be at par with bacteriological grade agar (Figure 2). Medium solidified with gelrite significantly increased the inter-nodal length (1.9 cm) followed by Agar PT (1.8 cm) and bacteriological agar (1.7 cm). Gelrite significantly increased the fresh (308.3 mg) as well as dry weight as compared to other solidifying agents (Table 2).

The accelerated growth in our study in the medium solidified with gelrite may be due to more availability of water in the media with gelrite, which was used in the lower concentration (Beruto et al., 1999). But Klimaszewska et al. (2000) reported that such effects were due to the physico-chemical characteristics of solidifying agent. The most prominent distinction among the solidifying agents which influences the *in vitro* growth characters is the water retention capacity of the gels and the availability of nutrients to the cultured tissue. Gelrite has been reported to yield better results than agar by many authors for regeneration and shoot multiplication (Henderson, 1987; Goldfarb et al., 1991; Van Ark et al., 1991; Welander and Maheswaran, 1992; Sharma et al., 2011). In addition to this, it was reported that agar from different sources contains various amounts of contaminants, whereas phytagel is free from phenolic compounds but has higher ash content than agar (Scherer et al., 1988). This may also be one of the reasons for reduced microplant growth with agar.

Table 2. Effect of different solidifying agents on morphological characters of potato microplants.

Characters/solidifying agent	S_1	S_2	S_3	S_4	$LSD_{0.05}$
Number of leaves	4.17[c]	4.45[ab]	4.22[bc]	4.56[a]	0.25
Inter-nodal length (cm)	1.77[bc]	1.82[b]	1.71[c]	1.91[a]	0.08
Number of roots	5.69	5.64	5.91	5.32	NS
Root length (cm)	5.71	5.99	5.94	5.83	NS
Fresh weight (mg)	221.72[c]	260.86[b]	223.26[c]	308.25[a]	21.09
Dry weight (mg)	16.43[c]	18.73[b]	16.04[c]	21.74[a]	1.90

S_1: Agar (PT); S_2: Agar (Bacteriological); S_3: Agar (Purified); S_4: Gelrite. Values are mean of four cultivars, three microplantlets and six replicates. Values superscripted with the same letter in each column are not significantly different on the basis of least significant difference (P≤0.05).

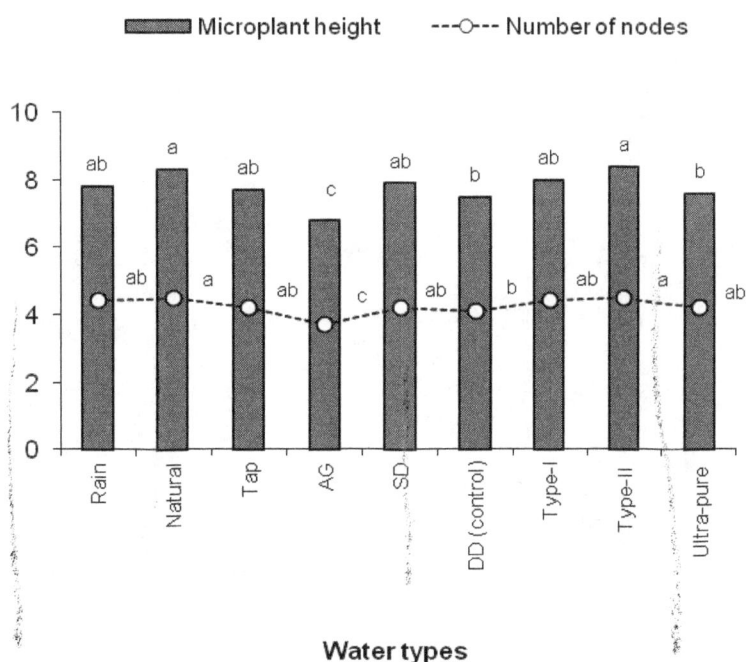

Figure 3. Effect of water types on micro-plantlet height and number of nodes.

Water types

Water sources significantly (P≤0.05) influenced the microplant height, number of leaves, number of nodes, inter nodal length, number of roots, root length and fresh as well as dry weight. Among different types of water, Type-II water recorded significantly maximum microplant height (8.4 cm) which was found to be at par with natural, rain, tap water, single distilled and Type-I water (Figure 2). Type-II recorded significantly maximum number of leaves (4.4) which was found to be at par with almost all other types of water apart from double distilled (standard) and aqua guard water. Type-II and natural water recorded significantly maximum number of nodes (4.5) which was found to be at par with rain, Type-I, tap, single distilled and ultra pure water (Figure 3).

Among different types of water, most of the water types resulted equal inter-nodal length. Ultra pure water recorded significantly maximum number of roots (7.1) in comparison to double distilled water (standard) whereas, aqua guard the minimum (4.4). Type-II water recorded significantly maximum fresh (230 mg) as well as dry weight (19.6 mg) whereas, rain water the minimum (Table 3). This may be due to the presence of bicarbonate and chlorides of calcium as well as magnesium in comparison with the distilled water where these elements are present in very trace amounts. Sunandakumari et al. (2004) also observed non significant differences in M. piperita cultured on media prepared with tap water + commercial sugar and double distilled water + tissue culture grade sucrose. However, we have not faced any problem of high rate of

Table 3. Effect of different water types on morphological characters of potato microplants.

Characters/water type	W_1	W_2	W_3	W_4	W_5	W_6	W_7	W_8	W_9	$LSD_{0.05}$
Number of leaves	4.30^{ab}	4.24^{ab}	4.05^{ab}	3.57^c	4.11^{ab}	4.00^b	4.20^{ab}	4.38^a	4.11^{ab}	0.36
Inter-nodal length (cm)	1.78^{ab}	1.73^{bc}	1.73^{abc}	1.65^c	1.81^{ab}	1.83^a	1.66^c	1.81^{ab}	1.82^{ab}	0.10
Number of roots	5.40^b	6.17^b	5.60^b	4.42^c	6.18^a	6.13^b	5.66^b	5.93^b	7.14^a	0.79
Root length (cm)	5.46	5.56	5.08	4.49	5.09	5.95	4.83	5.53	4.79	NS
Fresh Weight (mg)	193.4^d	218.2^{ab}	198.1^{cd}	192.5^d	202.2^{bcd}	210.1^{bcd}	216.1^{abc}	230.2^a	206.2^{bcd}	19.6
Dry Weight (mg)	15.88^d	18.69^{ab}	17.47^{bc}	18.13^{ab}	15.78^d	16.39^{cd}	18.71^{ab}	19.60^a	18.85^{ab}	1.57

W_1: Rain water; W_2: Natural water; W_3: Tap water; W_4: Aqua-guard water; W_5: Single distilled water; W_6: Double distilled water; W_7: Reverse osmosis /Type-I water; W_8: Electronically de-ionized/Type-II water; W_9: Ultra pure water. Values are mean of four cultivars, three microplantlets and six replicates. Values superscripted with the same letter in each column are not significantly different on the basis of least significant difference (P≤0.05).

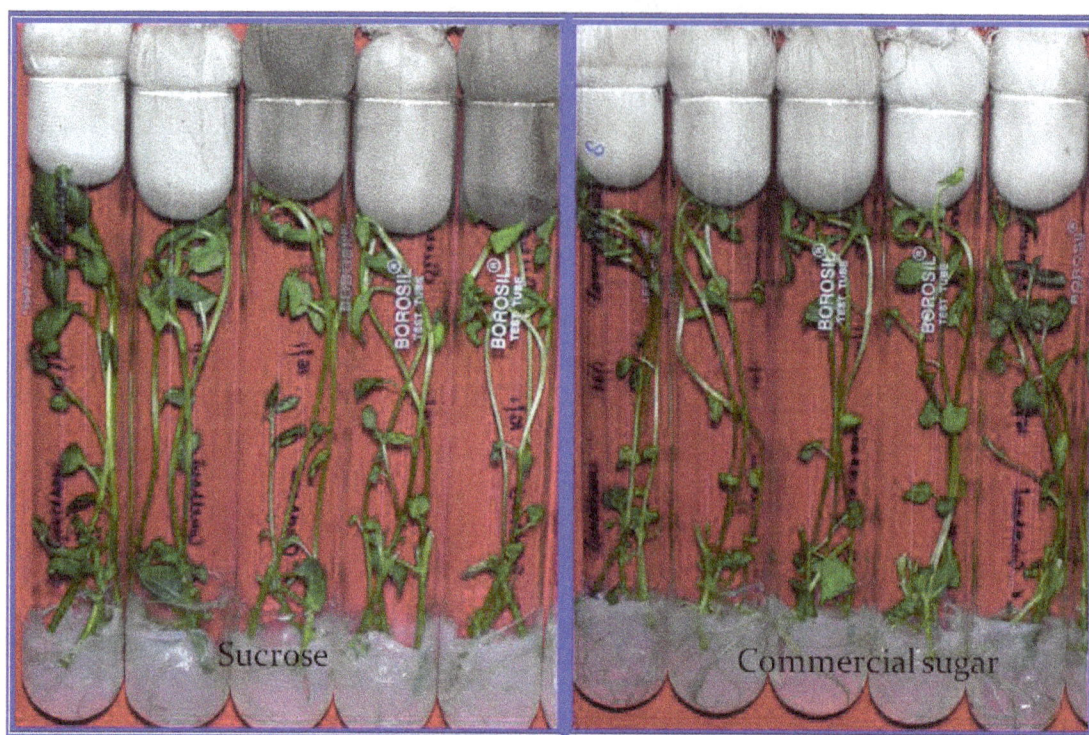

Figure 4. Growth of potato micro-plantlets on standard and low cost carbon source.

precipitation in the solution before or after adding agar or solidifying agent for preparation of medium as observed by Das and Gupta (2009) therefore, boiling of tap water was not required.

The cost of production for one litre of micro propagation medium using different components that was attributed to the cost of plantlet production was compared. The results stated that the cost of production of plantlets in micro propagation medium could be reduced from 59 to 68% by using different substitutes. The highest reduction of cost (68%) could be noticed when all the three components tap water, commercial sugar and agar (bacteriological) were together being substituted in place of double

distilled water, laboratory grade sucrose and agar (purified), respectively.

From the present study, it was inferred that the analytical grade (AR) sucrose can be successfully replaced by an ordinary commercial sugar, which is 5 to 6 times cheaper than the sucrose (Figure 4). Similarly, for preparation of media, we can use clean tap water instead of double distilled/ultrapure water which will reduce the investment on costly apparatus as well as on electricity (Figure 5). For solidifying the media, we can replace the Agar (purified) with agar ((bacteriological) or gelrite/phytagel, though the unit cost of gelrite/phytagel is more than agar but the quantity of gelrite/phytagel used

Figure 5. Growth of potato micro-plantlets on standard and low cost water source.

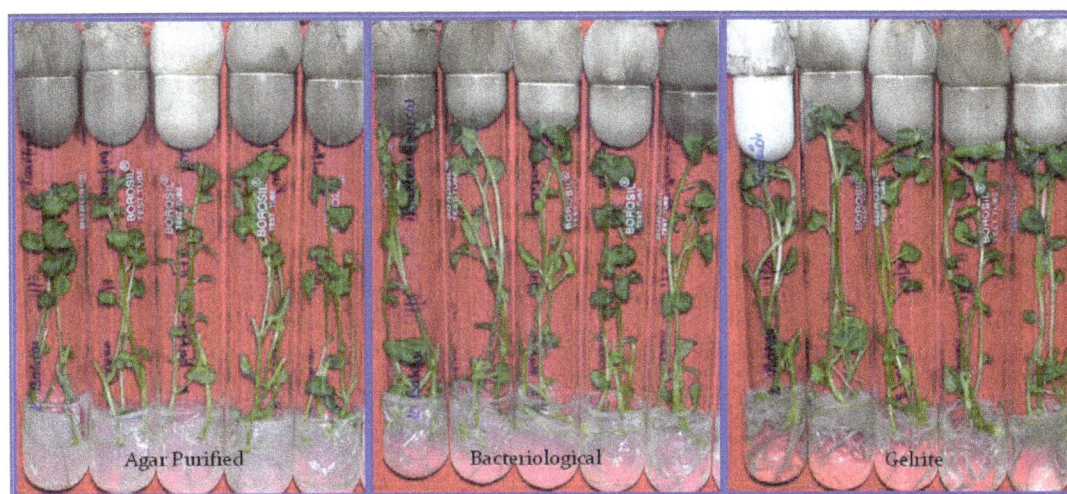

Figure 6. Growth of potato micro-plantlets on standard and low cost solidifying agents.

for solidifying unit quantity of media is much less (25%) and will lead to save 43 to 52% cost on solidifying agent (Figure 6). In short, easily available low cost alternatives presented in this work allow a low cost strategy for successful micropropagation of potato without compromising on quality of plants.

REFERENCES

Afrasiab H, Jafar R (2011). Effect of different media and solidifying agent on callogenesis and plant regeneration from different explants of rice (Oryza sativa L) varieties super basmati and IRRI-6. Pak. J. Biol. Sci. 43(1):487-501.

Beruto M, Beruto D, Debergh P (1999). Influence of agar on in vitro cultures: Physio-chemical properties of agar and agar gelled media. In Vitro Cell Dev-Pl 35:86-93.

Blanc G, Lardet L, Martin A, Jacob JL, Carron MP (2002). Differential carbohydrate metabolism conducts morphogenesis in embryogenic callus of Hevea brasiliensis (Mull. Arg.). J. Exp. Bot. 53:1453-1462.

Bridgen MP (1994). A review of plant embryo culture. Hort. Sci. 29:1243-1245.

Buah JN, Tachie-Menson JW, Addae G, Asare P (2011). Sugarcane Juice as an Alternative Carbon Source for in vitro Culture of Plantains and Bananas. Am. J. Food Technol. 6:685-694.

Das A, Gupta SN (2009). Use of low cost resources for banana micro propagation. Indian J. Hortic. 66(3):295-300.

Deb CR, Pongener A (2010). Search of alternative substratum for agar in plant tissue culture. Current Sci. India. 98:99-102.

Debergh PC (1983). Effects of agar brand and concentration on the tissue culture medium. Physiol. Plantarum 59:270-276.

Demo P, Kuria P, Nyenda AB, Kahangi EM (2008). Table sugar as an alternative low cost medium component for in vitro micro-propagation of potato (Solanum tuberosum L.). Afr. J. Biotechnol. 7:2578-2854.

Ebrahim MKH, Zingheim O, Veith R, Kassem EEA, Komor ME (1999). Sugar uptake and storage by sugarcane suspension cell at different temperatures and high sugar concentrations. J Plant Physiol. 154:610-616.

Ganapathi TR, Mohan JSS, Suprasanna P, Bapat VA, Rao PS (1995). A low cost strategy for in vitro propagation of banana. Curr. Sci. India 68: 646-650.

Goldfarb B, Howe GT, Bailey LM (1991). A liquid cytokinin pulse induces adventitious shoot formation from Douglas-fir cotyledons. Plant Cell Rep. 10:156-160.

Henderson JM (1987). The use of gelrite as a substitute for agar in medium for plant tissue culture. Ala. Agric. 2:5-6.

Henderson WE, Kinnersley AM (1988). Corn starch as an alternative gelling agent for plant tissue culture. Plant Cell Tiss. Org. 15:17-22.

Hossain MA, Hossain MT, Raihan, Ali M, and Mahbubur SM (2005). Effect of differenmt carbon sources on in vitro regeneration of Indian Pennywart (Centella asiatica L.). Pak. J. Biol. Sci. 8(7):963-965.

Kaur R, Gautam H, Sharma DR (2005). A low cost strategy for micropropagation of strawberry (Fragari ananassa) cv. Chandler. Proceedings of the VII International Symposium on Temperate Zone Fruits in the Tropics and Subtropics, 2005, Acta Horticult, pp. 129-133.

Klimaszewska K, Bernier CM, Cyr DR, Sutton BCS (2000). Influence of gelling agents on culture medium gel strength, water availability, tissue water potential, and maturation response in embryogenic cultures of Pinus strobes L. In Vitro Cell Dev-Pl 36:279-286.

Kuria P, Demo P, Nyende AB, Kahangi EM (2008). Cassava starch as an alternative cheap gelling agent for the in vitro micro-propagation of potato (Solanum tuberosum l.). Afr. J. Biotechnol. 7(3):301-307.

Murashige T, Shoog F (1962). A revised medium for rapid growth and bioassay with tobacco tissue culture. Physiol. Plantarum. 15:473-479.

Nagamori E, Kobayashi T (2001). Viscous additive improves micropropagation in liquid medium. J. Biosci. Bioengin. 91:283-287.

Raghu AV, Martin G, Priya V, Geetha SP, Balachandran I (2007). Low cost alternatives for the micropropagation of Centella asiatica. J. Plant Sci. 2:592-599.

Rashid H, Toriyama K, Qureshi A, Hinata K, Malik AK (2000). An improve method for shoot regeneration from calli of indica rice (Basmati). Pak. J. Biol. Sci. 3:2229-2231.

Riek JD, Piqueras A, Debergh PC (1997). Sucrose uptake and metabolism in a double layer system for micropropagation of Rosa multiflora. Plant Cell Tiss. Org. 47:269-278.

Savangikar VA (2002). Role of low cost options in tissue culture. In: Low cost options for tissue culture technology in developing countries. Proceedings of a technical meeting organized by the Joint FAO/IAEA Division of Nuclear techniques in food and agriculture, August 26-30, 2002, Vienna, IAEA, pp. 11-15.

Scherer PA, Muller E, Lippert H, Wolff G (1988). Multi-element analysis and gelrite impurities investigated by inductively coupled plasma emission spectrometry as well as physical properties of tissue culture medium prepared with agar or the gellan gum Gelrite. Acta Hortic. 226:655-658.

Sharma S, Venkatasalam EP, Patial R, Latawa J, Singh S (2011). Influence of gelling agents and nodes on the growth of potato microplant. Potato J. 38(1):41-46.

Sharifi A, Moshtaghi N, Bagheri A (2010). Agar alternatives for micropropagation of African violet (Saintpaulia ionantha). Afr. J. Biotechnol. 9(54):9199-9203.

Shimon KN, Mills D, Merchuk JC (2000). Sugar utilization and invertase activity in hairy root and cell suspension cultures of Symphytum officinale. Plant Cell Tiss. Org. 62:89-94.

Sima BD, Desjardins Y (2001).Sucrose supply enhances phosphoenolpyruvate carboxylase phosphorylation level in in vitro Solanum tuberosum. Plant Cell Tiss. Org. 67:235-242.

Sunandakumari C, Martin KP, Chithra M, Sini S, Madhusoodanan PV (2004). Rapid axillary bud proliferation and ex vitro rooting of herbal spice, Mentha piperita L. Indian J. Biotechnol. 3:108–112.

Van AF, Zaal MA, Cremers MJ (1991). Improvement of the tissue culture response of seed derived callus cultures of Poa protensis L.: effect of gelling agent and abcisic acid. Plant Cell Tiss. Org. 27:275-280.

Welander M, Maheswaran G (1992). Shoot regeneration from leaf nodes of dwarfing apple rootstocks. J. Plant Physiol. 140:223-228.

Yu CU, Joyce PJ, Cameron DC, McCown BH (2000). Sucrose utilization during potato microtuber growth in biorectors. Plant Cell Rep.. 19:407-413.

Zapata A (2001). Cost reduction in tissue culture of banana. Int. Atom Energy Labs. Agric. And Biotech. Lab Austria.

14

Inhibitory effects of *Acremonium* sp. on Fusarium wilt in bananas

Liu Yuelian and Lu Qingfang

Agricultural College, Guangdong Ocean University, Zhanjiang, Guangdong 524088, China.

An endophytic fungus, strain Q34, isolated from *Kandelia candel* (L.) Druce showed strong *in vitro* antagonistic activity toward *Fusarium oxysporum* f. sp. *cubense* (foc) race 4. The crude extract of the strain inhibited the growth of foc race 4 and caused conidial deformation. In gas chromatography-mass spectrometry (GC-MS) analysis, the crude extract showed 18 main peaks. Three substances from the main peaks showed strong inhibitory activity against the growth of foc race 4. Moreover, the strain Q34 reduced the incidence of Fusarium wilt in banana plantlets under greenhouse conditions and in the field. This strain was identified as *Acremonium* sp. on the basis of morphology. These results suggest that strain Q34 is potentially useful for the biological control of foc race 4.

Key words: *Fusarium oxysporum* f. sp. *cubense* race 4, *Acremonium* sp., biological control, Fusarium wilt.

INTRODUCTION

Widespread infection of banana (*Musa* spp.) by *Fusarium oxysporum* f. sp. *cubense* (E. F. Smith) Snyder and Hansen (foc) is increasingly affecting the production of this important fruit crop in tropical and subtropical regions. Among the 4 races of this species, foc race 4 is a particularly virulent pathogen that infects banana plants in all growth stages, from seedlings to fruit-bearing crops, often causing serious economic loss (Ploetz, 2006; Wu et al., 2010). Following visual detection of the infection, the most common method of controlling foc race 4 is removal of the infected banana plants and quarantine of the infested areas.

Currently, there are no effective methods to chemically control Fusarium wilt. For example, carbendazim, the most effective fungicide against other *F. oxysporum* strains, has been found to have relatively lower efficacy against foc race 4 (Buddenhagen, 2009; Cao et al., 2004; Ploetz, 2006; Wu et al., 2010; Sun et al., 2011). Moreover, its effectiveness against susceptible strains is often not as adequate as desired by growers, especially when the opportunity for early detection is missed and the infection is already widespread. To manage foc race 4 infection, researchers are increasingly focusing on biological control. For example, *Pseudomonas aeruginosa* FP10 reduces the vascular discoloration resulting from foc infection (Ayyadurai et al., 2006), and *Streptomyces noursei* Da07210 shows strong antagonism toward foc race 4 in plate bioassays (Wu et al., 2009). Moreover, *Pseudomonas* (Altinok et al., 2013), *Bacillus subtilis* (Sun et al., 2011), *Streptomyces* (Cao et al., 2005), Nonpathogenic *F. oxysporum* and *Pseudomonas fluorescens* (Belgrove et al., 2011), and bacteria in the rhizosphere (Li et al., 2012), were reported to have inhibitory effects against Fusarium wilt, encouraging researchers to continue searching for more effective strains.

In the current study, antagonistic strains were isolated from a variety of terrestrial environments and from marine environments Mangrove plants growing in specia saltwater environments, have potential of producing novel metabolite. This situation occurs to mangrove-derived endophytes too because of their co-evolution (Guo 2001; Wen et al., 2010). Here, we used *K. candel* (L.) Druce as endophytic fungal host, which grows in mangrove forests and is free from foc race 4 infection. To investigate endophytes derived from *K. candel*, a comprehensive isolation and screening program was initiated.

MATERIALS AND METHODS

Endophytic microbes and pathogenic fungi

Endophytic fungal strains were isolated from disease-free *K. candel* in Zhanjiang (Guangdong, China) using the modified method described by Hyde and Soytong (2008). Hyphal tips were transferred to new potato dextrose agar (PDA) dishes, and the dishes were incubated at 28°C for 7 days.

The phytopathogenic fungus foc race 4 was obtained from a stored culture from South China Agricultural University.

Screening of antagonistic strains

Pathogenic fungi were inoculated on PDA plates for 7 days at 28°C. The antagonistic activity of each endophytic fungal strain was determined by using the dual plate culture technique, with three replicates for each strain. Then, the following formula was used to assess the inhibition activity (Si et al., 2005):

Rate of inhibition (%) = [(average diameter of the control pathogen colony − average diameter of the test pathogen colony)/average diameter of the control pathogen colony] × 100

Identification of antagonistic strains

Identification of antagonistic strains to foc race 4 based on colony characteristics and morphological characteristics were performed under a light microscope (Olympus) referencing (Kirk et al., 2001).

Extraction of inhibitory compounds

The endophytic fungal strain showing antagonistic activity was grown in potato dextrose broth at 28°C with shaking at 120 rpm for 10 days. Next, 20 L of the culture was filtered and centrifuged at 8 kr min⁻¹ for 10 min at 4°C. The supernatant was extracted twice with an equal volume of ethyl ethanoate, and the ethyl ethanoate layer was subsequently evaporated under reduced pressure in a rotary evaporator to obtain the crude extract of this strain.

Inhibition assay using the crude extract

The hyphal extension-inhibition assay described by Roberts and Selitrennikoff (1986) was used for this experiment. In brief, agar plugs (5-mm diameter) of foc race 4 were obtained from a freshly growing colony (5 days) and placed at the center of PDA plates containing various concentrations (100, 50, 25, 12.5, and 6.25 µg mL⁻¹) of the crude extract; ethyl ethanoate was used as the control. After incubation at 28°C for 5 days, diameters of the fungal colonies were measured, and the fungal conidia were observed under a microscope (100 × magnification) to determine abnormal

morphology at 10 days after plating. The percentage of inhibition of mycelial growth was calculated using the aforementioned formula.

Gas chromatography-mass spectrometry analysis of the crude extract

The crude extract was dissolved in 1 mL of ethyl ethanoate (1:9, v/v). A high-performance gas chromatography-mass spectroscopy (GC-MS) system (GC6890-MS5973) with a mass-selective detector and electron impact ionization was used to analyze the crude extract. The following operating conditions were used: DB25 ms (30 m × 0.32 mm, 0.25-µm film thickness), carrier-gas flow rate of 1.0 mol min⁻¹, and split ratio of 50:1. The GC oven temperature was programmed from 40°C (5 min) to 280°C (5°C min⁻¹). The electron impact ionization conditions included 70 eV ion energy and a 50–650 amu mass range in full-scan acquisition mode. Compounds of the crude extract were identified using the Wiley mass spectral library and verified against reference compounds (Zhang et al., 2008).

Identification of the inhibitory compounds of the crude extract

Inhibitory compounds were identified, with a similarity index greater than 850, from a database search based on a comparison of the mass spectrum of the substance with GC-MS system databanks (Wiley 138 and NBS 75k Library). Each sample was tested twice. For each detected candidate inhibitory compound, antifungal activity was confirmed using the same pure commercial compound (analytical grade) instead of the crude extract.

The effects of the commercial compounds (100 µg mL⁻¹) on the mycelial growth of foc race 4 were evaluated in a hyphal extension-inhibition assay, and compounds with an inhibition rate of 100% in this assay were then selected for determining the IC_{50} by the FAO method (Georgopoulos and Dekker, 1982). Next, the effects of different concentrations of these compounds on the mycelial growth of foc race 4 were assessed by hyphal extension-inhibition assays. The compounds were dissolved in 99% methyl N-(1H-benzimidazol-2-yl)carbamate and were used as negative and positive controls, respectively.

Biological control assay under greenhouse conditions

Strain Q34 and foc race 4 were prepared as described above. The effectiveness of strain Q34 against foc race 4 on the plantlets of a Cavendish banana ('Yueke No. 1') was evaluated using the procedure described by Subramaniam et al. (2006) with minor modifications. Banana plantlets were grown in pathogen-free tissue culture until 3–4 fully expanded leaves had grown and healthy roots had formed. To facilitate infection, the roots were slightly bruised by removing the root ball. Next, plantlets were divided into six treatment groups, with three plants in each group and three replicates per plant. Plantlet roots were then treated for 30 min with (i) the fermented liquid of strain Q34, (ii) fermentation filtrate, (iii) medium only, (iv) methyl N-(1H-benzimidazol-2-yl)carbamate, (v) sterilized water (foc race 4 was later administered to i to v, as described below), and (vi) sterilized water (foc race 4 was not added, as described below). Plantlets were then placed in 250-mL plastic cups of sterile soil, with holes in the bottom of the cups. Three plantlets were planted in each cup. All plantlets were inoculated with 1 mL conidial suspension (10⁵ mL⁻¹) of foc race 4 near the region between the stems and roots, except for those treated with (vi). All cups were watered to saturation and placed in a greenhouse (28–32°C). The percent disease incidence was calculated based on the number of plantlets with chlorotic leaves for 25 days including whole 54 plantlets. Disease incidence (%) =

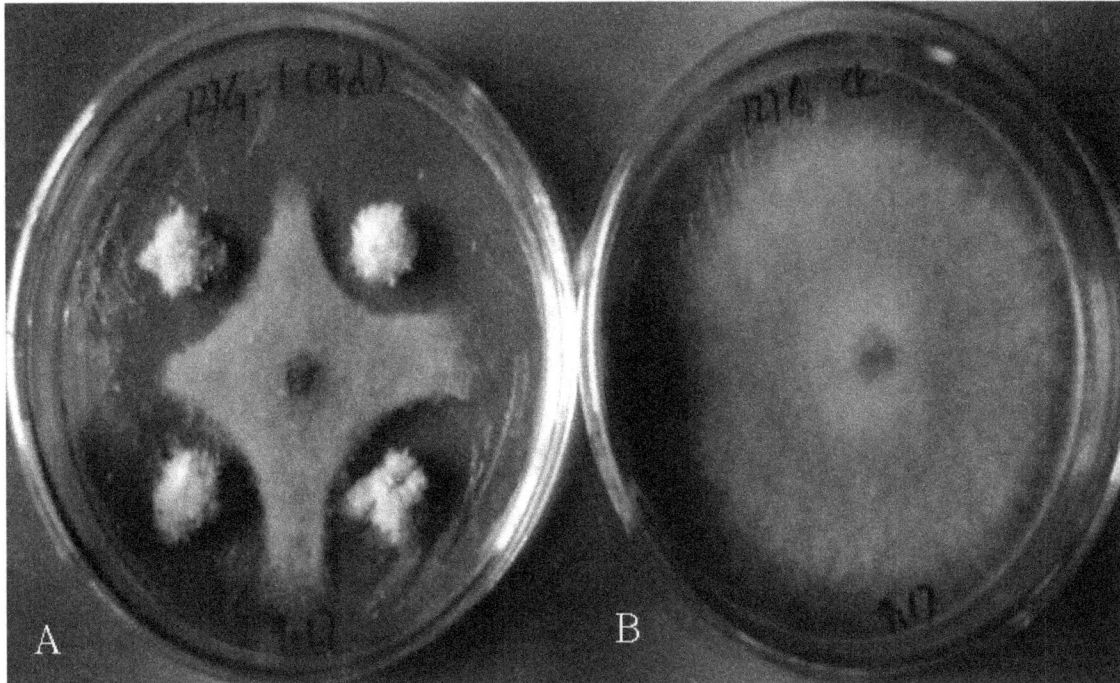

Figure 1. Q34 against Foc race 4 (4d) - A: Treatment; B: Control.

(number of plantlets with discolored leaves/total number of plantlets) × 100 (Subramaniam et al., 2006).

Biological control assay in field conditions

Field experiment was conducted at Zhanjiang, Guangdong province, China, latitude: 21°10'; longitude: 110°30 '; rainfall: 1100 mm; mean temperature: 23.2°C; RH: 80–95%. The crop was grown in the rainfed condition (soil pH: 5.4, soil type: clay, organic matter: 25.5%, P: 1.75 gkg^{-1}, N: 0.90 gkg^{-1}, K: 7.57 gkg^{-1}.The experiment was designed as follow: (A) the fermented liquid of strain Q34, (B); methyl N-(1H-benzimidazol-2-yl)carbamate, (C) sterilized water, (D). sterilized water. A,B and C were inoculated with conidial suspension (10^5 mL^{-1}) of Foc race 4, but not D. 100 mL conidial suspension, or 100 mL both the fermented liquid of strain Q34 (10 days), or 100 mL methyl N-(1H-benzimidazol-2-yl)carbamate (30 μgmL^{-1}) were dripped into each banana plantlet, near the region between the stems and the roots. Uniform tissue-cultured banana plantlets cv. "Yueke No. 1" with more than 4 new leaves and of 18 to 20 cm height was used in the experiments. In each treatment, there were 9 plants and 4 m × 2 m spacing was adopted. Disease severity was measured in the 30th, 50th, 70th and 90th day respectively. The disease incidence was calculated as the same as the assay in greenhouse.

Statistical analysis

Statistical analysis was conducted with the Statistical Analysis System (SAS Institute, Inc., Cary, NC, USA). LSD was performed using ANOVA in SAS in order to assess differences in mycelial growth rates and pathogenicities at a significance level of $p ≤ 0.05$. IC_{50} values were analyzed by ANOVA performed with the General Linear Model (GLM) of SAS.

RESULTS

Isolates and antagonistic activity

In total, 125 endophytic fungal strains were obtained from K. candel: 65, 53, and 7 strains were obtained from the stem, leaves, and roots, respectively. One strain isolated from the leaves of K. candel exhibited the strongest activity toward foc race 4, with an average inhibition rate of 72.83% ($P ≤ 0.05$; Figure 1). This strain was coded as Q34.

Strain Q34 was identified as Acremonium sp. (Figure 2), a fungus that is found in decomposed banana stud (Krishnaveni et al., 2009) and in banana cortex (Pocassangre et al., 2000), but not found in banana in China. Q34 colonies grew slowly and attained a diameter of 17 mm after 14 days of incubation at 25°C on PDA. At first, colonies appeared white; however, their color changed over time from pale, to yellow, and finally to brown. The conidiophores of Q34 were indistinct from the colorless hyphae of the vegetative mycelium. The conidiogenous cells arose from a single hypha: they were solitary, branchless, slender, hyaline-like, thin-walled, 10 to 20 μm long, 1.8 to 2.5 μm wide, and 0.5 to 1.0 μm wide at the apex. Conidia occurred in long chains; were hyaline-like, fusiform, and aseptate; and measured 3.6 to 4.8 × 0.9 to 1.4 μm. These morphological characteristics of strain Q34 were similar to those of Acremonium (J.C. Gilman and E.V. Abbott) W. Gams (1975). Strain Q34 is now stored in the Laboratory of the Agricultural College of

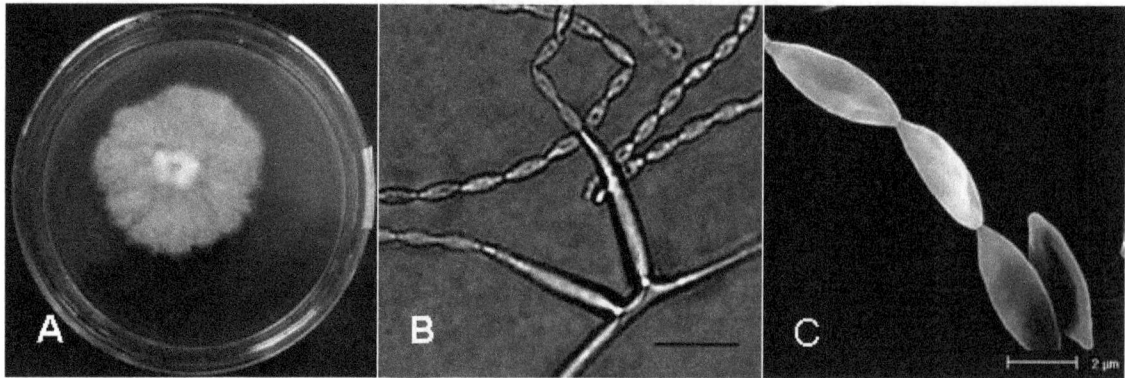

Figure 2. Morphology of Q34 - **A**: Colony; **B**: Conidiophores; **C**: Conidia Bar in **B** = 10 μm, Bar in **C** = 2 μm.

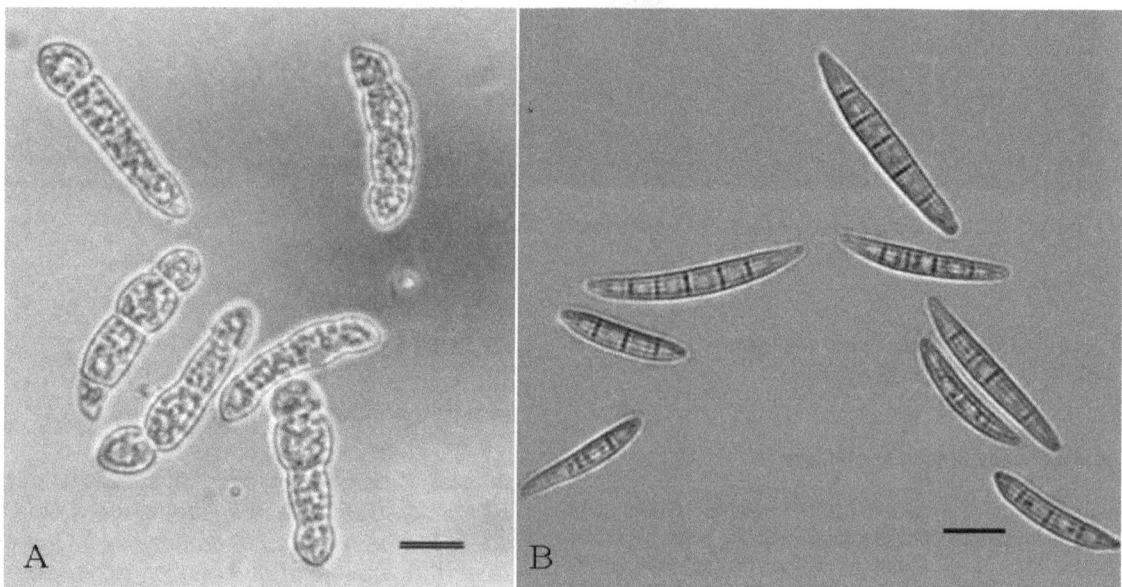

Figure 3. Effect of the crude extract of strain Q34 on conidia of Foc race 4 (5d) **A**: Crude extract; **B**: Sterilized water.

Guangdong Ocean University.

Effects of the crude extract on the growth of foc race 4

In the presence of the crude extract of strain Q34, the mycelial growth of foc race 4 was significantly inhibited (IC$_{50}$, 40.00 μg mL^{-1}; regression equation, $Y = 24.84 - 0.63X$; $R^2 = 0.87839$). Microscopic observation revealed that treatment with the extract caused distorted conidial growth of foc race 4 (Figure 3). After 4 h, the conidia appeared deformed and were unable to germinate.

Characteristics of the compounds derived from the crude extract

Next, we sought to identify the components of the crude

extract from strain Q34 using GC-MS analysis. GC-MS analysis revealed 18 compounds, as represented by major peaks, in the crude extract (Figure 4). Ten of the compounds with similarity indices of greater than 850 were sterilants, drug intermediates, or unknowns: 99% 2-phenylethanol; 100% 1,2,3,4-tetrahydronaphthalene; 95% 1,2,4,5-tetramethylbenzene; 100% 2,3-dihydro-4,7,dimethyl-1H-indene; 100% 1,1'-biphenyl; 100% benzyl benzene; 100% 9H-fluorene; 100% anthracene; 100% octadecane; and 100% icosane. Pretest results showed that 2-phenylethanol (peak 1), 2,3-dihydro-4,7,dimethyl-1H-indene (peak 4), and 1,1'-biphenyl (peak 5) displayed inhibition rates of 100% for the mycelial growth of foc race 4 (Figure 5).

Taken together, these results suggested that 3 compounds from strain Q34 were primarily responsible for inhibiting mycelial growth. Further studies showed that 2-phenylethanol significantly inhibited the mycelial

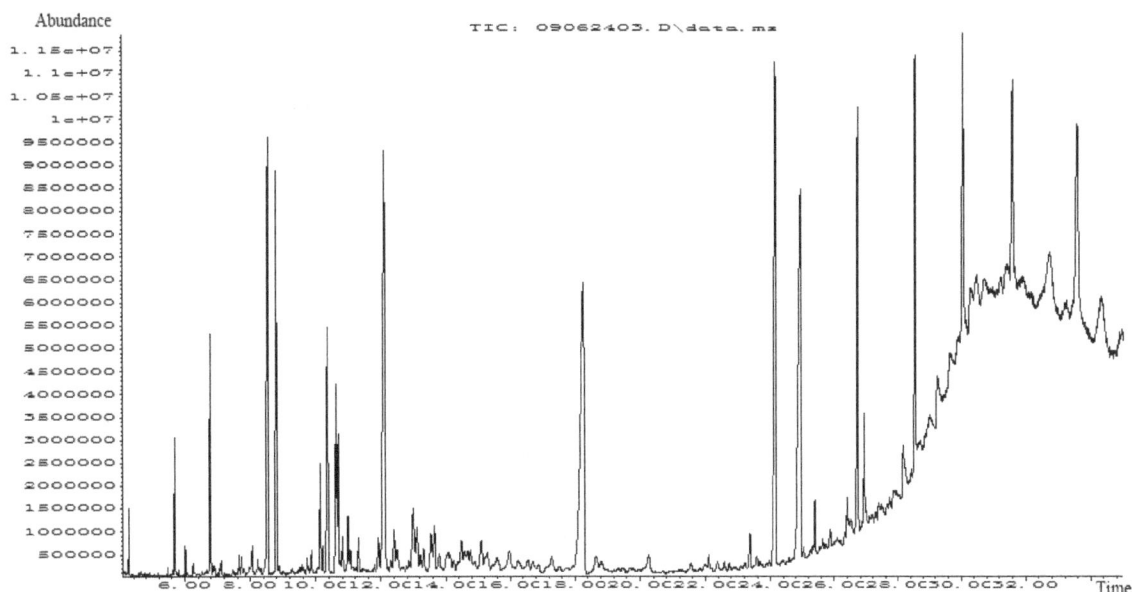

Figure 4. Abundance ion chromatography of chemical constituents in extract substances of Q34.

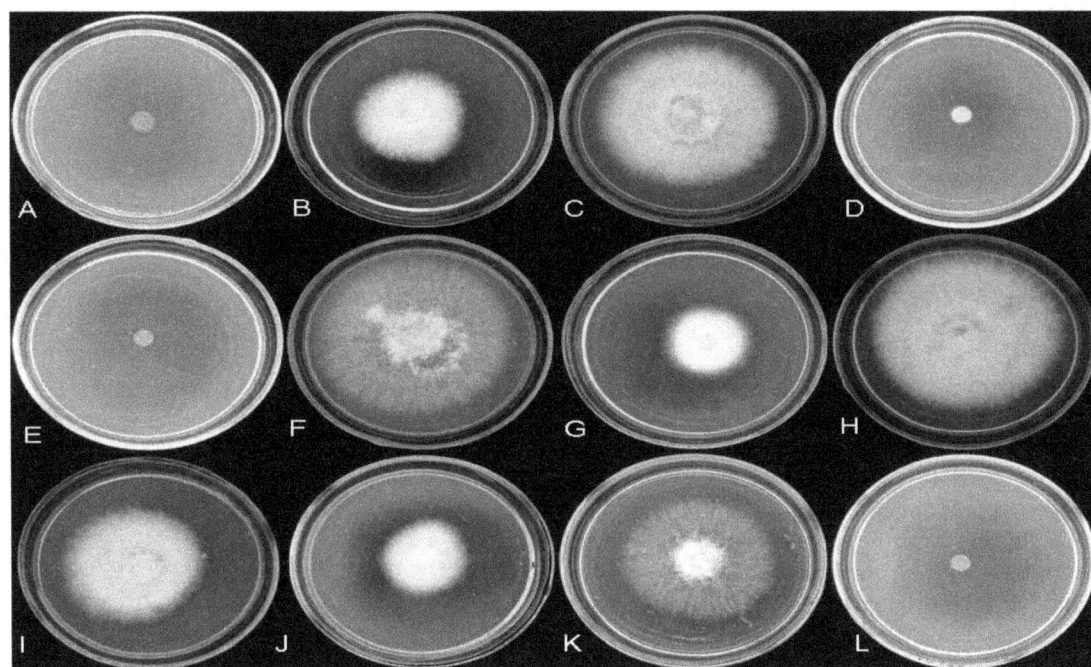

Figure 5. Effects of compounds at concentration 50 µgmL^{-1} on mycelia growth of Foc race 4 (after 5 days). A: phenylethyl alcohol; B: 1,2,3,4-etrahydronaphthalene; C: 1,2,4,5-tetramethylbenzene; D: 2,3-dihydro-4,7, dimethyl-1H-indene; E: biphenyl; F: diphenylmethane; G: fluorene; H: anthracene; I: octadecane; J: eicosane; K: ethyl acetate; L: methyl N-(1H-benzimidazol-2-yl)carbamate.

growth of foc race 4 (IC$_{50}$, 23.03 µg mL^{-1}) and was more effective than methyl N-(1H-benzimidazol-2-yl)carbamate (IC$_{50}$, 29.69 µg mL^{-1}). With an IC$_{50}$ value of 26.74 µg mL^{-1}, 1,1'-biphenyl was the second most effective compound and was also more effective than methyl N-(1H-benzimidazol-2-yl)carbamate. In contrast, 2,3-dihydro-4,7,dimethyl-1H-indene was inferior to methyl N-(1H-benzimidazol-2-yl)carbamate (IC$_{50}$, 40.36 µg mL^{-1}; Table 1). Moreover, 2-phenylethanol and 1,1'-biphenyl had short retention times (5.701 and 9.869 min,

Table 1. Effects of 4 compounds on mycelia growth of Foc race 4 (after 5 days).

Candidate of compounds	Concentration (μgmL^{-1})	Inhibition of micelial growth (%)	Regression equation	R^2	IC50 (μgmL^{-1})
Phenylethyl alcohol (1)	50	85.87±7.07[a]	y = 14.01 + 1.56 x	0.91044	23.03
	25	63.47±3.12[b]			
	12.5	39.93±2.04[c]			
	6.25	25.77±3.53[d]			
	3.125	16.37±3.12[d]			
1H-Indene,2,3-dihydro-4,7-dimethyl (4)	50	63.49±3.12[a]	y = 3.91 + 1.14x	0.97637	40.36
	25	28.15±1.18[b]			
	12.5	15.20±3.53[c]			
	6.25	12.84±1.18[c]			
	3.125	10.48±3.12[c]			
Biphenyl (5)	50	77.63±1.18[a]	y = 11.59 + 1.43x	0.90640	26.74
	25	56.24±2.36[b]			
	12.5	39.93±2.04[c]			
	6.25	15.19±2.04[d]			
	3.125	8.13±2.04[e]			
Methyl N-(1H-benzimidazol-2-yl)carbamate	50	77.62±1.18[a]	y = 0.67+ 1.66x	0.91518	29.69
	25	58.78±1.18[b]			
	12.5	14.02±2.36[c]			
	6.25	9.31±1.18[cd]			
	3.125	4.59±2.04[d]			

*: (P ≤ 0.05).

respectively) and more than 95% similarity to the standard. Therefore it was found to be three composts with antagonistic activity against foc.

Effects of strain Q34 on foc race 4 in banana plantlets grown under greenhouse conditions

Next, we wanted to determine whether treatment with strain Q34 extract could inhibit infection of foc race 4 in banana plants. As shown in Figures 6 and **7**, significant differences between the various treatments were observed. The fermented liquid and fermentation filtrate of strain Q34 inhibited infection by and growth of foc race 4 on banana plantlets (disease incidence, 31.81 and 43.60%, respectively; P ≤ 0.05; Figure 6). Plantlets treated with the fermented liquid showed disease symptoms on day 12 after inoculation, whereas those treated with medium only showed wilting on day 6 (Figure 7). In the latter case, the disease developed rapidly (disease incidence, 61.67%; P≤ ≤ 0.05).

Effects of fermented liquid of strain Q34 on Foc race 4 in banana plantlets in field

The practical application of strain Q34 was tested to

reveal that the strain Q34 controlled the disease by Foc race 4 in the field. The disease of the treatment (A), (B) and (C) were shown on the 30th day. Later, the disease incidence of the treatment (A) showed a peak in the 50th day, but declined in the 70th and 90th days, while the treatment (B) and (C)'s increased (Figures 8 and 9). These results showed that the fermented liquid of strain Q34 possessed inhibition on the development of Foc race 4.

DISCUSSION

In the current study, we sought to identify components isolated from *K. candel* that would inhibit the occurrence of Fusarium wilt in banana plants. We found that strain Q34, an endophytic fungal strain isolated from *K. candel*, had significant inhibitory activity toward foc race 4. In fact, Q34 showed stronger antagonistic activity toward foc race 4 than did cardenza, with an average inhibition rate of 72.83%. The crude extract of this strain was effective at inhibiting the growth of foc race 4 and caused distorted conidial growth. GC-MS analysis of the crude extract of Q34 revealed more than 1 compound with inhibitory activity against this pathogenic fungus. Moreover, in the biological control assay, the fermented liquid and fermentation filtrate of strain Q34 significantly inhibited

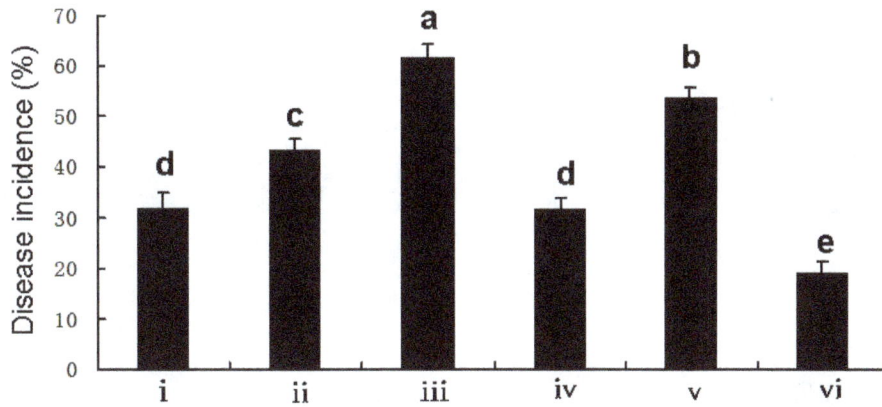

Figure 6. Disease incidence of different treatments on banana in the field (25 days). i: Fermented liquid of Strain Q34; ii: Fermentation filtrate; iii: Medium only; iv: Methyl N-(1H-benzimidazol-2-yl)carbamate; v: Sterilized water; vi: Sterilized water without Foc race 4.

Figure 7. Disease incidence of different treatments on banana in the field (25 days). i: Fermented liquid of Strain Q34; ii: Fermentation filtrate; iii: Medium only; iv: Methyl N-(1H-benzimidazol-2-yl)carbamate; v: Sterilized water; vi: Sterilized water without Foc race 4.

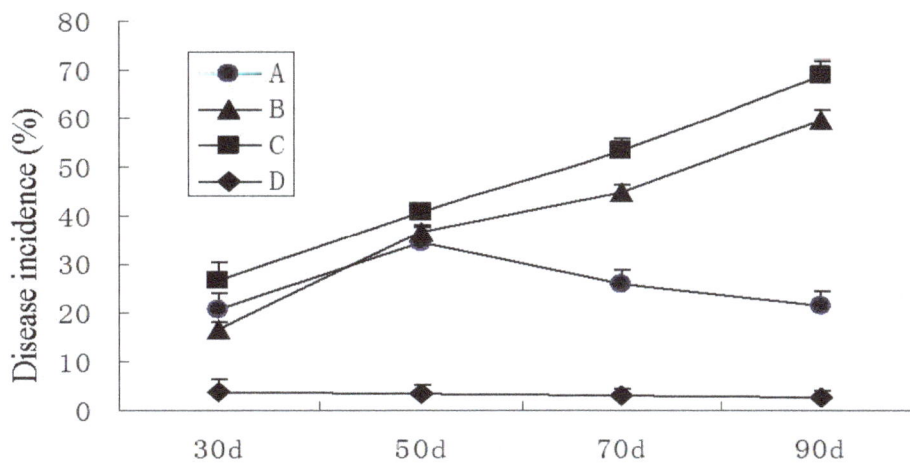

Figure 8. Disease incidence of different treatments on banana in the field (90 days). A: Fermented liquid of Strain Q34; B: Methyl N-(1H-benzimidazol-2-yl)carbamate; C: Sterilized water; D: Sterilized water without Foc race 4.

Figure 9. Effects of strain Q34 on Foc race 4 in banana plants grow in the field (90 days). A: Fermented liquid of Strain Q34; B: Methyl N-(1H-benzimidazol-2-yl)carbamate; C: Sterilized water; D: Sterilized water without Foc race 4.

infection by and growth of foc race 4 on banana plantlets. There appeared to be several compounds produced by strain Q34 that resulted in the inhibition of foc race 4.

In our experiment, we wounded the roots of plants by removing the root ball, allowing young plants to be easily infected by foc race 4. This method, combined with planting three banana plants per cup, which would cause nutrient depletion, resulted in the early appearance of symptoms of infection. Therefore, this method could be used for rapid screening of antagonistic strains inhibiting Fusarium wilt in an *in vivo* bioassay.

In the current study, we described novel structures of secondary metabolites produced by some endophytic fungi that have activity against foc, and this is becoming the focus of a new approaches for biological control (Backman and Sikora, 2008; Mejia et al., 2008). Previous studies have shown that endophytic fungi isolated from *K. candel* have antifungal activity. For example, a marine endophytic fungus (No. 1893) isolated from the dropper of *K. candel* produces two metabolites that exhibit significant activity against *Heliothis armigera* and *Sinergasilus* spp. (Chen et al., 2003). *Talaromyces* sp. ZH-154, another endophytic fungus isolated from the stem bark of the same host, contains 7-epiaustdiol and 8-O-methylepiaustdiol, which have been evaluated for their antimicrobial activities (Liu et al., 2010). In our study, only 10 of 18 compounds were tested; the anti-foc race 4 activities of the remaining 8 compounds are not known. In fact, these additional 8 compounds may represent novel antimicrobials. Indeed, with these novel resources from the mangrove, new species, new genes, and new compounds may be identified, allowing us to develop better tools with which to control foc race 4 infections.

In our study, Q34 was identified as *Acremonium* sp., some species of which are human pathogens. Previous reports have shown that the endophytic fungus *A. implicatum*, which forms beneficial symbiotic associations with *Brachiaria* spp., exhibits activity against *Drechslera* spp., species that cause leaf spots (Kelemu et al., 2001). From our study, we found that *Acremonium* sp. Q34 is similar to *A. implicatum* in morphology. While *A. implicatum* has not been implicated as a human pathogen (Gams, 1975), the potential for Q34 to behave as a human pathogen should be considered before its application as a biological control agent.

In the present study, we found that *Acremonium* strain Q34 inhibited the growth of foc race 4. This strain contains at least three compounds with inhibitory effects were found in the crude extract of Acremonium spp. Q34, which may explain its effectiveness against this pathogenic fungus. Antagonism involving a single substance is incomplete or weak, while a combination of two or more substances may enhance antagonism. To inhibit the disease process, a mixture of various fungicides, fungicides plus an antifungal strain, or a combination of two or more antifungal strains has been used in some studies (Chaves et al., 2009; Singh, 2010).

Antifungal substances have different modes of action. One mode may involve stimulation of some plant resistance mechanisms by antagonistic fungi, leading to activation of the corresponding signal transduction pathway. Another possible mode of action is antagonism between fungi, resulting in the inhibition of pathogen growth. Although we noted that strain Q34 inhibited the growth of foc race 4, the underlying mechanism is still unknown. Furthermore, *Acremonium* sp. Q34 could be colonized in banana with a long-term antifungal effect, while *Acremonium* sp. was found as endophytic fungi in banana cortex (Pocassangre et al., 2000). Further studies on these issues are in progress.

ACKNOWLEDGMENTS

We thank Prof. Jiang Zide (South China Agricultural University, Guangzhou, China) for the use of laboratory space and helpful suggestions. We also thank Dr Han Qiao Hu for writing assistance, associate professor Chang Hui Ye for help with statistical analyses, and the Editage Online Team for providing language help. This study was supported by the Science and Technology Project of Guangdong Province (No. 2012A020200009).

Abbreviations: **foc,** *Fusarium oxysporum* f. sp. *cubense*; **GC-MS,** gas chromatography-mass spectrometry; **IC$_{50}$,** median inhibitory concentration; **PDA,** potato dextrose agar; **FAO,** Food and Agricultural Organization.

REFERENCES

Altinok HH, Dikilitas M (2013). Potential of Pseudomonas and Bacillus isolates as biocontrol agents against Fusarium wil of eggplant. Biotechnol. Biotechnol. Equip. 27(4):3952-3958.

Ayyadurai N, Naik PR, Rao MS, Kumar RS, Samrat SK, Manohar M, Sakthivel N (2006). Isolation and characterization of a novel banana rhizosphere bacterium as fungal antagonist and microbial adjuvant in micropropagation of banana. J. Appl. Microbiol. 100:926-937.

Backman PA, Sikora RA (2008). Endophytes: an emerging tool for biological control. Biol. Control 46:1-3.

Belgrove A, Steinberg C, Viljoen A (2011). Evaluation of Nonpathogenic *Fusarium oxysporum* and *Pseudomonas fluorescens* for Panama DiseaseControl. Plant Dis. 95(8):951-959.

Buddenhagen I (2009). Understanding strain diversity in *Fusarium oxysporum* f. sp. *cubense* and history of introduction of 'tropical race 4' to better manage banana production. Acta Hortic. 828:193-204.

Cao LX, Qiu ZQ, Dai X, Tan HM, Lin YC, Zhou SN (2004). Isolation of endophytic actinomycetes from roots and leaves of banana (Musaacuminata) plants and their activities against *Fusarium oxysporum* f. sp. *cubense*. World J. Microbiol. Biotechnol. 20:501-504.

Cao LX, Qiu ZQ, You JL, Tan HM, Zhou SN (2005). Isolation and characterization of endophytic streptomycetes antagonists of fusarium wilt pathogen from surface-sterilized banana roots. FEMS Microbiol. Lett. 247:147-152.

Chaves NP, Pocasangre LE, Elango F, Rosales FE, Sikora R (2009). Combining endophytic fungi and bacteria for the biocontrol of Radopholus similis (Cobb) Thorne and for effects on plant growth. Sci. Hortic. 122:472-478.

Chen G, Lin Y, Lu W, Vrijmoed LLP, Gareth Jones EB (2003). Two new metabolites of a marine endophytic fungus (No. 1893) from an estuarine mangrove on the South China Sea coast. Tetrahedron 59:4907-4909.

Gams W (1975). *Cephalosporium*-like hyphomycetes: some tropical species. Trans. Br. Mycol. Soc. 64:389-404.

Georgopoulos SG, Dekker J (1982). Detection and measurement of fungicide resistance general principles. FAO Plant Prot. Bull. 30:39-42.

Guo LD (2001).Advances of researches on endophytic fungi. Mycosystrma 20(1):148-152.

Hyde KD, Soytong K (2008). The fungal endophyte dilemma. Fungal Divers. 3:163-173.

Kelemu S, White JF, Muñoz F, Takayama Y (2001). An endophyte of the tropical forage grass *Brachiaria brizantha*: Isolating, identifying, and characterizing the fungus, and determining its antimycotic properties. Can. J. Microbiol. 47:55-62.

Kirk PM, Cannon PF, David JC, Stalpers JA (2001). Ainsworth and Bisby's dictionary of the fungi.9th ed. CAB International, Egham. 1-655.

Krishnaveni R, Rathod V, Thakur MS, Neelgund YF, (2009). Transformation of L-tyrosine to L-dopa by a novel fungus, *Acremonium rutilum*, under submerged fermentation. Curr. Microbiol. 58:122-128.

Li P, Ma L, Feng YL, Mo MH, Yang FX, Dai HF, Zhao YX (2012). Diversity and chemotaxis of soil bacteria with antifungal activity against Fusarium wilt of banana. J. Ind. Microbiol. Biot. 39(10):1495-1505.

Liu F, Cai XL, Yang H, Xia XK, Guo ZY, Yuan J, Li MF, She ZG, Lin YC (2010). The bioactive metabolites of the mangrove endophytic fungus *Talaromyces* sp. ZH-154 isolated from *Kandelia candel* (L.) Druce. Planta Med. 776:185-186.

Mejia LC, Rojas EI, Maynard Z, Van Bael S, Arnold AE, Hebbar P, Samuels GJ, Robbins N, Herre EA (2008). Endophytic fungi as biocontrol agents of Theobroma cacao pathogens. Biol. Control 46:4-14.

Ploetz RC (2006). Fusarium wilt of banana is caused by several pathogens referred to as *Fusarium oxysporum* f. sp. *cubense*. Phytopathology 96:653-656.

Roberts WK, Selitrennikoff CP (1986). Isolation and partial characterization of two antifungal proteins from barley. Biochim. Biophys. Acta 880:161-170.

Si M, Xue Q, Yu B, Yuan H, Cai Y, Lai H, Chen Z (2005). Study on the antagonistic function of thirty-six strains' antagonistic actinomycete to four kinds of pathogenic fungi. J. Northwest. Sci-Tech. Univ. Agric. For. (Nat. Sci. Ed.) 33:55-58.

Singh AK (2010). Integrated management of *Cercospora* leaf spot of urd bean (*Vigna mungo*). J. Mycol. Plant Pathol. 40:595-596.

Subramaniam S, Maziah M, Sariah M, Puad MP, Xavier R (2006). Bioassay method for testing Fusarium wilt disease tolerance in transgenic banana. Sci. Hortic. 108:378-389.

Sun JB, Peng M, Wang YG, Zhao PJ, Xia QY (2011). Isolation and characterization of antagonistic bacteria against fusarium wilt and induction of defense related enzymes in banana. Afr. J. Microbiol. Res. 5(5):509-515.

Wen L, Chen G, She ZG, Yan CY, Cai JN, Mu L (2010). Two new paeciloxocins from a mangrove endophytic fungus *Paecilomyces* sp. Russ. Chem. Bull. 59:1656-1659.

Wu X, Huang H, Chen G, Sun Q, Peng J, Zhu J, Bao S (2009). A novel antibiotic produced by *Streptomyces noursei* Da07210. Antonie van Leeuwenhoek 96:109-112.

Wu YL, Yi GJ, Peng XX (2010). Rapid screening of *Musa* species for resistance to Fusarium wilt in an *in vitro* bioassay. Eur. J. Plant Pathol. 128:409-415.

Zhang CL, Zheng BQ, Lao JP, Mao LJ, Chen SY, Kubicek CP, Lin FC (2008). Clavatol and patulin formation as the antagonistic principle of *Aspergillus clavatonanicus*, an endophytic fungus of *Taxus mairei* Appl. Microbiol. Biotechnol. 78:833-840.

15

Development of hydraulic normal loading device for single wheel test rig

Satya Prakash Kumar, K. P. Pandey, Ranjeet Kumar and Man Singh

Agricultural and Food Engineering Department, IIT Kharagpur-721 302 WB, India.

A study was conducted to develop a hydraulic normal loading device for a single wheel test rig available in the Agricultural and Food Engineering Department of IIT Kharagpur. A hydraulic circuit was designed to apply additional normal load up to 2000 kg on wheel axle. The same circuit also can be used to remove 1350 kg load from the initial load. The circuit uses a fixed displacement pump, a 3-way-4 port-solenoid operated- spring centered directional control valve, a pressure relief valve and a double acting hydraulic cylinder. The normal load is sensed by a ring transducer and recorded in a data acquisition system. This device was used to study the effect of normal load on a radial-ply tire (14.9 to 28) under soft and hard soil condition. It was observed that, with hydraulic loading device the normal load on the wheel axle under dynamic condition was found to increase in the range of 1 to 5% for soft and hard bed soil surface.

Key words: Static normal loads, dynamic normal load, radial-ply tire, drawbar pull, slip.

INTRODUCTION

The prediction of tractive performance has been a major goal for many researchers. Research results show that, about 20 to 55% of the available tractor energy is wasted at the tire-soil interface. This energy is not only wasted, but wears the tire and compact the soil to a degree that may be detrimental to crop production (Burt and Bailey, 1982). Tractive performance is influenced by tire parameters, soil condition, implement type, and tractor configuration (Brixius, 1987). The tractive characteristics of a tire depend on tire geometry (width, diameter, section and height), tire type (radial, bias), lug design, inflation pressure, normal load on axle and soil type and conditions. A tire testing facility should have provisions to measure parameters such as pull, actual velocity, torque, axle rpm, tire sinkage, and dynamic normal load on tire.

The Agricultural and Food Engineering Department of IIT Kharagpur has an indoor soil bin to test the various sizes of traction tires used in tractors. The test tire is

loaded by putting dead weights on a platform attached to the test wheel. This is a very laborious and strenuous exercise, particularly when heavy loads are required for testing large tires. This operation may be facilitated by using a hydraulic loading device. Such a device would help in testing not only the large tires at higher loads but also the small tires where reduced loads are required.

Burt investigated the role of both dynamic load and slip (S) on tractive performance. At constant S, tractive efficiency (TE) increased with increases in dynamic load on compacted soil. On the soils with an uncompacted subsurface, TE decreased with increased dynamic load (Elwaleed et al., 2006). Wonderlich developed dynamic loading device for single wheel testing unity. The loading system consists of a hydraulic cylinder and an adjustable pressure reducing/relieving valve. The hydraulic cylinder is connected to the tractor's hydraulic couplings through a pressure reducing/relieving valve which keeps the loading constant over varying terrain conditions (Wonderlich and Goodall, 2007). The combined effect of normal load and inflation pressure is also significant on tractive performance of tires. A study was, therefore, undertaken to address this issue with the following objectives of design, development and evaluation of a hydraulic normal loading device for applying varying normal loads on single wheel test rig under different soil bed conditions.

Theoretical considerations

Traction performance parameters of radial and bias ply tires S, TE, coefficient of traction (COT) or net traction ratio (NTR). These parameters are defined as follows.

Slip (s)

When a tractor pulls a load, there is a reduction in distance traveled and/or speed that occurs because of flexing of the tractive device and shear within the soil. It is the ratio of decrease in the actual speed to the theoretical speed and is given by Kumar and Pandey (2012).

$$s = \frac{(V_t - V_a)}{V_a} \times 100 \qquad (1)$$

Where s = slip in percentage, V_t = theoretical velocity, m/s, and V_a = Actual velocity, m/s

Tractive efficiency (TE)

It is the ratio of drawbar power to the axle power and is given by Tiwari and Pandey (2010):

$$TE = \frac{[P \times V_a]}{[T \times \omega]} \times 100 \qquad (2)$$

Where T.E = Tractive efficiency, percent, P = Pull, Newton, T= Torque, Nm, and ω = Angular velocity, rad /sec.

Coefficient of traction (COT)

This can be defined as the ratio of pull to the normal load on the tyre and is given by:

$$C.O.T = \frac{P}{W} \qquad (3)$$

Where COT = Coefficient of traction, P = Pull, kg, W = Normal load on the tire, kg.

Relative deviation (RD)

It is a measure of how precise the average is, that is, how well the individual numbers agree with each other. The relative deviation (RD) is often times more convenient. It is expressed in percentage. (Kumar and Pandey, 2009):

$$RD = \frac{1}{N} \sum_{i=1}^{N} \frac{|P_i - O_i|}{|P_i|} \times 100 \qquad (4)$$

Where P is the predicted value, O is the observed value, and N the number of observations.

MATERIALS AND METHODS

Design of ring transducer normal load measuring

A proving ring with a maximum load bearing capacity of 3000 kg was selected for normal load measurement. The body of the proving ring was made of special steel, carefully forged to give maximum strength. Before mounting the strain gauges on proving ring, the surface of the ring was prepared carefully was rubbed by sand paper to remove paints and rust. The ring transducer was calibrated for load measurement using electronic pan balance (Figure 1) and found good linearity of applied load and output of ring transducer.

Design of hydraulic circuit for applying varying normal load

Due to sinkage of tire hydraulic load increases on the tire test rig that kept initial. During the whole procedure the system was running without any stoppage. The same procedure was followed for different hydraulic circuit with different valves that minimize increase in hydraulic normal load that kept on tire test rig. Double acting hydraulic cylinder of bore dia.80 mm, rod dia.45 mm and stroke length 250 mm was fabricated for applying different normal loads on tire test rig. The diameter of the piston rod is nearly half of the cylinder bore. Cylinder was capable of applying 2000 kg and lifting 1350 kg loads from tire test rig. The other components used in the hydraulic circuit were a pressure relief valve of size 10 lpm, a pump 12 lpm, a 3-postion-4 ports solenoid operated and spring centered DCV and a 10 L reservoir.

Developed hydraulic system was fitted over tire test rig to apply hydraulic normal load on radial ply tire for traction performance. Hydraulic loading circuit is shown in the Figure 2. This hydraulic system consisted of a 10 lpm size pressure relief valve, a hydraulic cylinder (area 50.24 cm²), a solenoid operated direction control valve, compensating valve, an accumulator, a ring transducer, an external gear pump, a 2 hp electric motor, an hydraulic oil reservoir with an oil filter. A pressure compensating set up had been adopted in order to prevent cavity in the cylinder. In the pressure compensation system two set of relief valve and check valves were placed between extend and retract end. If cylinder rod moves upward due to undulation on soil surface then cavity is created at the rod end side and excess pressure at bore end side, in this situation excess oil from bore end is diverted to the rod and through relief valve 2.

In another situation if we are operating the tire setup at reduced load than the weight of tire test setup, that is, we are taking out normal load and in this situation while in motion tire sink in the soil, in this case, vacuum is created at the bore end and excess pressure is created at rod end. In this situation excess pressure

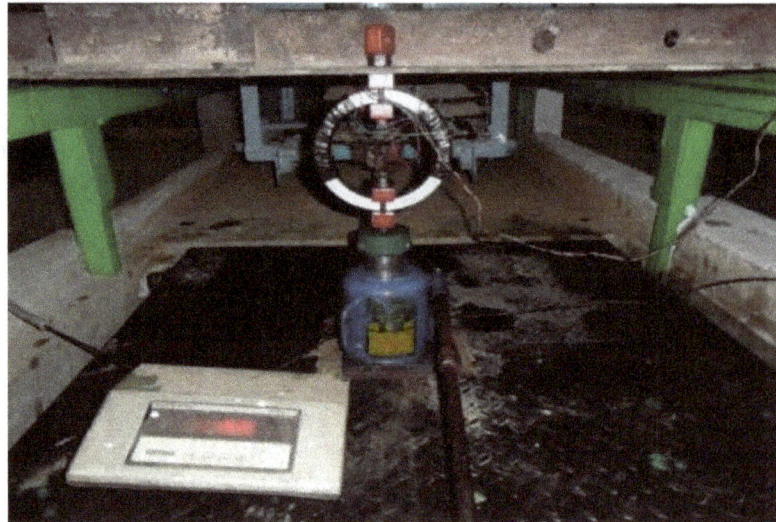

Figure 1. Calibration of ring transducer for measurement of normal load.

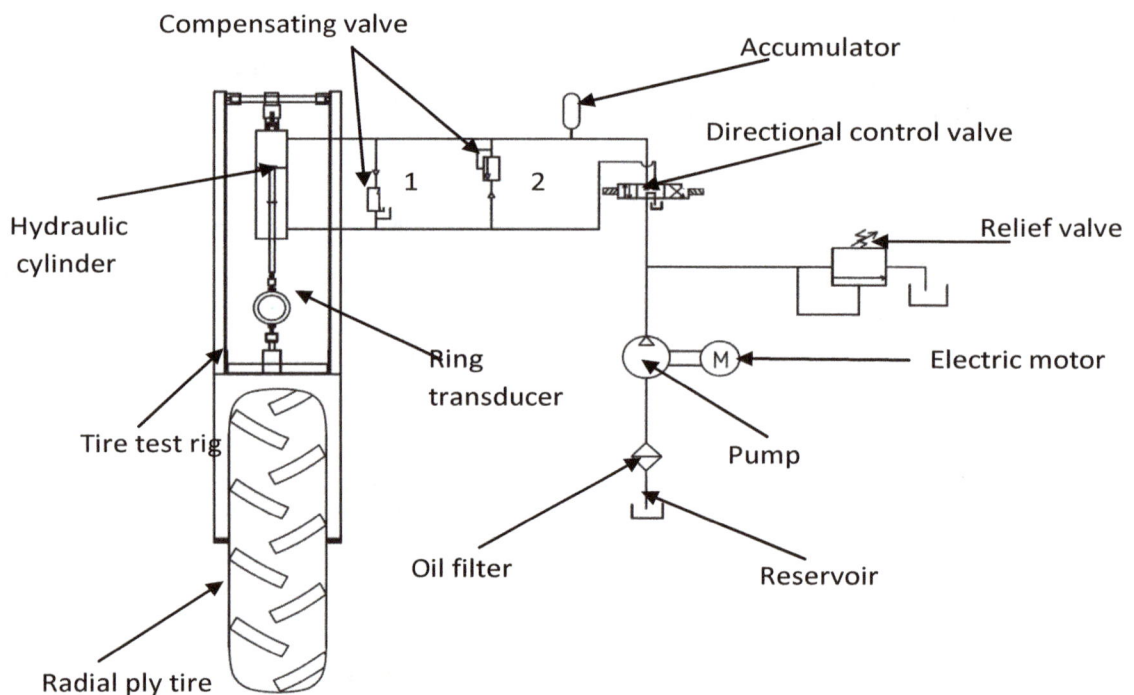

Figure 2. Hydraulic loading circuit on tire test rig.

from rod end is transferred to bore end through relief valve 1. An accumulator is used in this system to control sudden rise pressure in the system and thus, help in maintaining the applied normal load.

Experimental setup for validation of the designed hydraulic circuit and ring transducer under dynamic loading condition in soil bin

The experimental set-up consisted of a radial ply tire, tire test carriage, an indoor soil bin, a soil processing trolley, and a drawbar

pull loading device is shown in Figure 3. The different units of the experimental set-up are shown in Figure 4.

Test procedure

To investigate the effect of hydraulic normal loading behavior on soft and hard soil condition, first, soil bed was prepared with the help of soil processing trolley and cone index was measured with the help of hydraulic cone penetromete up to the soil compaction level of 600 to 1800 kpa. Normal load of 1150 kg (750 kg test rig

Figure 3. General view of the tire traction testing facility. 1= Drawbar pull loading device, 2= Radial ply tire, 3 = Indoor soil bin, 4 = Soil processing trolley, 5= Dead weight loading platform, 6 = Parallel bar linkage.

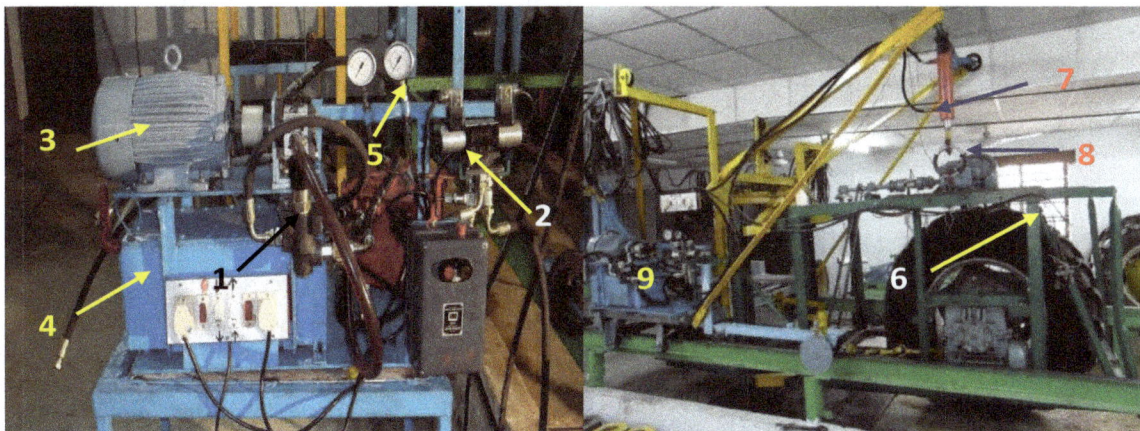

Figure 4. Constructional details of the tire test carriage along with designed hydraulic normal loading system . 1= Pressure relief valve, 2 = DCV 3.electric motor, 4= Reservoir 5.pressure gauge, 6 = Main frame, 7= Double acting hydraulic cylinder, 8 = Ring transducer, 9 = Hydraulic power system.

weight + 400 kg hydraulic weight) applied on the test tire (14.9 R 28) with the help of pressure relief valve. The tire was tested at two different bed conditions at different drawbar pulls until wheel indicate up to 15% S. Drawbar pull was applied with the help of drawbar loading device: S, drawbar pull and actual hydraulic normal load experience by the wheel were measured with the help of MGC plus data acquisition system. A program is written in mat lab to calculate parameters like average pull, normal load, actual and theoretical velocity. These data were used to determine the COT, S and TE and angular velocity of the wheel. The final results appeared on a matlab screen with graphical display and during each experiment the final values were saved in an excel file.

RESULTS AND DISCUSSION

Performance evaluation of develop hydraulic circuit in dynamic condition on different soil bed

A hydraulic normal load of 400 kg was set on wheel carriage by using the developed hydraulic circuit on soft and hard bed condition. The directional control valve was set in extending mode and the wheel was set in motion. During the entire process of run, the load fluctuated because of tire sinkage. The average dynamic normal load was found to be slightly more than the normal load that was kept initially on the tire. The normal loading behavior of the tire with drawbar pull for soft and hard bed condition is presented in Tables 1 and 2, respectively. The variation in dynamic load is shown in Figures 5 and 6 for soft and hard bed, respectively. The dynamic normal load on the radial tire was found to be 1 to 5% more than what was kept initially under static condition within 15% S on soft bed condition and 2 to 4.5% for hard bed condition.

Conclusions

A hydraulically varying normal load application system

Table 1. Increase in hydraulic normal load during dynamic condition of radial ply tire on soft bed.

Pull (kg)	Slip (%)	Initial applied hydraulic load (kg)	Average Hydraulic Load during motion (kg)	Increase in load (kg)	Increase in load w.r.t. static load (%)	Standard deviation	RD (%)	Tractive efficiency (TE) (%)
64.65	1.29	400.33	405.05	4.72	1.18	23.10	9.41	49.58
227.83	4.63	400.01	408.85	8.84	2.21	23.82	9.51	63.36
315.63	8.33	400.47	416.87	16.40	4.1	27.76	11.50	67.97
391.03	15.65	400.17	420.21	20.04	5.01	29.42	9.85	64.98

Table 2. Increase in hydraulic normal load during dynamic condition of radial ply tire on hard bed.

Pull (kg)	Slip (%)	Initial applied hydraulic load (kg)	Average hydraulic Load during motion (kg)	Increase in load (kg)	Increase in load w.r.t. static load (%)	Standard deviation	RD (%)	Tractive efficiency (TE) (%)
61.45	1.90	400.35	407.95	7.60	1.90	25.58	9.70	25.92
238.64	6.29	400.23	411.39	11.16	2.79	26.50	11.97	52.62
357.34	12.36	400.63	414.37	13.74	3.43	32.23	13.45	70.75
420.64	15.55	400.42	418.47	18.05	4.51	28.74	11.51	65.95

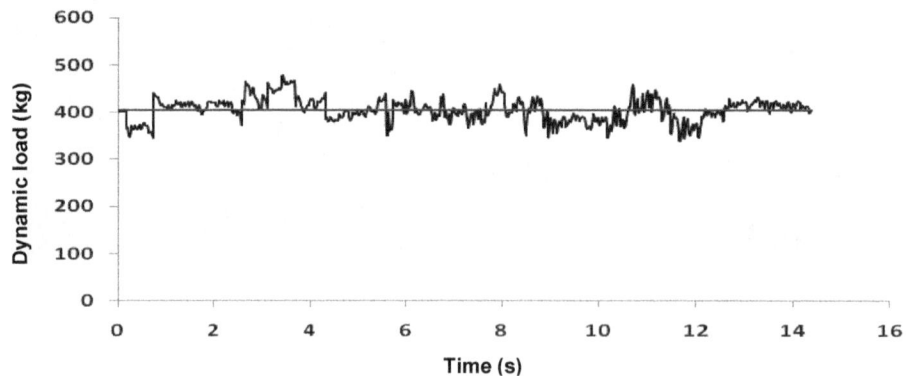

Figure 5. Variation in dynamic load with time for pull 64.65 kg on soft surface under dynamic condition.

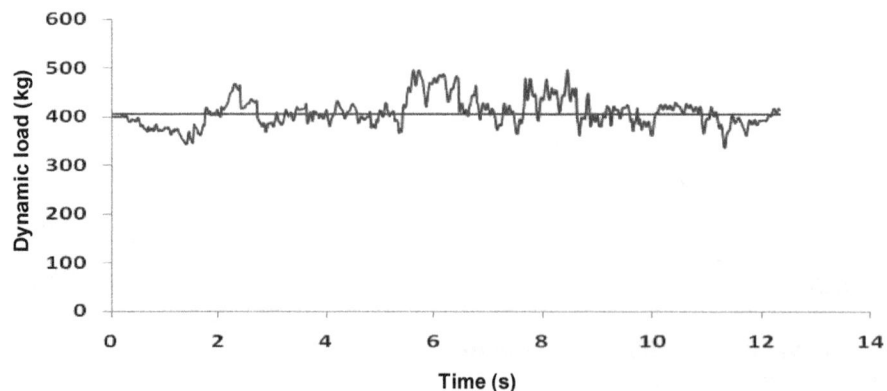

Figure 6. Variation in dynamic load with time for pull 61.45 kg on hard surface under dynamic condition.

instead of manual loading for traction studies of agricultural tires in a soil bin was designed and developed to reduce the drudgery of the labor. This system is simple and it has easy mode of operation. The hydraulically varying normal load application unit was tested rigorously and found satisfactory results. The designed and developed hydraulic loading device was capable of applying normal load of 2000 kg and lifting a maximum normal load of 1350 kg. The dynamic normal load on the radial tire was found to be 1 to 5% more than what was kept initially under static condition within 15 % S. The variation in soft bed condition was slightly more than that in hard bed condition.

REFERENCES

Burt EC, Bailey AC (1982). Load and inflation pressure effects on tires. Transactions of the ASAE 25(4):881-884.

Brixius WW (1987). Traction prediction equations for bias-ply tires. ASAE Paper No. 87-1622. ASAE St. Joseph, MI 49085-9659.

Elwaleed AK, Yahya A, Zohadie M, Ahmad D, Kheiralla AF (2006). Net traction ratio prediction for high-lug agricultural tire. J. Terramech. 43(2):119-139.

Kumar R, Pandey KP (2009). A program in visual basic for predicting haulage and field performance of 2WD tractors. J. Comput. Electr. Agric. 67:18–26.

Kumar A, Pandey KP (2012). A device to measure dynamic front wheel reaction to safeguard rearward overturning of agricultural tractors. J. Comput. Elect. Agric. 87:152–158.

Tiwari VK, Pandey KP (2010). A review on traction prediction equations. J. Terramech. 47:191–199.

Wonderlich G, Goodall A (2007). single wheel tester for small agricultural construction, and ATV tires. Biosystems and Agricultural Engineering Department University of Kentucky, pp. 7-21.

16

Isolation, characterization and screening of *Burkholderia caribensis* of rice agro-ecosystems of South Assam, India

Biswajit Deb Roy[1]*, Bibhas Deb[2] and Gauri Dutta Sharma[1]

[1]Microbiology Laboratory, Department of Life Science, Assam University, Silchar-788011, Assam, India.
[2]Microbiology Laboratory, Department of Botany, Gurucharan College, Silchar-788004, Assam, India.

Six isolates of *Burkholderia* were obtained from rhizosphere soils of rice grown in tropical lowlands of South Assam, India. Among the identified *Burkholderia* isolates, SDSA-I10/1 has shown higher nitrogen fixing potential and it was selected for 16S rDNA sequencing. The isolate SDSA-I10/1 showed highest resemblance to *Burkholderia caribensis* MWAP84 (Y17011) and hence, identified as *Burkholderia caribensis* strain SDSA-I10/1 (GU372342). Inoculation of this strain improved the growth and yield parameters of rice significantly over uninoculated control plants, thus it may be used as indigenous microbial inoculant for intensive rice cropping in tropical lowlands.

Key words: *Burkholderia caribensis* strain SDSA-I10/1, 16S rDNA sequence, biofertilizer, Diazotroph, nitrogenase activity, tropical lowland.

INTRODUCTION

The world population is increasing day by day but the expansion of land is limited. Moreover, almost half of the world's population is consuming rice (*Oryza sativa* L.) as the primary food grain, making it the most important food crop currently produced (Cottyn et al., 2001). Hence, to produce higher yields of rice, expensive nitrogenous fertilizers are commonly used. These are used to fulfill the nitrogen demand of rice that can be overcome partially by using biofertilizers when they are scientifically applied. Biofertilizer is important in crop farming systems because it is an inexpensive source of nitrogen for the higher yields of crops. This process diminishes the need for expensive chemical fertilizer. Thus the extensive use of biofertilizers would provide economic benefits to farmers improve the socio-economic condition of people and preserve natural resources.

Diverse diazotrophic and endophytic bacteria have been isolated from rice plants (Engelhard et al., 2000; Gyaneshwar et al., 2001). There is a possibility that all or some of these bacteria could be contributing to the N balance of wetland rice (Malarvizhi and Ladha, 1999; Nieuwenhove et al., 2000). Endophytic bacterial associations with rice are nonspecific and the size of the bacterial population in rice tissues is low (Malarvizhi and Ladha, 1999). The association of rhizospheric *B. vietnamiensis* (Gillis et al., 1995; Tran Van et al., 2000) explains well the biological nitrogen fixation (BNF) observed in certain rice genotypes. In rice, β-proteobacteria of the genus *Burkholderia*, for example, *Burkholderia brasilensis*, *B. vietnamensis* (Gillis et al., 1995) and *Burkholderia* spp. (Muthukumarasamy et al., 2007) have been reported in high numbers. These bacteria colonize the rice plants systemically, although the highest numbers were observed in the roots.

Some of the best studied diazotrophs for nitrogen fixation are *Burkholderia* spp., which fix N_2 when other sources of nitrogen are absent or at low levels (Dobₒereiner et al., 1993; Kirchhof et al., 1997). In Vietnam, rice grown inoculated with *B. vietnamiensis*

---
*Corresponding author. E-mail: b_deb_roy@sify.com

increased in yield 13 to 22% (Tran Van et al., 2000). An endophytic *Burkholderia* species isolated from rice plants in Brazil fixed 31% of the total nitrogen captured by the plant and resulted in a 69% increase in the rice biomass (Baldani et al., 2000). A yield increase of 54% was observed for the rice inoculated with *B. brasilensis* (Guimaraes et al., 2000). Bacteria belonging to the genus *Burkholderia*, which are very common in soil, water and associated with plants (McArthur et al., 1988), have a wide natural diversity, not only in taxonomy, but also in ecological features. In Brazil, *Burkholderia* species associated with rice plants could fix 31% of the total nitrogen captured by the plant (Baldani et al., 2000) and the inoculation of rice with this endophytic *Burkholderia* species led to a 69% increase in the rice biomass (Baldani et al., 2000). Rice varieties grown in low fertility soil in Vietnam inoculated with the *B. vietnamiensis* TVV75 strain gave yield increases of 13 to 22% (TranVan et al., 2000). Diazotrophs can affect plant growth directly by the synthesis of phytohormones and vitamins, inhibition of ethylene synthesis, improving nutrient uptake, enhancing stress resistance, solubilization of inorganic phosphate and mineralization of organic phosphate (Baldani et al., 2000). Indirectly, diazotrophs are able to decrease or prevent the deleterious effects of pathogenic microorganisms, mostly through the synthesis of antibiotics and/or fungicidal compounds, through competition for nutrients or by the induction of systemic resistance to pathogens (Baldani et al., 2000). In addition, they can affect the plant indirectly by interacting with other beneficial microorganisms.

In this study, we have isolated a diazotrophic bacterium, *Burkholderia caribensis* strain SDSA-I10/1from the rhizosphere of cultivated rice varieties grown in rainfed acidic lowlands of South Assam, India. The strain was identified using morphological, physiological, biochemical, and 16S rDNA nucleotide sequence analysis. Finally, the bacterial strain was used in plant inoculation experiment to demonstrate the potential to improve the growth performance and grain yield of rice.

MATERIALS AND METHODS

Assam, the eastern most state of the Indian sub-continent, extends from 22.19$'$ to 28.16$'$ North latitude and 89.42$'$ to 96.30$'$ East longitude. Southern Assam (Barak Valley) comprising three districts, namely Cachar, Karimganj and Hailakandi, is situated between longitude 92.15$''$ and 93.15$''$ East and latitude 24.8$''$ and 25.8$''$ North, covering an area of 6,941.2 sq. km. of land.

The soil samples were collected randomly from 6 locations of South Assam from the rhizosphere of rice growing in acidic lowlands at panicle initiation stage during July, 2009. Three parallel sampling lines were marked out at known distances (depending on the size of the selected rice growing field) from each other in each location. The first line was placed randomly and the others parallel to this line. Each line included four sampling areas (1 m^2) placed at regular distances from each other. Five rectangular soil cores (5 × 3.5 cm^2, 0 to 30 cm deep) were taken from each sampling area

involving rhizospheric zone of growing rice crop. The samples taken from the same line were combined and each pooled soil sample, henceforward consisting of 20 cores. The soil samples were placed in sterilized plastic bags, transferred to the laboratory within one day and stored at 4°C prior to isolation. Sterile gloves were used in the soil sampling, working tools were sterilized with ethanol and flamed, and further procedures were performed as aseptically as possible.

Shade-dried, ground and sieved rhizosphere soil samples (1 g) were used for tenfold serial dilution. 10 g soil from each sample was aseptically weighed and transferred to an Erlenmeyer flask with 90 ml sterilized distilled water, and were shaken for 30 min at about 150 rpm. Immediately after shaking, a series of tenfold dilutions of the suspension was made for each sample by pipetting 1 ml aliquot into 9 ml sterilized distilled water. Appropriate serial dilution (10^{-6}) of rhizosphere soil samples was inoculated was inoculated (1 ml) aseptically onto petriplates containing 20 ml melted agar medium at 30°C in triplicate for 5 to 7 days (Burbage and Sasserl, 1982). Quantitative enumeration of the strains from all three pooled soil samples of each location (n=3) was carried out by dilution plate count method in solid N-free modified PCAT medium (specific for isolation of *Burkholderia* sp.) without tryptamine (Estrada-de-los Santos et al., 2001). Colonies growing on the dishes were counted after incubation. Data from each of the three replicates were averaged for a location and expressed as cfu (colony forming units) per g of soil.

The sub-surface pellicles or the turbidity of the entire modified PCAT medium were considered presumptively positive for growth of *Burkholderia* strains. The pellicles/growth was sub-cultured in the same medium for acetylene reduction assay. This assay was carried out by injecting 10% (v/v) acetylene in the head space above the medium and incubated for 1 h (Hardy et al., 1973). Ethylene production was measured using a Systronics Gas Chromatograph with a Poropak Q column and a flame ionization detector connected to a chromatography data computer system. Nitrogenase positive pellicles were sub-cultured in fresh semisolid media and streaked onto solid plates supplemented with yeast extract (100 mg L^{-1}), incubated at 30°C for 5 days.

Colonies were further analysed for taxonomic identity. Identification of *Burkolderia* isolates was done by (a) cultural, (b) morphological, (c) biochemical and physiological study following Bergey's manual of determinative bacteriology (Hensyl, 1994). The isolates were studied following the methods of Cerney (1993) including morphological and physiological features. The physiological and biochemical activities of the isolates were tested through the methods as demonstrated by Collee and Miles (1989).

Genomic DNA was isolated by the method of Ausubel et al. (1987), except that the lysate was extracted twice with chloroform to remove residual phenol. The 16S rDNA gene was amplified using degenerative primers according to the direct sequencing method (Hiraishi, 1992). The 16S rDNA sequence of the isolate SDSA-I10/1 as determined in this work was edited manually and aligned with a selection of 16S rDNA sequences from members of the alpha, beta and gamma subclasses of the Proteobacteria. These were retrieved from the sequence collection of the Ribosomal Database Project (Maidak et al., 1996) and from GenBank. Non-resolved positions and gaps were removed prior to the phylogenetic analysis. The sequence was aligned using Clustal V software and homologies of sequences were determined using the basic alignment search BLAST against the NCBI database. Distance matrix was corrected for multiple base changes at single locations by the method of Jukes and Cantor (1969). Phylogenetic tree was constructed from evolutionary distance matrix by the neighbour-joining method of Saitou and Nei (1987) using NEIGHBOR, contained in the Phylogenetic Inference Package, PHYLIP 3.51 (Felsenstein, 1978). The phylogenetic tree was constructed using the 10 nearest neighbours and as well as other species available in the database. The analysis of distance matrix, similarity matrix and phylogenetic tree was done using the neighbor program included in Joe Felsenstein's Phylip 3.51c distribution. Parsimony analysis was performed by using DNAPARS. The results obtained by both NEIGHBOR and DNAPARS were subjected to bootstrap analysis by using SEQBOOT in sets of 1000 resamplings. The hierarchy view and the sequence were done using RDP release 10 (ribosomal database release 10 software).

A field experiment was set-up at Kalinjar (24°92′ N, 92°24′ E and 71 m above mean sea level) located 6 km away from Silchar, Assam, India. The field was not cultivated in the last year prior to this experiment. Climate of the region is sub-tropical humid and receives mean annual rainfall 2,431 mm and average rainy days 167 per annum. Total bright sunshine hour (BSSH) is 2,029 h against maximum possible BSSH of 4,242 h per year. Mean relative humidity is 82%. During experimental period in 2009, the mean maximum and minimum temperatures recorded during sali rice (winter, August to November) were 32.6 and 22.3°C, respectively. The length of crop growing period is >200 days in a year in this rice agro-ecological zone of Assam.

The experimental field was divided into two blocks (one for *Burkholderia* treatment and one for control). Within each block, five plots were the replicates, each with an area of 4×5 m^2. Each block was laterally isolated by polythene sheets embedded into the soil to a depth of 30 cm. The experiment was arranged as completely randomized block design.

The initial soil characteristics of the experimental field were determined. The clay loam inceptisol had the following properties: sand 36%, silt 14%, clay 49%, pH (1:2, soil/water) 4.90, total organic C 0.38%, total N 32.5 kg ha^{-1}, total P 27.4 kg ha^{-1}, total K 35.7 kg ha^{-1}, organic matter 0.48%, and electrical conductivity 0.94 dsm^{-1}. The soil of the experimental plot was analyzed as per the standard methods of Jackson (1973).

The plant inoculation experiment was undertaken in sali (winter) cropping season (August to December) and the crop was successfully harvested. Three rice seedlings together (25 days old) were transplanted in the puddled plots at spacing 30×10 cm (between rows × between plants). The two treatments were: *Burkholderia caribensis* or T_1 and control or T_2. The standard inoculum of the strain was prepared by growing the bacteria in 250 ml glucose-peptone broth for 72 h at 28 ± 2°C on a shaker. The cells in active growth stage were harvested by centrifugation at 8000 rpm for 10 min and resuspended the pellet in sterile distilled water to attain a concentration of 10^8 cfu ml^{-1}. Then sterilized charcoal powder was mixed with the aqueous suspension of the diazotroph strain as described by Thakuria et al. (2004) to prepare carrier based inoculant. Plant infection of the strain was done by seedling root dip method. For this purpose, an aqueous slurry of charcoal powder based inoculants of the strain was made in shallow plastic tubs. Twenty five day old seedlings of rice were uprooted carefully and the root portions of the uprooted seedlings were dipped into the aqueous slurry for 12 h to ensure maximum contact of the strain on the root surface. The seedlings treated with aqueous slurry of sterilized charcoal powder but devoid of the *Burkholderia* strain were used as a control. Five replicates were maintained for each treatment. The cfu count of the biofertilizer strain was 10^8 colony forming units (cfu) g^{-1} charcoal. The charcoal based inoculants were applied (at 4 kg ha^{-1}) to rice seedlings by root-dip technique.

At the harvesting stage, the plants were uprooted with intact roots, washed thoroughly to remove the adhered soil and taken for analysis of growth and yield parameters. Plant height (cm plant^{-1}), shoot length (cm plant^{-1}) and root length (cm plant^{-1}) were measured: the average height of five randomly chosen plants from each plot was measured from ground level to the panicle tip. Number of tillers/plant was counted in five randomly chosen plants (average) in each plot. The fresh weight (g plant^{-1}) and 100-grain weight (g plant^{-1}) were analysed from five randomly taken samples of each plot. Plant dry weight (g plant^{-1}) was recorded after drying

In an oven for 1 day at 70°C.

The N-content of shoot was estimated by micro-kjeldahl method (Fallik et al., 1988) and the chlorophyll content of leaves was estimated by Arnon method (1949). Protein content of grains was determined by Lowry's et al. (1951) method. Nitrate reductase (NR) activity was assayed by soaking the roots in KNO$_3$ solution for 48 h. NR was recorded as reduction of KNO$_3$ to KNO$_2$ per hour of reaction at 30°C in terms of concentration (μM NO$_2$ h^{-1}g^{-1} fresh wt.) of the latter (Hewitt and Nicholas, 1964).

Summary statistics were used to obtain the mean, standard error and percent increase over control (Snedecor and Cochran, 1967). The least significant difference (LSD) was calculated following the method of Misra and Misra (1983). The nucleotide sequence of the 16S rDNA gene of the strain SDSA-I10/1 investigated in this experiment was deposited in the NCBI GenBank through the GenBank submission tool BankIt (http://www.ncbi.nlm.nih.gov) under accession number GU372342.

RESULTS

Altogether, six isolates of indigenous *Burkholderia* sp. were obtained from the acidic rice rhizosphere soils of South Assam, India. The data in Table 1 show the cfu count of *Burkholderia* isolates in rice rhizosphere soils at panicle initiation stage in the six locations of South Assam. The isolates of *Burkholderia* showed the highest cfu count in the rhizosphere of acidic submerged rice fields of Cachar district. The isolate SDSA-I10/1 showed highest cell count in the rice fields of Cachar district. The cell count of the isolate SDSA-I10/4 was higher in the lowland rice fields of Karimganj district and that of the isolate SDSA-I10/5 was higher in the lowland rice agro-ecosystems of Hailakandi district. The population of the isolates differed significantly not only between the locations but between the different isolates as well. On an average, SDSA-I10/1 population was highest followed by SDSA-I10/4 and SDSA-I10/5 which are the dominant *Burkholderia* strains in the rice rhizosphere soils of South Assam.

The pure cultures of the isolates of *Burkholderia* were tested for the ability to fix atmospheric N$_2$ by acetylene reduction technique. The greater the activity of nitrogenase enzyme the greater is the ability of atmospheric N$_2$-fixation. The maximum nitrogenase activity was observed in SDSA-I10/1 culture (402.3 nM C$_2$H$_4$ hr^{-1}ml^{-1} culture). Other isolates also showed good N$_2$-fixation ability as characterized by ARA activity > 300 nM C$_2$H$_4$ h^{-1}ml^{-1} culture (Table 2).

Burkholderia isolates have been identified as per following characteristics: round, entire, convex or raised, smooth, opaque, orange or brown colony, cell rod or oval shaped and occur singly or in pairs, gram negative, cyst and extra-cellular PHB granules present, catalase positive, non sporulating, motile, aerobic or occasional anaerobic, optimum growth temperature 25 to 30°C, pH 3.0 to 5.7, growth at maximum 2.5% NaCl, hydrolyzed urea and nitrate, fermentative, oxidase positive, and acid production from dextrose, galactose, inositol, arabinose, mannitol and sucrose. According to Bargey's Manual of

Table 1. Quantitative enumeration of *Burkholderia* isolates in the rhizosphere of cultivated rice varieties at panicle initiation stage grown under acidic lowland ecosystems.

District	Location	Isolate	Rice variety	*cfu g^{-1} dry soil ($\times 10^6$)
Cachar	Salchapra	SDSA-I10/1	Ranjit	51 ± 4.32
	Narsingpur	SDSA-I10/2	Bahadur	28 ± 3.1
Karimganj	Kaliganj	SDSA-I10/3	Mahsuri	21 ± 1.32
	Patharkandi	SDSA-I10/4	Ranjit	47 ± 2.82
Hailakandi	Lalabazar	SDSA-I10/5	Ranjit	43 ± 1.75
	Panchgram	SDSA-I10/6	Ranjit	17 ± 1.25

± SE, * cfu g^{-1} dry soil is the average of all the locations of the district which was again average of the three pooled soil samples of each location in the district.

Table 2. Estimation of N_2-fixing potential (nitrogenase activity) of *Burkholderia caribensis* isolates by ARA.

Isolate/ Strain	*Nitrogenase activity (nM C_2H_2 h^{-1} ml^{-1} culture)
SDSA-I10/1	402.3
SDSA-I10/2	352.4
SDSA-I10/3	371.7
SDSA-I10/4	312.4
SDSA-I10/5	302.1
SDSA-I10/6	338.5
LSD at 5% significant level	12.25

*Nitrogenase activity is the average of three replicates.

Systemic Bacteriology, the biochemical and carbohydrate fermentation tests indicated that the characters represented by the isolates were similar to *Burkholderia* spp. (Table 3).

Out of six isolates, the one showing higher N_2-fixation rate (SDSA-I10/1) as envisaged by its higher nitrogenase activity was analyzed for 16S rDNA gene sequence due to feasibility of experimentation. The sequence of the 16S rDNA gene of the isolate investigated in this work has been deposited in GenBank as described in the methods section. The 16S rDNA gene sequence of the isolate SDSA-I10/1 showed close resemblance to *B. caribensis*; MWAP84; Y17011 and *B. caryophylli* (Figure 1). The length of 16S rDNA sequence was approximately 1119bp and showed a high similarity to *B. caribensis* (99.3%) MWAP84 (Y17011). The phylogenetic analysis of the isolate and the type strain of the genus *Burkholderia*, formed one compact cluster with *B. caribensis* str. MWAP84, str. TFD2, *B. caribensis* str. MWAP64, *B. caribensis* str. MWAP71, and *B. caryophylli* str. MCII-8 (Figure 1). 16S rDNA sequence analysis confirmed that this isolate had a 99.3% sequence similarity with *B. caribensis* MWAP84 (Y17011).

The isolate SDSA-I10/1 has showed highest N_2-fixing ability in comparison to other isolates of *Burkholderia* and strain identification of this isolate was carried out on the basis of morphological, physiological, and biochemical

study coupled with 16S rDNA gene sequencing. Therefore, this strain was selected for bioinoculation experiment. Since rice is mostly grown in sali (winter) crop during August to December under acidic rainfed lowland ecosystems of South Assam, the bioinoculation effect of this strain in N_2-assimilation of rice was observed on winter cropping season. The popular variety of winter rice in South Assam is *cv.* Ranjit which was selected for bioinoculation experiment.

The data in Table 4 revealed that treatment with the *Burkholderia* strain SDSA-I10/1 significantly increased the growth and yield of rice over untreated control plants in winter season. Almost 13% increase in plant height over uninoculated control was recorded. The shoot length of plants was improved by 6% following inoculation with SDSA-10/1 strain. Root length of rice was improved by 18%. Number of tillers was increased by 29%, fresh and dry biomass by 17% and 34% respectively, and weight of grains by 28% followed by inoculation of winter rice *cv.* Ranjit with *Burkholderia caribensis* strain SDSA-I10/1 in acidic flooded lowlands of South Assam. Inoculation of *B. caribensis* treatment resulted in maximum increase in weight of grains in autumn season among all the treatments.

38% increase in the plant chlorophyll$_a$ content was reported at the harvesting stage of winter rice. The nitrogen content of shoot was improved by 31% following

Table 3. Morphological, physiological and biochemical characteristics of Burkholderia isolates.

Tests	Isolates					
	SDSA-I10/1	SDSA-I10/2	SDSA-I10/3	SDSA-I10/4	SDSA-I10/5	SDSA-I10/6
Colony morphology	Round, entire, highly raised, opaque, orange colony	Round, entire convex, opaque, colony	Round, entire raised, opaque orange colony	Round, entire highly raised, colourless, colony	Round, entire, convex, colourless colony	Round, irregular Convex, opaque, light brown colony
Cell shape	Oval rod	Oval	Oval rod	Rod	Rod	Spherical
Arrangement	Single	Single	Paired	Single	Single	Single
Gram's reaction	-	-	-	-	-	-
Motility	+	+	+	+	+	+
Spore	Absent	Absent	Absent	Absent	Absent	Absent
PHB granule	Present	Present	Absent	Present	Present	Present
Cyst	Present	Present	Present	Present	Present	Absent
Optimum growth temperature (°C)	25-30°C	25-30°C	25-30°C	25-30°C	25-35°C	25-30°C
Optimum pH	3.0-5.7	3.5-5.7	3.0-6.4	3.5-5.7	4.0-5.7	4.0-5.7
Anaerobic growth	±	±	±	±	±	±
NaCl tolerance (%)	2.5	2.5	2.5	2.5	2.5	2.5
Gas production from glucose	-	-	-	-	+	-
Urea hydrolysis	±	±	±	±	±	±
Starch hydrolysis	-	-	-	-	-	-
Nitrate reduction	±	±	±	±	±	±
Catalase	+	+	+	+	+	+
O/F test	F	F	F	F	F	F
Cytochrome oxidase	±	±	±	±	±	±
H_2S production	-	-	-	-	-	-
Methyl red test	-	-	-	-	-	-
Voges Proskauer test	-	-	-	-	-	-
Fructose	-	-	-	-	-	-
Galactose	+	+	+	+	+	+
Inositol	+	+	+	+	+	+
Dextrose	+	-	+	+	+	+
Cellobiose	-	-	-	-	-	-
Arabinose	+	+	+	+	-	+
Mannitol	+	+	+	+	+	+
Sorbitol	-	-	-	-	-	-
Sucrose	±	±	±	±	±	-
Xylose	-	-	-	-	-	-

+, Tests positive; ±, tests variable; -, tests negative.

inoculation with *B. carinbensis* strain SDSA-I10/1. The protein content of grains was also increased over the uninoculated plants. 25% increase in protein content of matured grains was observed in the plant inoculation test. The NR activity of rice roots was improved by 16% over uninoculated control in winter crop (Table 5).

DISCUSSION

Six isolates of N₂-fixing heterotrophic *Burkholderia* sp.

were observed in the rhizosphere region of cultivated rice varieties grown in the tropical rainfed lowlands of South Assam, India. The occurrence of N₂-fixing *B. caribensis* strain in the root region of rice grown in acidic rainfed submerged lowlands of South Assam, India is reported for the first time in this present investigation. Enumeration, isolation and identification of diazotrophs from rhizosphere soil of Korean rice varieties revealed three groups of N₂-fixing bacteria belonging to the genera *Azospirillum*, *Burkholderia* and *Gluconacetobacter* (Kang, 2006). Agronomically important paddy rice varieties in

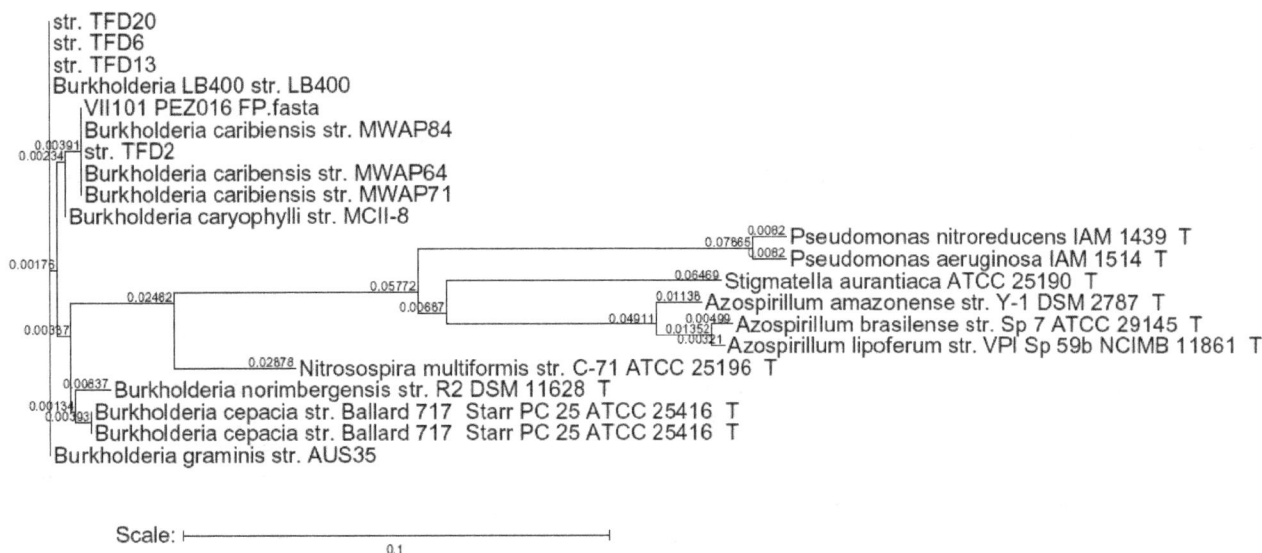

Figure 1. Phylogenetic tree based on 16S rDNA gene sequence comparison showing the position of *B. caribensis* strain SDSA-I10/1 (VII101_PEZ016_FP.fasta) and other related strains of the family Burkholderiaceae.

Table 4. Effect of *B. caribensis* strain SDSA-I10/1 on the growth and yield of winter (sali) rice cv. Ranjit in the absence chemical fertilizers.

Treatment	Plant height (cm)	Shoot length (cm)	Root length (cm)	No. of tillers/ plant	Plant fresh weight (g)	Weight of 100 grains/ plant (g)	Dry/ biomass plant (g)
B. caribensis, SDSA-I10/1 (T1)	125.3 (113)	100.3 (106)	25.0 (118)	17.3 (129)	137.3 (117)	2.66 (128)	44.05 (134)
Control (T2)	110.4 (100)	94.2 (100)	16.2 (100)	13.4 (100)	117.1 (100)	2.08 (100)	32.82 (100)
LSD (P=0.05)	9.45	7.13	4.47	2.99	7.22	0.43	3.77

*Each variable is the average of five replicates; figures in bracket indicate percent increase over control; LSD (P<0.05) = Least significant difference at 5% level of significance.

Table 5. Effect of *B. caribensis* strain SDSA-I10/1 on the biochemical parameters of winter (sali) rice cv. Ranjit in absence of chemical fertilizers.

Treatment	Chlorophylla content of leaves (mg g^{-1} fr wt.)	Nitrogen content of shoot (mg g^{-1} dry wt.)	Protein content of root grains (%)	Nitrate reductase activity (µM NO$_2$ h^{-1}g^{-1} fr wt.)
B. caribensis, SDSA-I10/1(T1)	1.08 (138)	0.59 (131)	9.73(125)	9.78(116)
Control (T2)	0.78 (100)	0.45 (100)	7.80 (100)	8.42 (100)
LSD (P=0.05)	0.06	0.05	0.28	1.08

*Each variable is the average of five replicates; figures in bracket indicate percent increase over control; LSD (P<0.05) =Least significant difference at 5% level of significance.

Korea harbour different diazotrophic isolates such as *Azospirillum* spp., *Herbaspirillum* spp., *Burkholderia* spp. and *Gluconacetobacter* spp. (Kang, 2006). The identification experiments revealed the association of N$_2$ fixing *Herbaspirillum* spp., *Gluconacetobacter diazotrophicus* and *Burkholderia* spp. in Korean rice varieties for the first time apart from the association of *Azospirillum* spp. and *Pseudomomnas* spp.

This present work revealed that the number of cultivable N$_2$-fixing *B. caribensis* in the root region of rice at panicle initiation stage in most of the sampling sites of South Assam was over 10^6 cfu g^{-1}dry soil confirming the number found by Watanabe et al. (1978) in paddy soil at the International Rice Research Institute. This may be due to the fact that the number of N$_2$ fixers is strongly governed by soil organic matter content (Xie et al., 2003) and rice agro-ecosystem soils of South Assam are rich in organic matter. The population count of the isolate SDSA-I10/1 was highest followed by SDSA-I10/4 in the acidic rice field soils of South Assam. Watanabe and Barraquio (1979) revealed that nitrogen-fixing bacteria are present in greater numbers in the roots of wetland rice. All the isolates are more or less acid tolerant and therefore they are prevalent in the acidic rice agro-ecosystems of South Assam. All the six isolates formed the typical orange yellow sub surface pellicles on semisolid selective modified PCAT medium which showed the micro aerobic nature of the organism (Cavalcante and Dobereiner, 1988). Rennic and Vose (1983) used single nitrogen free medium for isolating nitrogen-fixing bacteria and showed that at the higher dilutions 75% of the isolates exhibited acetylene reduction. The results confirmed that the isolates were of Burkholderia spp. The studied isolates used galactose, inocitol, dextrose, arabinose, mannitol, and sucrose like that of recently reported Burkholderia phytofirmans (Sessitsch et al., 2005).

Recently studies have been carried out in different parts of the world to investigate the role of diazotrophic inoculation in crop growth and yield to develop alternative sources of chemical fertilizers. Nitrogenase activity of maize rhizospheric bacteria was detected in 19 isolates ranging from 21.8 to 3624 nM C_2H_2 h^{-1} ml^{-1} culture (Naureen et al., 2005). In this present study, the nitrogenase activity range (302.1 to 402.3 nM C_2H_4 h^{-1} ml^{-1} culture) of the isolates fell within that of rhizospheric bacteria as detected by Naureen et al. (2005). Reis et al. (2004) reported variation in nitrogen fixing efficiency in different strains under different conditions. In this present study, six isolates of Burkholderia sp. from rhizosphere of rice were designated as SDSA-I10/1, SDSA -I10/2, SDSA-I10/3, SDSA-I10/4, SDSA-I10/5, and SDSA-I10/6 respectively. The isolates showed appreciable amount of nitrogenase activity and the strain SDSA-I10/1 showed highest ARA and hence the strain was selected for 16S rDNA sequencing and plant inoculation test. Bacteria were enumerated by cfu count and the isolates were identified based on morphological, physiological, and biochemical characteristics. This work has led to the identification of a novel strain of culturable N_2-fixing heterotrophic B. caribensis SDSA-I10/1 of the class β-proteobacteria. The isolate SDSA-I10/1 has the highest 16S rDNA similarity to the members of the family Burkholderiaceae and are in the same clade of B. caribensis (99.3%). This present investigation has led to the molecular characterization of one indigenous diazotroph strain, B. caribensis, GU372342 from acidic lowland rice agro-ecosystem of South Assam, India which has shown higher N_2-fixing capacity and can be used as microbial fertilizer for rice.

One of the probable causes of poor rice production may be due to low (<50%) agronomic use efficiency of nitrogenous fertilizers in acidic rice field soils under high rainfall (>2000 mm) conditions (Ladha et al., 1998).

Therefore, most of the rice field soils of South Assam are deficient in N and the farmers are using excessive amount of chemical N fertilizers. Since the long term use of chemical nitrogenous fertilizers depletes the soil organic matter and poses a threat to the survival of indigenous soil micro-flora, biological nitrogen fixation (BNF) technology can play an important role in substituting the use of chemical N fertilizers in rice cultivation.

In this present study, gnotobiotic assays were conducted to test the inoculation effect of B. caribensis (GU372342) on biological and biochemical parameters of rice. Seedlings treated with the strain showed a considerable increase in plant height, shoot length, root length, number of tillers, fresh weight, grain weight, dry biomass, chlorophyll$_a$ content, N and protein content, and NR activity compared with the control. In a similar study conducted under gnotobiotic conditions, Baldani et al. (2000) reported that the inoculation of rice with Herbaspirillum seropedicae enhanced plant dry weight by 71.5%. These results could be explained by the fact that inoculated plants absorbed more nutrients which reflect on growth activity, nitrogenous compound assimilation, forming more growth substances (IAA), more cell division and enlargement, more forming of tissues and organs. However, any practical application of these results should be preceded by further evaluation under field conditions. Besides exploring the potential for BNF and other promising PGP functions carried out by free-living diazotroph strains, it is also important to ensure that the bacteria are well adapted to environmental conditions before they are utilized as inoculant strains. The isolated free-living and/or associative B. caribensis strain could be very useful in the formulation of new microbial inoculants and could be applied most profitably to enhance the growth and yield of rice in tropical rainfed acidic lowlands.

Bacteria such as Pseudomonas spp., Burkholderia caryophylli, Achromobacter piechaudii were shown to lower the endogenous ethylene level in planta by producing a degradative enzyme 1-aminocyclopropane- 1-carboxylic acid (ACC)-deaminase (Shaharoona et al., 2007). The effects of ACC-deaminase producing rhizobacteria on plants included increased root growth, and improved tolerance of salt and water stress (Shaharoona et al., 2007).

Elcoka et al. (2008) hypothesized that microbial inoculants will replace mineral fertilizer; rather many studies, for example Adesemoye et al. (2009), have shown that microbial inoculants are good and reliable supplements to fertilizer. This present experiment revealed that inoculation of winter rice with acid tolerant B. caribensis strain improved plant height, shoot length, root length, number of tillers, fresh weight, weight of grains, dry biomass, chlorophyll content, N-content, protein content and NR activity by almost 20 to 30% over control and this is in conformity with the findings of Deb Roy et al. (2009) who studied the inoculation effect of three native diazotroph strains of South Assam namely,

Azotobacter chroococcum, Azospirillum amazonense and *Beijerinckia indica* on the growth and yield of summer (ahu) rice *cv.* IR-36 grown in acidic flooded lowland ecosystem. There were repeated beneficial effects on rice plants inoculated with a *Burkholderia vietnamiensis* strain on early and late components in low fertility sulfate acid soil of Vietnam (Tran Van et al., 2000). The nitrogen fixation by the inoculated strain per se seems not to be the sole cause of increased plant growth as indicated from the present data on percent N content of inoculated samples. It is presumed that the bacterium used for inoculation in this present study appears to act as that of plant growth promoting rhizobacteria (PGPR). It would have effected plant growth by the synthesis of phytohormones and vitamins, inhibition of plant ethylene synthesis and improved nutrient uptake as reported recently (Dobbelaere et al., 2003).

Conclusion

The isolates of *Burkholderia* spp. are indigenous and best suited to the ecophysiological condition of lowland rice agro-ecosystems of South Assam which have showed higher N_2 fixation rate and one of the isolates SDSA-I10/1 (GU372342) might be used as efficient microbial fertilizer strain for growing rice, the major staple food of the people in tropical rainfed acidic lowlands.

ACKNOWLEDGEMENTS

Authors are thankful to Enzene Biosciences Pvt. Ltd., Bangalore and IIT, Guwahati for providing the facility of DNA sequencing and gas chromatography respectively.

REFERENCES

Adesemoye AO, Torbert, HA, Kloepper JW (2009). Enhanced plant nutrient use efficiency with PGPR and AMF in an integrated nutrient management system. Can. J. Microbiol. 54:876–886.

Arnon DI (1949). Copper enzymes in isolated chloroplasts: polyphenol oxidase in *Beta vulgaris*. Plant Physiol. 24:1-25.

Ausubel FM, Brent R, Kingston RE, More DD, Seidman JG, Smith JA, Struhl K (1987). Current protocols in Molecular Biology. New York: Jon Wiley and Sons. pp. 12-19.

Baldani VLD, Baldani JI, Dobereiner J (2000). Inoculation of rice plants with the endophytic diazotrophs *Herbaspirillum seropedicae* and *Burkholderia spp.* Biol. Fertil. Soils 30:485-491.

Burbage DA, Sasser M (1982). A medium selective for *Pseudomonas cepacia*. Phytopathol. Abst. 72:706.

Cavalcante VA, Dobereiner J (1988). A new acid-tolerant bacterium associated with sugarcane. Plant Soil 108:23–31.

Cerney G (1993). Method for the distinction of gram negative and gram positive bacteria. Eur. J. Appl. Microbiol. 3:223-225.

Collee JG, Miles RS (1989). Tests for identification of bacteria. In:Mackie and Maccartney (eds), Practical Medical Microbiology. Churchill Livingstone, London, pp. 141-160.

Cottyn B, Regalado E, Lanoot B, De Cleene M, Mew TW, Swings J (2001). Bacterial populations associated with rice seed in the tropical environment. Phytopathology 91:282-292.

Deb Roy B, Deb B, Sharma GD (2009). Diversity of soil diazotrophs in acid stress rice agro-ecosystems of Barak Valley of Assam, India. J. Pure and Appl. Microbiol. 3(1):117-124.

Dobbelaere S, Vanderleyden J, Okon Y (2003). Plant growth-promoting effects of diazotrophs in the rhizosphere. Crit. Rev. Plant Sci. 22:107-149.

Dobereiner J, Reis VM, Paula MA, Olivares FL (1993). Endophytic diazotrophs in sugar cane, cereals and tuber plants. In: Palacios R, Mora J, Newton WE (eds), New horizons in nitrogen fixation. Current Plant Science and Biotechnology in Agriculture, Dordrecht: Kluwer 17:671-676.

Elcoka E, Kantar F, Sahin F (2008). Influence of nitrogen fixing and phosphorus solubilizing bacteria on the nodulation, plant growth, and yield of chickpea. J. Plant Nutr. 31:157–171.

Engelhard M, Hurek T, Reinhold-Hurek B (2000). Preferential occurrence of diazotrophic endophytes, *Azoarcus* in wild rice species and land races of *Oryza sativa* in comparison with modern races. Environ. Microbiol. 2:13–141.

Estrada-de-los Santos P, Bustillos-Cristales MR, Caballero-Mellado J (2001). *Burkholderia*, a genus rich in plant associated nitrogen fixers with wide environmental and geographic distribution. Appl. Environ. Microbiol. 67:2790–2798.

Fallik E, Okon Y, Fisher M (1988). Growth response of maize roots to *Azospirillum* inoculation: effect of soil organic matter content, number of rhizosphere bacteria and timing of inoculation. Soil Biol. Biochem. 20:45.

Felsenstein J (1978). Cases in which parsimony or compatibility methods will be positively misleading. Syst. Zool. 27:401-410.

Gillis M, Tran Van V, Bardin R, Goor M, Hebbar P, Williems A, Segers P, Kersters K (1995). Polyphasic taxonomy in the genus *Burkholderia* leading to an emended description of the genus and proposition of *Burkholderia vietnamiensis* sp. nov. for N_2-fixing isolates from rice in Vietnam. Int. J. Syst. Bacteriol. 45:274–289.

Guimaraes SL, Silva RA, Baldani JI, Baldani VLD, Dobereiner J (2000). Effects of the inoculation of endophytic diazotrophic bacteria on grain yield of two rice varieties (guarani and CNA 8305) grown under field conditions. In: Pedrosa FO, Hungria M, Yates G, Newton WE (eds) Nitrogen fixation: from molecules to crop productivity. Current Plant Sciences and Biotechnology in Agriculture, 38, Dordrecht: Kluwer. p. 431.

Gyaneshwar P, James EK, Mathan N, Reddy PM, Reinhold-Hurek B, Ladha JK (2001). Endophytic colonization of rice by a diazotrophic strain of *Serratia marcescens*. J. Bacteriol. 183:2634–2645.

Hardy RWF, Burns RC, Holsten RD (1973). Application of the C_2H_4 assay for measurement of nitrogen fixation. Soil Biol. Biochem. 24:47-81.

Hensyl WR (1994). Bergey's Manual of Determinative Bacteriology. 9th Edn. Baltimore, Williams and Wilkins.

Hewitt EJ, Nicholas JD (1964). Enzymes if inorganic nitrogen metabolism. In: Paech K, Tracy MV (eds), Modern Methods of Plant Analysis VII. Springer Verlag, Berlin. pp. 67-172.

Hiraishi A (1992). Direct automated sequencing of 16S rDNA amplified by polymerase chain reaction from bacterial cultures without DNA purification. Lett. Appl. Microbiol.15:210–213.

Jackson ML (1973). Soil Chemical Analysis, Prentice Hall (India) Pvt. Ltd, New Delhi. pp. 239-241.

Jukes TH, Cantor CR (1969). Evolution of protein molecules. In: Munro HN (eds), Mammalian Protein Metabolism. New York: Academic press. pp. 21–132.

Kang UG (2006). Enumeration, isolation and identification of diazotrophic bacterial species from paddy rice in Korea. Paper presented in 18th World Congress of soil Science. pp. 140-146.

Kirchhof G, Reis VM, Baldani JI, Eckert B, Dobereiner J, Hartmann A (1997). Occurrence, physiological and molecular analysis of endophytic diazotrophic bacteria in graminaceous energy plants. Plant. Soil. 194:45-50.

Ladha JK, Kirk GJD, Bennett J, Peng S, Reddy PM, Singh O (1998). Opportunities for increased nitrogen-use efficiency from improved lowland rice germplasm. Field Crops Res. 56:41–71.

Lowry OH, Rosenbrough NJ, Farr AL, Randall RJ (1951). Protein measurement with follin phenol reagent. J. Biol. Chem. 193: 265-275.

Maidak BL, Olsen GJ, Larsen N, Overbeek R, McCaughey MI, Woese CR (1996). The ribosomal database project (RDP). Nucleic Acids

Res. 24:82-85.

Malarvizhi P, Ladha JK (1999). Influence of available N and rice genotype on associative nitrogen fixation. Soil Sci. Soc. Am. J. 63:93-99.

McArthur JV, Kovacic DA, Smith MH (1988). Genetic diversity in natural populations of a soil bacterium across a landscape gradient. Proc. Natl. Acad. Sci. USA 85:9621-9624.

Misra BN, Misra MK (1983). Introductory Practical Biostatistics. Naya Prokash, Calcutta, pp. 118-130.

Muthukumarasamy R, Kang UG, Park KD, Jeon WT, Park CY, Cho YS, Kwon SW, Song J, Roh DH, Revathi G (2007). Enumeration, isolation and identification of diazotrophs from Korean wetland rice varieties grown with long-term application of N and compost and their short-term inoculation effect on rice plants. J. Appl. Microbiol. 102:981–991.

Naureen KM, Khan MA, Ahmed MS (2005). Characterization and screening of bacteria from rhizosphere of maize in Indonesian and Pakistani Soils. J. Basic Microbiol. 45(6):447-459.

Nieuwenhove VC, Holm IV, Kulasooriya SA, Vlassak K (2000). Establishment of *Azorhizobium caulinodans* in the rhizosphere of wetland rice (*Oryza sativa* L.). Biol. Fertil. Soils. 34:143-149.

Reis VM, Estrada-de-los Santos P, Tenorio-Salgado J, Vogel M, Stoffels M, Guyon S, Mavingui P, Baldani VLD (2004). *Burkholderia tropica* sp. nov., a novel nitrogen fixing, plant-associated bacterium. Int. J. Syst. Evol. Microbiol. 54:2155-2162.

Rennic RJ, Vose PV (1983). ^{15}N-isotope dilution of quantity dinitrogen fixation associated with Canadian and Brazillian wheat. Can. J. Bot. 61:967-971.

Saitou N, Nei M (1987). The neighbor-joining method: a new method for reconstructing phylogenetic trees. Mol. Biol. Evol. 4:406-425.

Sessitsch A, Coenye T, Sturz AV, Vandamme P, Ait Barka E, Salles JF, Van Elsa JD, Faure D (2005). *Burkholderia phytofirmans* sp. nov., a novel plant-associated bacterium with plant-beneficial properties. Int. J. Syst. Evol. Microbiol. 55:1187–1192.

Shaharoona B, Jamro GM, Zahir ZA, Arshad M, Memon MS (2007). Effectiveness of various *Pseudomonas* spp. and *Burkholderia caryophylli* containing ACC-deaminase for improving growth and yield of wheat (*Triticum aestivum* L.). J. Microbiol. Biotechnol. 17:1300-1307.

Snedecor GW, Cochran G (1967). Statistical methods. Iowa State University, U.S.A. Oxford and I.B.H. Publishing Co., New Delhi, pp. 350-372.

Thakuria D, Talukdar NC, Goswami C, Hazarika S, Bora RC, Khan MR (2004). Characterization and screening of bacteria from rhizosphere of rice grown in acidic soils of Assam. Curr. Sci. 86(7):978-985.

Tran Van V, Berge O, Ngo KS, Balandreau J, Heulin T (2000). Reproducible beneficial effects of rice inoculation with a strain of *Burkholderia vietnamiensis* on early and late yield components in low fertility sulphate acid soils of Vietnam. Plant Soil 218:273–284.

Watanabe I, Lee KK, Alimagne BV (1978). Seasonal change of N_2-fixing rate in rice field assayed by in situ acetylene reduction technique. In: Experiments in long-term fertility plots. Soil Sci. Plant Nutr. 4:1-13.

Watanabe I, Barraquio W (1979). Low levels of fixed nitrogen required for isolation of free-living N_2-fixing organisms from rice roots. Nature (London) 277:565–566.

Xie GH, Cai MY, Tao GC, Steinberger Y (2003). Cultivableheterotrophic N_2-fixing bacterial diversity in rice fields in the Yangtze river plain. Biol. Fertil. Soils 37:29-38.

17

Ergonomic intervention in sugarcane harvesting knives

R. Thiyagarajan[1]*, K. Kathirvel[2] and Jayashree G. C.[3]

[1]Agricultural Engineering College and Research Institute, Tamil Nadu Agricultural University, Kumulur, Tamil Nadu, India.
[2]Agricultural Engineering College and Research Institute, Tamil Nadu Agricultural University, Coimbatore, India.
[3]Indian institute of Crop Processing Technology, Tanjavur, Tamil Nadu, India.

Ergonomic evaluation of farm tools is necessary to improve the fit between the physical demands of the tools and the worker who perform the work. In spite of improved farm mechnization, the use of the hand tools is inevitable in certain agricultural operations like sugarcane harvesting. Commonly used and high energy demanding tools like sugarcane harvesting knives of various models available in India were selected to assess the ergonomic suitablity. Ten individuals were selected for the investigation based on the age and fitness. They were screened for normal health through medical investigations. Four models of sugarcane harvesting knives were selected for ergonomical evlauation. The parameters used for the ergonomical evaluation of screened sugarcane harvesting knives include heart rate and oxygen consumption rate, energy cost of operation, acceptable work load, over all discomfort rate and body part discomfort score. The maximum aerobic capacity of the selected ten individuals varied from 1.84 to 2.19 L min^{-1} for sugarcane harvesting. The heart rate and oxygen consumption rate of the sugarcane knives varied from 132.55 to 138 beats min^{-1} and 1.171 to 1.253 L min^{-1}, respectively. The energy cost of sugarcane harvesting knives, varied from 24.45 to 26.16 kJ min^{-1} respectively. The values of percent maximum aerobic capacity (VO_2 maximum) and work pulse for sugarcane harvesting knives were much higher than that of the acceptable workload (AWL), limits of 35%. Based on the analysis of results, the sugar cane harvesting knife (H1) ranked as I in terms of minimum value of heat rate (132.55 beats min^{-1}), energy cost of work (24.45 KJ min^{-1}), acceptable work load (58.14%), over all discomfort rate (moderate discomfort) and Body part discomfort score (29.39) when compared with other three models (H2, H3 and H4) of sugarcane harvesting knives.

Key words: Sugarcane harvesting, ergonomics, heart rate, oxygen consumption.

INTRODUCTION

Harvesting of sugarcane crop is an important agricultural operation in which, it is estimated that less than 20% of the world's more than 100 million tonnes of sugarcane is harvested mechanically (Anonymous, 2005). Harvesting systems vary widely and the choice of one system over another will depend on labour availability, labour cost, topography and climatic conditions. Sugarcane harvesting is highly labour intensive operation requiring around 1200 labour hour per heactare (Anonymous, 2005). For most of the sugarcane crops in India harvesting is done manually using locally made small hand tools such as knives. The farmers generally use old designs made of iron for harvesting sugarcane. The sugarcane harvesting operation involves the unit operations including, cutting the sugarcane, detrashing the cane, detopping and carrying to farm shed for manufacturing of jaggery or loading the bunch of sugarcane in truck for transport to sugar mills.

Agricultural ergonomics emerges as a potential discipline for whole ranging application in farming methods and practices. This discipline specifies application of those work sciences relating human performance to the improvement of work system in farming activity. It encompasses the persons, the jobs, the tools and equipment, the work place and space, and the working environment. Most designers of agricultural equipment concentrate to improve efficiency and durability, but none seem to give importance to the operators' comfort. Hence there is an urgent need to critically analyse these agricultural tools/equipment for their ergonomics in order to improve man-machine system efficiency without sacrificing performance. This would greatly help the researchers to appropriately design simple and labour effective gadgets considering ergonomic requirements. In view of the sugarcane harvesting operation, it is desirable to ergonomically eva-luate the available sugarcane harvesting knives to assess their suitability for farm workers for reduced drudgery and adequate comfort with the following specific objectives:

i. To measure the physiological cost (Heart rate and Oygen consumption rate) of the individuals while performing sugarcane harvesting operation with selected knives.
ii. To classify the workload in terms of energy cost of performing these operations.
iii. To assess the overall discomfort and body part discomfort rating of the subjects in the operation of selected sugarcane harvesting knives.

Srivastava et al. (1962) investigated the energy requirements, harvest rate and efficiency of grain harvesting equipment and concluded that there was a difference in the energy expenditure of different persons using the same harvesting equipment under similar condition. Sanders and McCormick (1993) stated that the linear relationship between heart rate and oxygen consumption is different for different people. So they suggested calibration of each person to determine the relationship between heart rate and oxygen consumption. They reported that heart rate is the best used as a predictor of oxygen consumption when moderate to heavy work is performed. They also stated that heart rate continuously sampled over a work day or task, is useful as a general indicator of physiological stress without reference to oxygen consumption or energy expenditure.

Kroemer et al. (1997) stated that heart rate and oxygen consumption have a linear relationship. They found that the relationship may change within one person with training, and it differs from individual to another. They inferred that heart rate measurements could be substituted for measurement of metabolic processes, particularly for oxygen consumption, since it could be performed easily. Bimla et al. (2002) investigated the efficiency of sickles in wheat harvesting. They found that the average heart rate was 110 and 107 to 109 beats min^{-1} for existing and improved sickles respectively and the corresponding average energy expenditure was 9.6 and 8.3 to 9.5 kJ min^{-1}. Maximal oxygen uptake, heart rate and muscle strength decreases significantly with old age (Astrand et al., 1965; Astrand and Rodahl, 1986).

The maximum strength or power can be expected from the age group of 25 to 35 years (Grandjean, 1982; Gite and Singh, 1997; Umrikar et al., 2004). Maximum muscle strength and at the same time the cross-sectional area of muscle is also greatest for this age group (Mc Ardle et al., 1994; Nigg and Herzog, 1999).

Other studies have researched the effect of different cane knives on performance. A study by de L Smit et al. (2001) compared short handled and long handled curved knives, and found the only difference in output to be the cutters' perception of exertion − indicating that the choice of knife and its usefulness depends on the preferences of the cane cutter. Nonetheless, other studies (Brooks, 1983) found some productivity enhancement with modified knife types.

MATERIALS AND METHODS

Sugarcane harvesting knives used in different regions of India procured and selected for ergonomical evaluations are shown in Figure 1 and Table 1. Selection of individuals plays a vital role in conducting the ergonomic investigations. The subject should be physically and medically fit to undergo the trials (Seidel et al., 1980). There should not be any major illness and handicaps and also they should be a true representative of the user population in operation of the selected sugarcane harvesting knives. Age and medical fitness is the main criteria for the selection of subjects. The medical and bio- clinical investigations like Electro Cardio Graph (ECG), blood pressure and bio-clinical analysis were conducted to assess the medical fitness of selected ten individuals participated in the investigation. Hence from the available workers, ten male workers in the age group 25 to 35 years were chosen considering their experience in the operation of the selected sugarcane harvesting knives. The characteristics of individuals are furnished in Table 2. A preliminary study was conducted with the selected 10 individuals for screening the selected ten models of knives. The criteria used for screening include: Over all discomfort rate (ODR) and body part discomfort score (BPDS), field capacity; subjective feedback, configuration similarity and versatility.

Ergonomical evaluation of screened sugarcane harvesting knives

Ergonomical evaluation was conducted with the screened sugarcane harvesting knives for assessing their suitability with the ten selected individuals. The evaluation was carried out in terms of heart rate and oxygen consumption rate, energy cost of operation, acceptable work load (AWL), over all discomfort rating (ODR) and body part discomfort score (BPDS).

Heart rate and oxygen consumption

Heart rate and oxygen consumption rate are the pertinent parameters for assessing the human energy required for performing various types of operation (Curteon, 1947). All the ten individuals were calibrated in the laboratory condition by indirect assessment of oxygen uptake. Oxygen consumption was measured by using the computerized ambulatory metabolic measurement system (Metamax-II) while running on the computerized treadmill (Viasys LE 200CE model). The corresponding heart rate was recorded using Polar Vantage NV computerized heart rate monitor (S 810i) at the submaximal loads. The maximum heart rate of all the selected individuals was computed using the equation proposed by

Figure 1. Different sugarcane harvesting knives selected for ergonomical evaluation.

Table 1. Specifications of ten models of sugarcane harvesting knives.

Model No.	Weight (g)	Thickness (mm)	Material of the blade	Diameter of the hand grip (mm)	Thickness of the cutting edge (mm)	Effective cutting length of the blade (mm)	Concavity (mm)	Material of the handle	Over all dimensions, L × B (mm)
01	550	4	Mild steel	32	1.02	160	58	Wood	347 × 160
02	550	4	Mild steel	31	1.02	160	58	Wood	357 × 150
03	600	4	Mild steel	36	1.02	203	40	Wood	380 × 150
04	550	4	Mild steel	36	1.02	133	86	Wood	375 × 160
05	450	4	Mild steel	33	1.02	180	34	Bamboo stick	357 × 98
06	700	6	Mild steel	34	1.02	220	-	Wood	413 × 74
07	500	3	Mild steel	34	1.02	155	49	Wood	380 × 122
08	500	1.5	Mild steel	168 × 62 × 20	1.02	225	67	Plastic	494 × 85
09	500	1.5	Mild steel	168 × 62 × 20	1.02	225	67	Plastic	494 × 85
010	400	1.5	Mild steel	155 × 60 × 22	1.02	215	-	Plastic	505 × 100

Astrand (1960) and the arrived values of maximum aerobic capacity (VO$_2$ maximum) for all the individuals. Because of the advantages of the indirect assessment of oxygen uptake, during the operation of each of the selected knives, only the heart rate of the subject performing the task was noted. The procedure adopted for each operation is explained subsequently.

Energy cost of operation

The recorded heart rate values from the computerized heart rate monitor were transferred to the computer through the interface in all the above cases. From the down loaded data, the values of heart rate at resting level and 6th to 15th minute of operation were taken for

calculating the physiological responses of the subjects (Tewari and Gite, 1998). The heart rate increases rapidly in the beginning of an exercise and reaches a steady state by the end of sixth minute (Davies and Harris 1964). The stabilized values of heart rate for each subject from 6th to 15th minute of operation were used to calculate the mean value for all the selected sugarcane harvesting knives.

Table 2. Characteristics of selected individuals for sugarcane harvesting.

S/N	Individuals	Age (Year)	Weight (kg)	Height (cm)	Experience (Year)
1	I	25	57	157	> 5
2	II	26	61	159	> 5
3	III	29	53	161	> 5
4	IV	27	65	158	> 5
5	V	24	58	153	> 5
6	VI	32	54	164	> 5
7	VII	31	62	163	> 5
8	VIII	29	58	159	> 5
9	IX	28	64	156	> 5
10	X	26	49	156	> 5

From the values of heart rate (HR) observed during the trials, the corresponding values of oxygen consumption rate (VO_2) of the subjects for all the screened sugarcane harvesting knives were predicted from the calibration chart of the each subject. The energy costs of operation of the screened sugarcane harvesting knives were computed by multiplying the oxygen consumed by the subject during the trial period with the calorific value of oxygen as 20.88 kJ l^{-1} (Nag et al., 1980) for all the individuals. The values of heart rate, oxygen consumption and the energy expenditure for all the subjects were averaged to get the mean values for all the screened sugarcane harvesting knives.

Grading energy cost of work

Measured physiological demands are evaluated against various criteria to determine whether the physical demand of a certain task is excessive, and whether the worker performing the task may suffer from physical fatigue. The energy cost of individuals for screened sugarcane harvesting operation thus obtained was graded as per tentative classification of strains in different types of jobs according to the young Indian male workers given in ICMR report (Sen, 1969).

Acceptable workload (AWL)

During any physical activity, there is increase in physiological parameters depending upon the workload, and the maximum values, which could be attained in normal healthy individuals, will be up to VO_2 max. However at this extreme workload, a person can work only for a few seconds. The acceptable workload (AWL) for Indian workers was the work consuming 35% of VO_2 max (Saha et al., 1979). To ascertain whether the operation of all the screened sugarcane harvesting knives is within the acceptable workload (AWL), the VO_2 maximum for each treatment was computed and recorded. The acceptable workloads for extended periods as 33% of maximal aerobic capacity for an 8-hour shift and 28% for 12-hour sift (NIOSH, 1981). Kodak (1986) later confirmed this.

Overall discomfort rating (ODR)

For the assessment of overall discomfort rating a 10 - point psychophysical rating scale (0 - no discomfort, 10 - extreme discomfort) was used which is an adoption of Corlett and Bishop (1976) technique. A scale of 70 cm length was fabricated having 0 to 10 digits marked on it equidistantly. A moveable pointer was provided to indicate the rating. At the end of each trial subjects was asked to indicate their overall discomfort rating on the scale. The overall discomfort ratings given by each of the ten individuals are added and averaged to get the mean rating.

Body part discomfort score (BPDS)

To measure localized discomfort, Corlett and Bishop (1976) technique was used. In this technique the subject's body is divided into 27 regions. A body mapping similar to that of body mapping was made with thermocoal to have a real and meaningful rating of the perceived exertion of the subject. The subject was asked to mention all body parts with discomfort, starting with the worst, the second worst and so on until all parts have been mentioned (Lusted et al., 1994). The subject was asked to fix the pin on the body part in the order of one pin for maximum pain, two pins for next maximum pain and so on (Legg and Mohanty, 1985).

Field evaluation of screened sugarcane harvesting knives

The experiment was conducted in sugarcane fields located at sarvanampatti village of Tamil Nadu. Field experiments were conducted with selected sugarcane harvesting knives during the month of February and March 2006. The temperature and relative humidity varied from 32 to 36°C and 28 to 64%, respectively during the period of evaluation. The field selected for trail was planted with CO 86032 variety of sugarcane. The individuals were given information about the experimental requirements so as to enlist their full cooperation. The trail was conducted between 7.30 AM to 5.00 PM. They were given rest for 30 min before starting the trial. After rest period of half an hour, individuals performed the harvesting operation shown in Figure 2. Trials with duration of 20 min were conducted for all ten individuals. The data for heart rate was recorded using computerized heart rate monitor. The same procedure was repeated for all the individuals and also for screened harvesting knives with three replications.

RESULTS AND DISCUSSION

Ten models of sugarcane harvesting knives are screened in to four for ergonomic evaluation according to the feedback of the sugarcane cutters, BPDS, ODR values and field capacity were presented in Table 3. The sugarcane cutters are mainly preferred to use the Gobichettypalayam model (01), Dharmapuri model (02), Cuddalore models (03 and 04) and Kallakuruchi model (07). Model 02 (Dharmapuri model) is selected from the

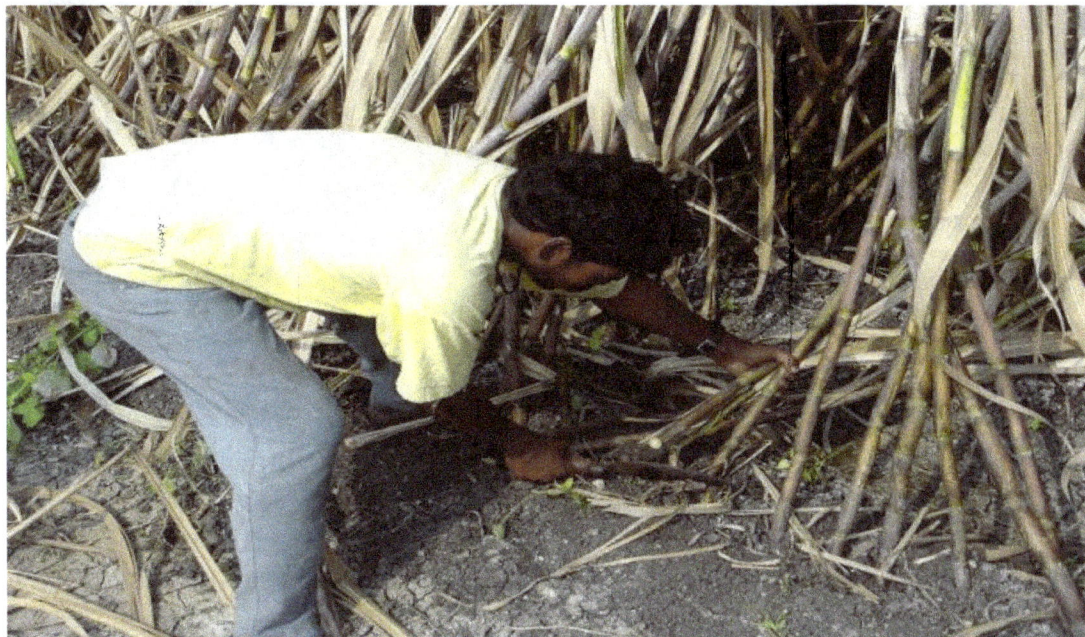

Figure 2. View of sugarcane harvesting knife in operation.

Table 3. Criteria used for screening of harvesting knives.

S/N	Sugarcane harvesting knives	Field capacity (Ha h^{-1})	ODR values	BPDS values
1	Gobichettipalayam model 01	0.00201	5.51	30.10
2	Dharmapuri model 02*	0.00209	5.67	29.39
3	Cuddalore model 03*	0.00173	6.04	29.65
4	Cuddalore model 04*	0.00177	6.19	32.92
5	Melalathur model 05	0.00129	5.82	29.37
6	Coimbatore model 06	0.00135	6.95	46.32
7	Kallakuruchi model 07*	0.00185	5.71	28.21
8	Pune model (right) 08	0.00101	6.17	36.02
9	Pune model (left) 09	0.00100	6.29	37.13
10	Pune model 10	0.00110	6.38	31.41

*Screened models of knives for ergonomical evaluation.

two models (01 and 02) because the both the models have the same similarities. The following four harvesting knives (H1, H2, H3 and H4) were screened for ergonomic evaluation owing to their suitability, increased comfort, user friendly, versatility and field capacity. The screened harvesting knives (H1, H2, H3 and H4) are shown in Figure 3.

The physiological response of the individuals with respect to time for the operation of the four screened sugarcane harvesting knives (H1, H2, and H3 and H4) is depicted in Figures 4 and 5. From the graph it is observed that the heart rate of the individuals increased steeply from the beginning of the operation and stabilized in the range of 120 to 145 beats min^{-1} after 6th minute of operation. The readings for the individuals from 6th to

15th minute were considered for the calculation of the heart beat rate for screened models of sugarcane harvesting knives. It is also observed that there existed a difference in the heart rate among the subjects using the same tools under the same conditions due to difference in subject's age, weight and stature. The mean values of heart rate of all the selected individuals and the corresponding oxygen consumption value are furnished in Table 4.

The mean value of energy expenditure of subjects for operation with sugarcane harvesting knives varied from 24.45 to 26.16 kJ min^{-1}. The values are in close agreement with the value of 24.58 kJ min^{-1} reported by Yadav and Srivastava (1984) for sugarcane harvesting knives. Performing the sugarcane harvesting operation in bending

Figure 3. Screened sugarcane harvesting knives.

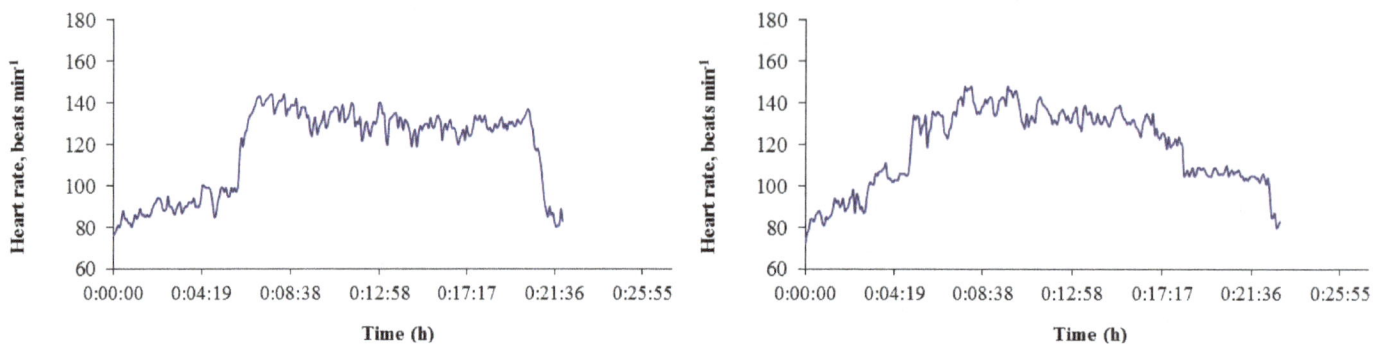

Figure 4. Heart rate response of the individual during the operation of sugarcane harvesting with H1 and H2 model.

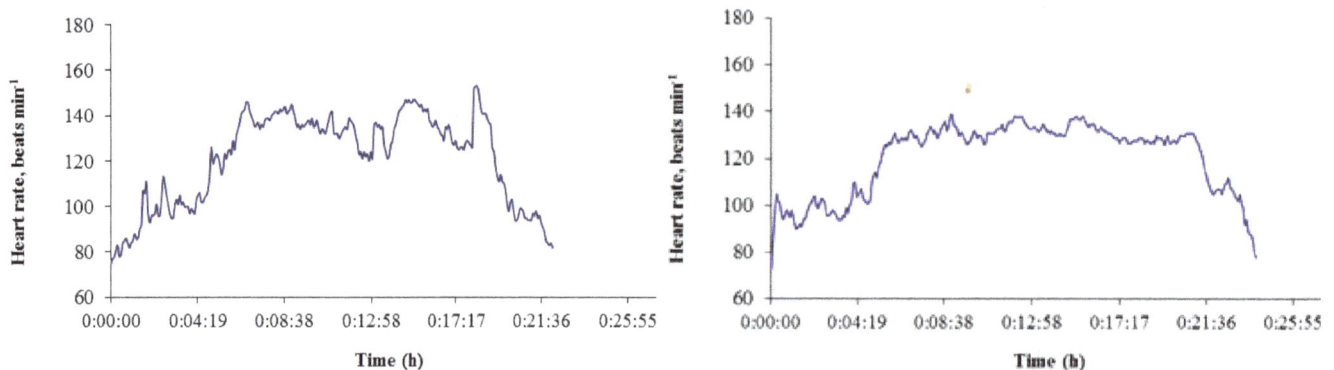

Figure 5. Heart rate response of the individuals during the operation of sugarcane harvesting with H3 and H4 model.

posture, holding the sugarcane crop in one hand and forcing the knives in sideward for harvesting the cane, these operations are graded as "heavy". The order of ranking of the harvesting knives based on energy cost was H1, H4, H3 and H2. The mean values of oxygen consumption rate (OCR) in terms of percent VO_2 max for screened sugarcane harvesting knives were 58.14 to 62.21%, respectively. These values were much higher than that of the AWL limit of 35% of VO_2 max indicating that all the screened sugarcane harvesting knives could not be operated continuously for 8 h without frequent rest-pauses.

From the rating of perceived exertion of the subjects, the ODR scale for screened sugarcane harvesting knives was "moderate discomfort" for models H1, H3 and H4 and more than moderate for model H2. Based on the

Table 4. Average physiological responses of individuals for operation of screened sugarcane harvesting knives.

Tools	Heart rate (beats min^{-1})	VO$_2$ (L min^{-1})	Energy expenditure (kJ min^{-1})	Energy grade of work	Rank
H1	132.55	1.171	24.45	Heavy	I
H2	137.41	1.253	26.16	Heavy	IV
H3	138.00	1.252	26.14	Heavy	III
H4	133.17	1.187	24.78	Heavy	II

Table 5. Average body part discomfort score for screened knives.

Screened knives	Body part experiencing pain	Score
H1	Moderate pain in shoulders, palms, elbow, wrist, mid back	29.39
H2	Moderate pain in shoulders, palms, elbow, wrist, mid back	31.65
H3	Moderate pain in shoulders, palms, elbow, wrist, mid back	32.92
H4	Moderate pain in shoulders, palms, elbow, wrist, mid back	28.21

Corlett and Bishop (1976) regional discomfort scale, the mean values of body part discomfort score for all the screened sugarcane harvesting knives were presented in Table 5.

The majority of discomfort experienced by the workers was in the right shoulders, palm, mid back and right elbow for all the individuals. This discomfort experienced by the individual's subjects was mainly due to the frequent sideward force given for cutting the sugarcane, a rough handle grip and weight of the knife.

Conclusion

Based on the analysis of the results the following conclusions are drawn.The selected ten individuals were calibrated in the laboratory by indirect assessment of oxygen uptake. The relation ship between the heart rate and the oxygen consumption was found to be linear for all the subjects. The maximum aerobic capacity of the selected ten subjects for sugarcane harvesting knives varied from 1.84 to 2.19 l min^{-1}. For sugarcane harvesting operation with H1, H2, H3 and H4 knives, the mean value of heart rate was 132.55, 137.41, 138 and 133.17 beats min^{-1} respectively and the corresponding oxygen consumption value was 1.171, 1.253, 1.252 and 1.187 L min^{-1}, respectively. From the mean value of oxygen consumption, the energy expenditure for screened sugarcane harvesting knives H1, H2, H3, H4 was computed as 24.45, 26.16, 26.14 and 24.78 kJ min^{-1}, respectively. The operation was graded as "heavy". The energy expenditure for screened sugarcane harvesting operation indicated that the energy cost of work was the highest for H1 followed by H4, H3 and H2. The oxygen consumption rate in terms of VO2 max for screened sugarcane harvesting knives H1, H2, H3 and H4 varied from 58.14 to 62.21%. The minimum over all discomfort

rate value of 5.67 for sugarcane harvesting knives (H1) indicated that model H1 exerted relatively lesser fatigue on the operator when compared with other three models. The majority of discomfort experienced by the workers in the operation of sugarcane harvesting H1, H2, H3, and H4 was in the right shoulder, wrist, elbow and mid back due to more weight and frequent sideward force given for cutting the sugarcane, a rough handle grip of the knife.

RECOMMENDATIONS

From the results, sugarcane harvesting knife (H1) is registered the lowest value of ergonomical evaluational parametersis taken and necessary ergonomics refinements like redesigning the knife (weight may be reduced by replacing the carbon steel instead of mild steel and width may also be reduced, a fine gripness may be provided in the handle of knife to reduce the palm itching, blister or welts) should be carried out for enhancing the comfort of the operator without jeoparadizing the efficiecy of the sugarcane harvesting knife.

REFERENCES

Anonymous (2005). Cotton crop improvement. Crop Production guide. Published by Department of Agriculture, Government of Tamil Nadu, Chennai.

Astrand I (1960). Aerobic work capacity in men and women. Acta Physiologica Scandinavica 4:1691-1692.

Astrand PO, Ekblon B, Messin R, Stallin B, Stenberg J (1965). Intra-Arterial blood pressure during exercise with different muscle group. J. Appl. Physiol. 20:253-256.

Astrand PO, Rodahl K (1986). A Textbook of work physiology. New York, Mc. Graw Hill.

Bimla S, Gandhi, Dilbaghi M (2002). Efficiency testing of sickles in wheat harvesting. An ergonomic study. Department of Family Resource Management, College of Home Science, CCSHAU, Hisar.

Brooks RC (1983). Improving cane cutters' productivity. Proc. South Afr. Sug. Technologist Assoc. 57:165-166.

Corlett EN, Bishop RP (1976). A technique for assessing postural discomfort. Ergonomics 19:175-182.

Curteon TK (1947). Physical fitness appraisal and guidance. The C.V.Mosby Co., St Louis.

Davies CTM, Harris EA (1964). Heart rate during transition from rest to exercise in relation to exercise tolerance. Journal of applied Physiology. 19(5):857-862.

de L Smit M, Coetsee MF, Davies SHE (2001). Energy expenditure and economy of movement of sugarcane cutters in burnt and green cane. Proc South Afr. Sug. Technologist Assoc. 75:46-50.

Gite LP, Singh G (1997). Ergonomics in agricultural and allied activities in India. Central Institute of Agricultural Engineering, Bhopal.

Grandjean E (1982). Fitting the task to the man – An ergonomic approach. Taylor and Francis Ltd., London.

Kodak, Eastman Company (1986). Ergonomic design for people at work, vol.2. New York: Van Nostrand Reinhold.

Kroemer KHE, Kroemer HJ, Kroemer KE (1997). Engineering physiology. Van Nostrand Reinhold. New York. pp. 193-228.

Legg SJ, Mahanty A (1985). Comparison of five modes of carrying a load close to the trunk. Ergonomics 28(12):1653-1660.

Lusted M, Healey S, Mandryk JA (1994). Evaluation of the seating of Qantas flight deck crew. Appl. Ergonomics 25:275-282.

Mc Ardle DE, Katch FI, Katch VL (1994). Essentials of exercise physiology. Lea &Febiger, A waverly company, Philadelphia: p.508.

Nag PK, Sebastian NC, Malvanker MG (1980). Occupational workload of Indian Agricultural workers, Ergonomics 23(2):91-102.

Nigg BM, Herzog W (1999). Bio mechanics of the musculo skeletal system. 2nd Edn., John Wiley and Sons, New York. p. 156.

NIOSH (National Institute for Occupational Safety and Health) (1981). Work practices guide for manual lifting, DHHS/NIOSH Publication #81-122. Washington D.C: Government Printing Office.

Saha PN, Datta SR, Banerjee PK, Narayane G (1979). An acceptable workload from a modified scale of perceived exertion, Ergonomics 37:485-491.

Sanders, McCormic (1993). Human Factors in Engineering and Design. Seventh Ed., McGraw Hill, Inc, New York.

Seidel H, Bastek R, Brauer D, Buchholz CH, Meister A, Metz AM, Rothe R (1980). On human response to prolonged repeated whole body vibration. Ergonomics 23(30):191-211.

Sen RN (1969). Tentative classification of strains in different types of jobs according to the physiological responses of young Indian workers in comfortable climates. ICMR report, Indian Council of Medical Research, New Delhi.

Srivastava K, Johnson AC, Zahradnik JW (1962). Energy requirements, harvest rates and efficiency compared for several hand grain harvesting implements. Paper presented at the Annual Meeting of ASAE at May Flower Hotel, Washington, D.C.

Tewari PS, Gite LP (1998). Human energy expenditure during power tiller operation. Paper presented at XXXIII annual convention of ISAE held at CIAE, Bhopal, Sept. pp. 21-23.

Umrikar SH, Zend JP, Upadhyay RK, Murali D (2004). Health status of farm women. Paper presented at National Conference on Humanizing Work and Work Environment, National Institute of Industrial Engineering, Mumbai, April. pp. 22-24.

Yadav RNS, Srivastava NSL (1984). Sugarcane harvesting knives used in India. The Indian Sugar Crops J. July-September. pp. 5-8.

18

Evaluation of strategies for developing the agriculture technology in the science and technology parks of Iran from faculty member aspect

Hanieh Davodi[1], Tahmasb Maghsoudi[2], Hossien Shabanali Fami[3] and Khalil Kalantari[3]

[1]Rural Development, Tehran University, Iran.
[2]Department of Agricultural Management, Shoushtar Branch, Islamic Azad University, Iran.
[3]Economics and Agricultural Development, Tehran University, Tehran, Iran.

Technology can be developed through supportive policies, suitable rules, and efficient institutional structure. Science and technology parks have the necessary potential for the technology development. Regional varieties and different economical-social fields need different strategies to reach the new agriculture technologies. This study purpose is identifying strategies for developing the agriculture technology in the science and technology parks. Based on paradigm and objective, this study is quantitative and applicatory, respectively, which was carried out by correlation method. Statistical population of the study was 200 of faculty members of Tehran province's agriculture faculties; Using Cochran formula, 90 ones were estimated as sample size, and samples were selected by random sampling. Questionnaire was the main study tool; its validity and reliability were determined by the expert panel and Cronbach alpha coefficient (0.87), respectively. The most suitable strategies were developing the technology in universities, increasing the national investment in research and the development of higher educational centers, reinforcing the relation between government and industry, universities, and society, taking incentives into account for the active researches in the field of technology development and new products, and more financial supports for thesis to complete and commercializing the studies` results. Results of factor analysis showed that 6 factors, totally, explains about 83.9% of the variance related to the agriculture technology development strategies. These factors are development of technology –oriented studies (19.5%), development of infrastructures (17.34%), improvement of the policies and programs (16.83%) information and communications (13.32%), incentives and supports (13.32%), and technology transfer (7.41%).

Key words: Technology, science and technology parks, strategies, factor analysis.

INTRODUCTION

Nowadays, agriculture sector is the dominant sector of the developing countries` national economy. These countries` economical growth and development is closely related to their agriculture sector's overall development. In the developed capitalist countries, passage from the traditional economy to the monetary one was accompanied by the technical developments and the raise of agriculture sector's productivity (Anonymous, 2002).

Scientific progresses and technological innovations of

the 20th century have resulted considerable achievements in producing the agriculture products. Agricultural productivity growth has made it possible for different countries to increase their income, participate in the global market, resolve the hunger crisis, and improve their citizens` quality of life (Stone, 2005). Science and technology, by developing and implementing supportive policies, can increase the suitable set of laws and efficient institutional framework and stimulate the economical growth in each country (Derjer, 2004). Consequently, the sustainable economical development would be achieved through technological dynamic changes supported by the effective and efficient innovation systems (Berg, 2005).

Science and technology, by developing and implementing supportive policies, can increase the suitable set of laws and efficient institutional framework and stimulate the economical growth in each country (McGann, 2007). Governments and commercial units, which are unable to innovate or use the scientific and technologic achievements, would be failed and crossed out from the competence circle (Cassiman and Veugelers, 2006). Thus, sustainable economical development will be resulted through dynamic technological changes supported by efficient innovation systems (Morin and Rafferty, 2005). Technological innovations smooth the progress of achieving unrivaled values such as new method introduction to carry out tasks, services, and etc (Porezat, 2010). Technology development includes planning, management, and execution of investigative activities in order to develop, evaluate, well-match, and test the higher technologies (Swanson et al., 1997). This process can be realized as a technologic system including all persons, groups, organizations, and institutes that are involved with the production, development and extending the new and the available technologies (Kaimowite, 1991).

Evaluating the innovation and technology development theories, linear innovation models such as entrepreneurship, vertical integration model, and technological cooperation model can be named, which, according to the criticism on the linear innovation models, nonlinear models in the form of institutional variety were introduced. The institutional variety model is the newest organizational form of knowledge production. The interaction of university, institution, and government in the science and technology development is considered in this model (Saljoghi, 2005), and, through this, science and technology parks are the most important institutes of this interaction.

Innovation system thinking represents a significant change from the conventional linear approach to research and development. It provides analytical framework that explore complex relationships among heterogeneous agents, social and economic institutions, and endogenously determined technological and institutional opportunities. It demonstrates the importance of studying innovation as a process in which knowledge is accumulated and applied by heterogeneous agents, through complex interactions that are conditioned by social and economic institutions (Agwu et al., 2008).

According to Tugrul and Ajit (2002) it is not a simple aggregation of organizations as portrayed by some views, but a group of agents who operate like an invisible orchestra characterized by coherence, harmony and synergy. It is an interactive learning process in which enterprises/agents in interactions with each other, supported by organizations and institutions play key roles in bringing new products, new processes and new forms of organizations into social and economic use (Francis, 2006).

Business today is based on the principle that there are different ways for technology acquisition. Therefore, identification of technology acquisition strategies, strengths and weaknesses and the implications of investment-oriented attitude are of importance.

Sun et al. (2007) believe that technology acquisition include three stages, Scanning technology (including identification Potential technologies), the choice of technology (technology assessment based on decision criteria), and internal technology.

According to Naito (1998) from the perspective of environmental resources looks to the issue of technology acquisition. The business requirements from the perspective of technology and human resources, information, and physical capital have been discussed. Each of these components of his vision and technique appropriate mechanism to implement the appropriate technology they need. The above definitions point to the three essential elements of innovation system namely:

1. The organizations and individuals involved in generating, diffusing, adapting and using knowledge.
2. The interactive learning that occurs when organizations engage in generating, diffusing, adapting and using new knowledge and the way in which this leads to innovation (new products, processes or services).
3. The institutions (rules, norms, conventions, regulations, traditions) that govern how these interactions and processes occur. The concept of innovation system is built on several assumptions and integrates current trends in development in the analytical framework. They include the followings:

a. Innovation takes place everywhere in the society and therefore bringing the diffuse element of a knowledge system and connecting them around common goals should promote economic development.
b. Innovation is an interactive process and is embedded in the prevailing economic structure and this determines what is to be learnt and where innovation is going to take place.
c. Innovation includes development, adaptation, imitation and the subsequent adoption of technology or application of new knowledge.

d. Innovation takes place where there is continuous learning and opportunity to learn is a function of the intensity of interactions among agents.

e. Heterogeneous agents are involved in innovation process, and formal research is a part of the whole innovation processes.

f. Linkages and/or interaction among components of the system (knowledge generating, transfer and using agents) are as important as direct investment in R and D.

g. Institutional context rather than technological change drives socio-economic development.

h. In addition to technical change and novelty, innovation includes institutional, organizational and managerial knowledge (Agwu et al., 2008).

Spielmen (2005) reported that analysis of innovation system may focus on the study of the system at different spatial (local, regional, national) at different sectoral levels (agriculture, pharmacy) in relation to a given technological set (biotechnology, ICTs), focus on the material (particular goods or services) and temporary dimension that studies how relationships among agents change over time as result of knowledge flow.

Analytical dimension at national level is referred to as national innovation system. It is that set of distinct, institutions which jointly and individually contributes to the development and diffusion of new technologies and which provides the framework within which government forms and implements policies to influence the innovation process. Metcalfe (1995) defined it as a system of interconnected institutions to create, store and transfer the knowledge, skills and artifacts which define new technologies. The element of nationality, according to Metcalfe (1995) follows not only from the domain of technology policy but from elements of shared language and culture which bind the system together, and form the focus of other policies, laws and regulations that condition the innovative environment.

Technology innovations include the invention phase (the ideas and devices, technical innovation), innovation (including the first practical or commercial use of the invention) and releases (if that technology adoption by others, the technology reaches the publication stage) is. The most important indicator of technological innovation is the number of patents registered some measure of technology development including basic infrastructure, information technology, energy status (total domestic energy consumption, the importance of energy as a strategic factor), the rate of investment on research and development, environment and technology (basic research, and ...), patents and intellectual property (ouch right Tra proportion of the population, the value of exported technology, etc.) is (Tsai and Wang, 2008).

Science and technology parks try to do the local development, industrial reconstruction encouragement, and facilitation of the industrial and commercial innovations and its activities reinforce the regional economical bases through improving the technological and scientific culture and providing wealth and employment, which, according to the different issues and tendencies, this role might be different, too (Farjadi and Riahi, 2007).

Science and Technology Park is a based –on – ownership-based development in a high quality park-like physical environment. It takes advantage from the accession of spiritual assets, suitable infrastructures and strategic policies and supports the technology based companies and NGOs in a managed environment; thus, the technology development and economical growth are facilitated (Eom and Lee, 2010).

Research parks, making the quick development of research centers and the research cycle (from universities to industries) completion possible by providing centralized supportive rules and sponsorship services despite of having different names such as technology park, techno police, and research town, are all participated on three main objectives of research cycle completion from university to the industries, acceleration of technology transmittal, support the new research centers and companies, and helping them to grow and succeeed and commercializing the results of researches (Frenz and Ietto-Gillies, 2009).

Science and technology park affect the technology development by creating research structures in the universities and accelerating the shift from individual researches to the organized ones, commercializing the results of researches and contribution to provide research credits, facilitating the presence of industries and research companies near the universities and developing their cooperation, developing the spin-off companies and creating the knowledge based technologies (Cassiman and Veugelers,2006).

Problem statement

Agriculture sector of Iran had 11.2% of GDP share. 22.8% share of the country's employment .Of the country's, 3.47 million producing units, about 86.7% are units smaller than 10 ha and the average size of the smaller than 10 ha units has reduced during the last 30 years. In addition, average land size has reduced from 6.05 to 5.07 hectares during this period, too (Iran Statistic Center, 2010).

The country's rural population was 30.1% of the total population in 2006, which, compared with the 39.5% rate in 1996, has reduced. The reduction process of the rural population in Iran has a massage that it is necessary to find a suitable environment to counter the reduction of foodstuff. Per capita of access to land in Iran is little (0.7 ha), while it is 59.6 ha in the developed countries. Senility of the population working in the agriculture sector is one of the most important issues of this sector, too, population working in Iran's agriculture sector, compared with those of other sectors, are older; therefore, if the process of senility continues, the situation of the agriculture would be critical in future. The ratio of the more than 60 years old employed population was 21.1% in 2006, while this ratio for the industry and services sectors was 3.9 and 5.2%, respectively. The most important results of senility of agriculture sector's population are reduction of productivity, jeopardizing the food security, country's dependence to the foodstuff, and the need of import from foreign countries (Iran statistic center, 2010).

The necessity of productivity improvement is come out as a basic strategy based on the population unlimited growth and its need of food, land per capita reduction for each farmer and the expansion of yeoman ship in the country, life economy's situation, high water use in the agriculture sector, unemployment rate increase among the sector's workers, senility, immigration increase from villages to cities; agriculture technology is one of the most important factors improving the productivity level. These technologies can increase the productivity rate of the production factors. Agriculture technologies are a suitable area for increasing the foodstuff production in a Medium-term period. It is clear that the development of technology in the agriculture sector is not random or spontaneous, but needs a provident and comprehensive attitude toward the executive, investigational, planning, and policymaking fields. Science and technology parks have a suitable institutional structure in order to develop the agriculture technologies, but because of the short-term existence, the development of agriculture technology materials and techniques are not comprehensible in these parks. So, this study is aimed at identifying the schematization and political strategies of agriculture technology development in the science and technology parks in Iran.

Research objectives

This study is generally aimed to evaluate the strategies of the agriculture technology development in the science and technology parks of Iran; specific objective are also included as follow:

1. Prioritizing the strategies of the agriculture technology development in the science and technology parks, and
2. Identifying the strategies of agriculture technology development in the science and technology parks

MATERIALS AND METHODS

Based on paradigm and objective, this study is quantitative and applicatory, respectively, which was carried out by correlation method. It is also an investigational study due to its extension. Because this study was carried out in a specific time, it is a single cross-sectional study. Statistical population of the study was 200 of faculty members of Tehran University's Pardis agriculture faculty;

Using Cochran formula, 90 ones were estimated as sample size, and samples were selected by random sampling. Questionnaire was the main study tool; its validity was determined by the experts' panel at the external, content, and instrument validity types. Reliability of the study was determined by Cronbach's coefficient alpha indicating the suitable reliability of the study tool. To evaluate the strategies, Likert scale with 28 items was used in the fields related to the groups. Change coefficient and factor analysis were used for analysis. Agricultural Technology Development in the park with a variety of strategies in knowledge-based organizations began and various strategies were studied. Because of the uncertain aspects of exploration strategies in order to identify factors in the development of agricultural technology park was used. Hence, exploratory factor analysis was used to categorize strategies. Of solutions within an enterprise scale solutions and communication between departments and external solutions were evaluated. Their suitability was assessed in the five-level Likert type of very good, good, ineffective, inadequate and was very inappropriate

RESEARCH FINDINGS AND DISCUSSION

Individual and professional characteristics of the faculty members

Average age was 46.65. 95.6% of them had PhD degree, 32.2% were assistant professors, and 32.2% were lecturers. Respondents had 14.59 years of work experience. 64.4% of them were managers who, averagely, had 7.74 years of managerial experience. The numbers of their articles were 23.12 research articles, 8/4 completed research projects, and 28.9 guidance paper indicating the activity of the faculty members and their adequate experience to answer the questionnaire questions.

Technology development strategies

Evaluating these strategies, 28 ones were selected; the faculty members were asked to evaluate each item's importance degree and rate of suitability at the current conditions of the country. They selected the strategies of improving the technology development infrastructures in the university (such as workshops, labs, etc), increasing the national investment in the field of research and development in the higher education centers, reinforcing the relation between industry, universities, society, and government, considering some incentives for the active researches in the field of technology development and new products, and more financial support for thesis in order to complete and commercialize the research results as the most suitable strategies (Table 1).

Identification of the technology development strategies in the agriculture sector

Factor analysis was used to evaluate the strategies of agriculture technology development. Data evaluation showed that these data are suitable for the factor

Table 1. Evaluation and prioritization of the agriculture technology development strategies.

Agriculture technology development strategies	Mean	SD	CV	Rank
Improving the technology development infrastructures in the university (such as workshops, labs etc)	8.9	1.40	0.157	1
Increasing the national investment in the field of research and development in the higher education centers	8.3	1.46	0.175	2
Reinforcing the relation between industry, universities, society, and government	8.8	1.71	0.195	3.5
Considering some incentives for the active researches in the field of technology development and new products	8.4	1.63	0.195	3.5
More financial support for thesis in order to complete and commercialize the research results	8.3	1.71	0.206	5
Developing researches on the new technologies(such as bio, nano ,and etc) in the universities	7.4	1.81	0.244	6
Consult and legal supports for faculty members in order to commercialize the research results	8.1	2.10	0.258	7
Designating resources to the groups based on the priorities of that group's technology development plan	6.9	1.86	0.269	8
Making the researches practical in order to resolve the agriculture economical ,social, technical problems	7.9	2.18	0.277	9
Using integrated strategies of a simultaneous technology importation and develop it inside the country	7.9	2.21	0.280	10
Technology transfer from all technological countries	7.1	2.03	0.285	11
More dependence to the countrywide scientific and technical abilities in order to increase the power of national competence in the international environment	7	2	0.287	12
Developing and facilitating the relations between groups by carrying out interdisciplinary research	7.6	2.02	0.289	13
Revising the research regulations to facilitate the technology development process (e.g. the possibility of university students` teamwork)	7.6	2.31	0.304	14
The possibility of carrying out the thesis jointly by a number of educational group (teamwork)	7.8	2.43	0.313	15
Merging the programs of technology development with the five- year plan of agriculture development in a national level	7	2.28	0.324	16
Determination of supportive criterions ,regulations, and rules to attract and the imported technologics as well as making them compatible	6.4	2.08	0.326	17
Modification of research instructions to facilitate the technology development	7.1	2.30	0.327	18
Developing the needed human resources in the field of technology development (education)	6.6	2.37	0.357	19
Institutionalization of the national program of technology development in the country's agriculture research centers	6.9	2.57	0.372	20
Providing the necessary advance educational environments for working on the new technologies	6.8	2.58	0.378	21
Improvement of the groups and faculties relationships with the science and technology parks	6.9	2.80	0.408	22
More information concerning the news of technology developments for the faculty members	6	2.60	0.436	23
Establishing research and development centers in the industrial companies and relating them with the universities and research centers	6.7	2.97	0.441	24
University science and technology parks development and reinforcement	6.1	2.96	0.485	25

Table 1. Contd.

Establishing a policy and strategy making center for the technological development	6.2	3.04	0.487	26
Distribution of new resource list (current awareness newsletter)	5.2	2.55	0.493	27
Designating a part of faculty members` activity time to the fields of technology development	5.9	2.93	0.498	28

The most inappropriate strategy 0 = the most appropriate strategy.

Table 2. Extracted factors with the eigenvalue, variance percent, cumulative variance percent.

Rank	Factors	Eigenvalue	Variance	Cumulative
1	technology-centered research development	4.46	19.5	19.5
2	Infrastructure development	4.85	17.34	36.84
3	improvement of policies and plans	4.71	16.83	53.68
4	Information and communications	3.73	13.32	67
5	Incentives and supports	2.65	9.49	67.49
6	Technology transmittal	2.07	7.41	83.90

analysis. KMO coefficient was 0.724 and the Bartlett`s test was significant, too (chi-square = 3880.94, df = 378, sig =0.00). Confirming the suitability of the data, factor analysis was carried out by the main item analysis method and using the varimax rotation; factors with eigenvalue (>1) were evaluated as the factor (Table 2).

Results showed that 6 factors, totally, explained about 83.9% of the variance related to the strategies of the agriculture technology development. The first factor with eigenvalue of 5.46 explains about 19% of the variance related to the strategies. According to the placing situation of the variables in the factor, this factor was named the factor of technology-centered research development. The second factor, infrastructures development, explains about 17.34% of the variance related to the agriculture development strategies. The third factor, improvement of policies and plans, with eigenvalue of 4.71 explains about 16.38% of the variance related to the agriculture development strategies. The fourth factor, information and communications, with eigenvalue of 3.713 explain about 13.32% of the variance related to the agriculture development strategies. The fifth factor, incentives and supports, with explains about 9.49% of the variance related to the agriculture development strategies. The sixth factor, technology transmittal, explains about 7.41% of the variance related to the agriculture development strategies (Table 3).

CONCLUSION AND SUGGESTIONS

The most important strategies of agriculture technology development in the country with emphasizing on the role of science and technology parks are technology-centered researches development, infrastructures development, improvement of policies and plans, information and communications, incentives and supports, and technology transmittal.

Sahebkar (2002), concerning the technology-centered researches development, believed that the development of technology in the agriculture sector is not random and spontaneous, but needs a complete and forecasting view for the fields of execution, schematization, policy making and investigation. Creating the national charter of technology is a necessary groundwork in order to institutionalize the research. The most important criterions of the national charter of technology are conceptualization, value orientation, future orientation, landscape creation, oral representation of the interaction between culture and technology, paradigm orientation, leadership orientation, and comprehensive orientation.

Results related to the technology development strategy confirms the results gained by Porsolymanian (2006), in which different supports, infrastructures improvement ,considering the social ,cultural, and political capitals are emphasized. In addition Hellsten (2007) believed that when a suitable economical ecosystem is created, a dynamic cycle should also be created, in which synergy between parameters and stimulus would be able to create a sustainable economical value. Research and technology infrastructure is the firm faith and belief of the country's macro management to the issue of science and technology.

Unsuitable infrastructure of Iran agriculture and its macro management believe in the industrial sector development as the first priority than the agriculture one need a change of the management's attitude toward the agriculture technology and that we are able to provide our own requirements. Improper structure of technology in making a relation between research and development centers and agriculture sector in addition to the inefficiency of the extension sector need a revision in the agriculture technology structure in the country.

Table 3. Coefficients related to each factor and coefficients gained from the rotation matrix

Factor	Variables inserted in each factor	Factor coefficient
Technology-centered research development	Making the researches practical in order to resolve the agriculture economical, social, technical problems	0.842
	reinforcing the relation between industry, universities, society, and government	0.752
	Institutionalization of national program of technology development in the country's agriculture research centers	0.677
	Modification of research instructions to facilitate the technology development	0.622
	Establishing research and development centers in the industrial companies and relating them with the universities and research centers	0.748
	Developing researches on the new technologies(such as bio, Nano ,and etc) in the universities	0.576
	increasing the national investment in the field of research and development in the higher education centers	0.795
	Providing the necessary advance educational environments (like universities) for working on the new technologies	0.675
Infrastructure development	University science and technology parks development and reinforcement	0.760
	Developing the needed human resources in the field of technology development (education)	0.579
	Improving the technology development infrastructures in the university (such as workshops, labs, etc)	0.835
	Establishing a policy and strategy making center for the technological development	0.581
Improvement of policies and plans	Revising the research regulations to facilitate the technology development process (e.g. the possibility of university students` teamwork)	0.858
	Merging the programs of technology development with the five -year plan of agriculture development in a national level	0.606
	More information concerning the news of technology developments for the faculty members	0.575
	Developing and facilitating the relations between groups by carrying out interdisciplinary research	0.736
Information and communications	Improvement of the groups and faculties relationships with the science and technology parks	0.645
	The possibility of carrying out the thesis jointly by a number of educational group (teamwork)	0.899
	Distribution of new resource list (current awareness newsletter)	0.781
	more financial support for thesis in order to complete and commercialize the research results	0.559
Incentives and supports	considering some incentives for the active researches in the field of technology development and new products	0.863
	Consult and legal supports for faculty members in order to commercialize the research results	0.807
	Determination of supportive criterions ,regulations ,and rules to attract and the imported technologies as well as making them compatible	0.683

Table 3. Contd.

Technology transmittal	Technology transfer from all technological countries	0.601
	Using integrated strategies of a simultaneous transfer of technology from out of the country and develop it inside it	0.781

One other important strategy is the development of information and communication. Hashemi and Asghari (2006) believed that the most important information services which should be provided for the science and technology parks are the accessibility to intranet and internet, library and library resources of all centers having agreement with, and software.

Study's results indicated a technology import (at the current situation) in the country, which is because of the lack of structures, infrastructure and necessities for a short-term agriculture technology development in the country; but, in long-term, compilation of technology importation and in-country production should be considered. Based on this strategy, Morin and Rafferty (2005) believed that the technology development in the country should occur through both technology transmittal and technology making. Wherever there is ability, technology can be made, and, wherever there is not, technology development should be through importation. According to the results, the following suggestions are dedicated to improve the process of developing the agriculture technology in the science and technology parks:

1. Identification of the technology requirements for the country `s different exploitation systems in order to route the applicatory and technology-centered researches by the governmental financial supports, which should be increased in the agriculture research sector;
2. Developing the agriculture research infrastructure with an emphasize on the development of both participated technologies and agriculture systems;
3. Improvement of policies and plans in order to prioritize the agriculture development by changing the strategies from industrial development to agriculture development;
4. Improvement of agriculture information system to develop the communications between the agriculture sector and research centers
5. More supports for agriculture researches and research centers in addition to make financial and emotional incentives for the researches in order to make an agriculture research tendency, and
6. Importing the technology for overseas into the country to meet the short-term technological needs and effort to make these technologies compatible with the agricultural conditions of the country's regions.

REFERENCES

Agwu AE, Dimelu MU, Madukwe MC (2008). Innovation system approach to agricultural development: Policy implications for agricultural extension delivery in Nigeria. Afr. J. Biotechnol. 7(11):1604-1611. Available online at http://www.academicjournals.org/AJB.

Anonymous A (2002). Agriculture Incubator Center, Developed and Operated by: Threshold to Maine Rc & D Area Inc.

Berg D (2005). Technology management: Brand value and the technology sector. 14th International Conference on Management of Technology, Vienna, Austria. 13-15 June. pp. 38-40.

Cassiman B, Veugelers R (2006). In search of complementarily in innovation strategy: internal R&D and external knowledge acquisition. Manage. Sci. 52:68-82.

Eom B, Lee K (2010). Determinants of industry-academy linkages and, their impact on firm performance: the case of Korea as a latecomer in knowledge industrialization. Res. Policy 39:625-639.

Farjadi GH, Riahi P (2007). Local market of Iran's science and technology parks. J. Res. Plann. Higher Edu.132(44):49-21.

Francis J (2006). National Innovation System Relevance for Development. Training of Trainers Workshop for ACP Experts on Agricultural Science, Technology and innovation (ASTI) system 2nd-3rd October.

Frenz M, Ietto-Gillies G (2009). The impact on innovation performance of different sources of knowledge: evidence from the UK Community Innovation Survey." Res. Policy 38:1125-1135.

Hashemi H, Asghari H (2006). The Role of Technological Entrepreneurs in Technology Transfer Process as Intelligent Carriers. IASP Asian Divisions Conference, ASPA 10th Annual Conference, 3rd Iranian National Conference on Science and Technology Parks, Isfahan, Iran.

Hellsten E (2007). The European nanotechnology strategy: environmental and health aspects. Available in: http://www.oeaw.ac.at/ita/nano07/Hellsten.pdf.

Iran Statistic Center (2010). Selected statistics of rural and agricultural situation in Iran. Available in: http://www.amar.org.ir/.

Kaimowite D (1991). The evaluation of links between research and education. Holt, Rinehart and Winston, Inc. Orlando, Florida, pp. 75-76.

McGann J (2007). Survey of Thins Tanks, a Summary Report, Foreign Policy Research Institute, USA, pp. 11-12.

Metcalfe JF (1995). The economic foundations of technology policy. In P. Stoneman (Ed.), Handbook of the Economics of Innovation and Technological Change: Oxford: Blackwell. pp. 409-512.

Morin J, Rafferty PJ (2005). Six key role in the management of technology resources. Translation: Department of Industrial Management Technology Management. Tadbir. Fifteen years. In Persian, P. 145.

Naito Y (1998). System Innovation: Technology Transfer. Enterprise Diagnosis 10:60.

Porezat A (2010). Explain barriers to entrepreneurship and the commercialization of university students in Tehran University. J. Sci. Technol. Policy. In Persian, 2:65-77.

Porsolymanian F (2006). The role of science and Technology Park to develop industries technology. J. Technol. In Persian, 9:49.

Sahebkar KM (2002). National Technology Charter (definition and criteria) technology analysts network. In Persian. pp 25-35.

Saljoghi KH (2005). Knowledge-based economy. The third and the tenth international congress. Congress. Cooperation of government, university and industry. J. Technol. In Persian. 5:45-56

Spielman DJ (2005). Innovation Systems Perspectives on Developing-Country Agriculture: A Critical Review. ISNAR Discussion 1:10-40.

Stone D (2005). Think Tanks and Policy Advice in Countries in Transition. Presented in Asian Development Bank Institute

Symposium: How to Strengthen Policy- Oriented Research and Training in Vietnam. Asian Development Bank Institute, Vietnam, pp. 2-5.

Sun YL, Yonglong W, Tieyu M and Guizhen H. (2007). Pattern of patent-based environmental technology innovation in China, Technological Forecasting & Social Change, [Available online at www. Sciencedirect.com].

Swanson BE, Bentz R, Sofranko A (1997). Improving agricultural extension. a reference manual was prepared under a contract between FAO and the International Program for Agricultural Knowledge System (INTERPAKS), College of Agricultural, Consumer, and Environmental Sciences, University of Illinois at Urbana-Champaign, United States.

Tsai KH, Wang J (2008). External technology acquisition and firm performance: A longitudinal study", J. Bus. Venture 23:91–112.

Tugrul T, Ajit M (2002). The cotton supply chain in Azerbaijan, ISNAR, the Hague, Netherlands, pp. 13-17.

19

Production, utilization and acceptability of organic fertilizers using palms and shea tree as sources of biomass

P. O. Oviasogie, J. O. Odewale, N. O. Aisueni, E. I. Eguagie, G. Brown and E. Okoh-Oboh

Date Palm and Shea Tree Research and Development Department, Nigerian Institute for Oil palm Research, P. M. B. 1030 Benin City, Edo State, Nigeria.

This document contains relevant information on biomass generation, utilization for the production of organic fertilizers and their alternative uses with respect to The Nigerian Institute for Oil Plam Research (NIFOR) mandate crops. The production of organic amendments using empty fruit bunches from the Oil palm *Elaeis guineensis* and its utilization in soil fertility management is well established. On an average, for every tonne of fresh fruit bunches (FFB) processed wastes of 230 to 250 kg empty fruit bunch (EFB) and 130 to 150 kg of fiber is produced. This large amount of biomass/waste can be successfully converted to fertilizer through the process of composting and can then be ploughed back into the soil. The palm oil mill effluent (POME) is also available, which in more recent times is being used as a soil conditioner and a source of bio fertilizers. Composted coir pith obtained from coconut *Cocos nucifera* husk is found to be rich in plant nutrients with nitrogen, phosphorus and potassium percentage values of 1.24, 0.06 and 1.20, respectively among other required plant nutrient. Percentage nutrient values for composted date palm *Phoenix dactylifera* biomass are also available in this paper. These mandate crop posses several alternative uses which makes them useful to the food and feed industry as well as oleo chemical, fuel, building and construction, pharmaceutical and confectionary industries. The information presented strongly suggests and confirms that the various biomass and residue produced by the palms and shea tree industry can be fully harnessed for organic fertilizer (manure) production even for commercial purposes by and for farmers in order to increase crop production.

Key words: Organic fertilizer, composting, biomass, alternative uses.

INTRODUCTION

Fertilizers are the most important input that ensures optimum crop production and are broadly classified into chemical and organic fertilizers. The use of organic fertilizer dates back to mans early farming activities. Modern research on the use of organic manure for the production of arables in Nigeria is shown in the pioneer work of Hartley and Greenwood (1933). The availability and affordability of chemical and high analysis fertilizers to the average farmer is less than optimal hence the need to source for locally available compostable organic materials. More recently, organic farming has become preferable especially for vegetable production particularly in the advanced countries. Research has shown that organically produced foods/ crops are healthier for consumption and safer for the environment. Hence, there is a rapidly growing demand for organic fertilizers that have levels of macro and micro nutrients which are comparable to that of inorganic fertilizers when applied at the same dosages and intensity.

The development of organic fertilizer entirely from naturally derived sources with the required specifications is for purpose of satisfying this growing demand in the agricultural industry. Organic fertilizers are sources of soil nutrients produced from organic materials usually of plant and /or animal origin. These fertilizers are manufactured using organic substances which are bio-degradable. These organic substances are further decomposed and broken into smaller and soluble particles by numerous microorganisms. Some advantages of organic fertilizers include: Replenish the soil, keeps soil friable, promotes beneficial soil life, increases crop yield and grows larger plant, prevents hardpan, benefit environment by recycling and reducing waste, minimize green house gas emission, protects certain crops from disease, are safe and cheaper than chemical fertilizers as it can be locally prepared.

Bio fertilizers simply mean "living fertilizers". Unlike the organic fertilizers, bio fertilizers are a mixture of beneficial microbes such as bacteria and fungi which have been reported to enhance the yield of crops. These microbes include soil inoculants which are preparations of the nitrogen fixing bacterium rhizobium for inoculating legume seeds (Odeyemi, 1991). The use of vesicular arbuscular mycorrhiza (VAM) has been shown to increase crop yield when used with woody leguminous prunnings (Rafiu-Adio et al., 2000). Some benefits of bio fertilizers include: Increase crop yield by 20 to 30%, replace chemical N and P by 25%, stimulates plant growth, activate soil biologically (Toyota and Kuninaga, 2006), restores natural fertility and provides protection against drought and some soil borne diseases. The following are types of bio fertilizers: For nitrogen supply; *Rhizobium* for legume crops, *Azotobacter*/*Azospirillum* for non legume crops, *Acetobacter* for sugarcane only and BGA and Azolla for low land paddy. For Phosphorous: *Phosphatika* for all crops to be applied with *Rhizobium*, *Azotobacter*, *Azospirillum* and *Acetobacter*; For enriched compost, cellulolytic fungal culture and *Phospholika* and *Azotobacter* culture (Singh and Amberger, 1991).

Organic manures/fertilizers are obtained from the composting/processing of organic materials. These materials could be: Plant material, animal material, urban/industrial material (Municipal Wastes), night soil, mineral organic materials and bacterial/ fungal materials (Chude and Uzoigwe, 2001).

NIFOR MANDATE CROPS–SOURCE OF BIOMASS FOR ORGANIC FERTILIZER PRODUCTION

The Nigerian Institute for Oil Plam Research (NIFOR) has five mandate crops. These include: Oil palm (*Elaeis guineensis*), Coconut palm (*Cocos nucifera*), Date Palm (*Phoenix dactylifera*), Raphia Palm (*Raphia hookeri*) and the Shea tree (*Vitellaria paradoxa*). These tree crops produce biomass which can be used for the production of organic fertilizer as well as other alternative uses. This document shows the availability of biomass of these mandate crops and possible uses as organic fertilizer as well as alternative uses.

Oil palm (*E. guineensis*)

Cultivation of the oil palm (*E. guineensis* Jacq.) has expanded tremendously in recent years such that it is now second only to soybean as a major source of the world supply of oils and fats. From its home in West Africa, the oil palm (*E. guineensis* Jacq.) has spread throughout the tropics and is now grown in 16 or more countries (Basri et al., 2003). However, the major centre of production is in South East Asia (SEA) with Malaysia and Indonesia together accounting for around 83% of world palm oil production in 2001. Presently, Southeast Asia is the dominant region of production with Malaysia being the leading producer and exporter of palm oil. In spite of the huge production, the oil consists of only about 10% of the total biomass produced in the plantation. The remainder consists of huge amount of oil palm wastes (biomass) such as oil palm shells, mesocarp fibers, empty fruit bunch (from the mills), oil palm fronds and oil palm trunk (Plate 1). At the palm oil mill, the sterilized fresh fruit bunches (Plate 2) goes through a threshing process to separate the fruit from the nuts. The emptied fruit bunch mainly consists of a main stalk (20 to 25%) and numerous spikelets (75 to 80%) with sharp spines at their tips. For the years of 2007 to 2020, it is projected that an average of 2.856 million tones (dry basis) of empty fruit bunches will be produced per year (ADB, 2006). Oil palm fronds are collected during pruning and replanting activities. The availability of fronds during the pruning activity was calculated using an estimate of 10.4 tonnes ha^{-1}, which currently gives an average of 6.97 million tonnes per year. Meanwhile, it was estimated that an average of 54.43 million tonnes per year of oil palm fronds will be available during the replanting process in the years 2007 to 2020. The Oil palm tree normally passes their economic age, on an average after 25 years. During replanting, the bole length of felled palm trunk is in the range of 7 to 13 m, with a diameter of 45 to 65 cm, measured at breast height. The area due for replanting has to be multiplied with the average number of 134 palms ha^{-1}, or in volume of 1.638 m^{3}ha^{-1}. About 53.87% (dry weight) of fiber bundles can be extracted from a trunk, with the remaining parts being the bark and parenchyma tissues which contribute to 14.45 and 31.68% of the dry weight of the trunk respectively (ADB, 2006).

In the processing of palm oil, the main by-product and wastes produced are empty fruit bunches (EFB), palm oil

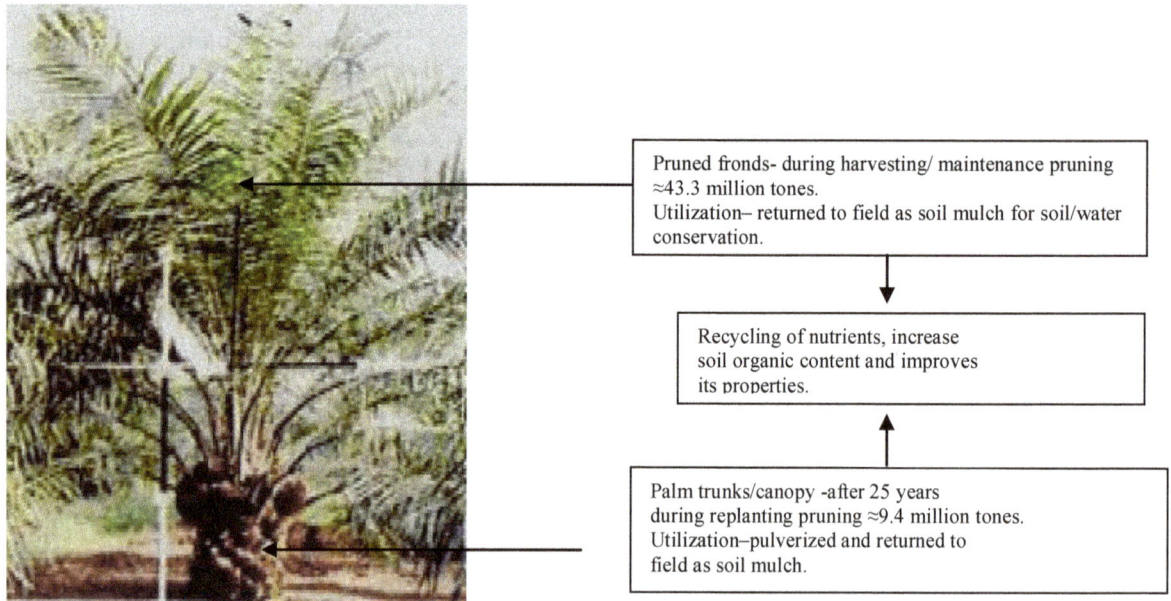

Plate 1. An oil palm showing output of recyclable biomass of oil palm in Malaysia, 2006.

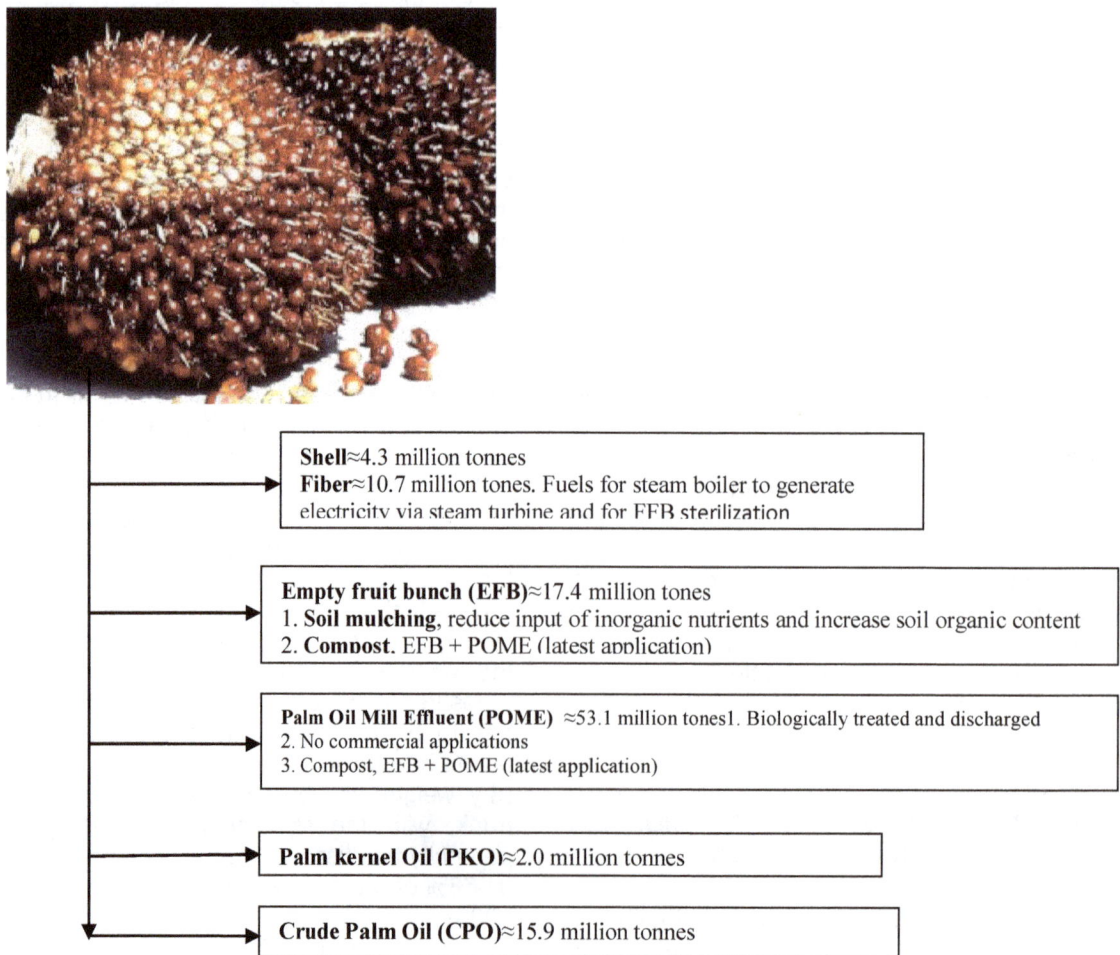

Plate 2. Total fresh fruit bunch processed≈79.3 million tones.

Figure 1. Flow chart of the MDF manufacturing and samples of MDF.

Plate 3. Plywood from oil palm trunk.

mill effluent (POME), mesocarp fiber and palm kernel shell/ cake residue (Yusoff, 2006).

On an average, for every tonne of fresh fruit bunches (FFB) processed, wastes of 230 to 250 kg empty fruit bunch (EFB), 130 to 150 kg of fiber, 60 to 65 kg of shell and 55 to 60 kg of kernel are produced (Karen, 2008). Currently, 15 million tonnes of EFB are generated annually from Malaysian palm oil industry (Rahman et al., 2006). Therefore, EFB as lignocelluloses, afford a renewable biomass and low-cost raw material for the production of high valued products such as polylactate, ethanol and compost.

The following are alternative uses of oil palm biomass: Medium density fiberboard (MDF) production. Malaysia is one of the top five exporters of MDF in the world. Our total production capacity exceeds one million cubic meters per year. Presently, most of the major producers use rubberwoods as the raw material. With the decline in the price of latex, interest in rubber is only about 1.2 to 1.4 million hectares, which is approximately half of its original size slightly over a decade ago. As such, it is envisaged that the supply of rubberwood for MDF production would not be sufficient to cater for the huge demand. The oil palm biomass has been tested in the laboratory and pilot scale at MPOB and elsewhere to be suitable as substitutes for rubberwood as raw materials for MDF production (Figure 1).

Plywood is also produced from oil palm trunks (Plate 3). A study was made to manufacture oil palm plywood consisting of 100% oil palm for the core, and the tropical hardwood for face and back veneer. Oil palm trunks used for manufacturing plywood were collected during replanting. Specific cutter gap and rotation during the peeling process are vital for processing the oil palm trunk, due to its inconsistence of vascular bundles arrangement. In general, the mechanical strength properties of oil palm trunk plywood meet the strength requirements as stipulated in the Japanese - Standard Method (JAS 233:2003).

Manufacturing of flat particleboard from oil palm biomass was envisaged a long time ago and the research started over 20 years ago with a few companies attempting to manufacture particleboard especially from EFB. Such end products are school and office desks and chairs, table tops and cabinets. The specialty of the flat particleboard produced from EFB using this process is that it has high screw withdrawal strength that is essential for the manufacture of quality furniture. Oil palm is also used as a delicacy meal. Palm heart has become an exotic mass product mainly used as salad ingredient especially in Europe. Normally, in our country, the source of oil palm heart is from coconut tree, but now oil palm heart has been introduced, which can be collected during replanting of oil palm tree. Oil palm hearts are slender, ivory-colored, delicately flavored, quite expensive and normally can be recovered at around 12 to 15 kg. A technology on mechanizing the extraction of the oil palm hearts had been developed, which consist of a mobile tractor system and attached with an auger. Activated carbon and advanced carbon products are also obtained from the oil palm tree (Plate 4). Oil palm biomass contains about 40 to 45% (wet basis) of carbon content, which is suitable for the preparation of carbon products.

Density =   0.95 - 1.30 g /cm^3
Porosity surface area (S_{BET}) =   680 - 1123 m^2/g
Micropore surface area (S_{MICRO}) =   360 - 1016 m^2/g

Plate 4. High porosity carbon powder and carbon pellets from EFB and the porosity characteristics.

Plate 5. A coconut palm, Nigerian Institute for Oil Palm Research (NIFOR), Nigeria.

The research and development for the preparation of advanced carbon products from oil palm empty fruit bunch (EFB) had been embarked since 2003, and with recent technology, MPOB successfully developed a process on preparation of carbon pre-cursors for the production of carbon electrode, electrical carbon brushes and Molecular Sieve Carbon (MSC) for gas filtration from EFB. Chemical treatment, carbonization and physical activation processes were applied to the carbon pre-cursors for the preparation of high porosity activated carbon powder and carbon pellets.

Coconut palm (*C. nucifera*)

Available coconut palm biomass include: Palm fronds, trunk, husk, coconut shell. The fronds and husk are easily compostable biomass of the coconut palm and have been used in the production of organic manure. The largest by products of coconut is coconut husk from which coir fibre is extracted. This extraction process generates a large quantity of dusty material called coir dust or coir pith. Large quantity of coir waste of about 7.5

million tones is available annually form coir industries in India (Plate 14). In Tamil Nadu state alone 5 lakh tons of coir dust is available. Coir pith has gained importance owing to its properties for use as a growth medium in Horticulture. Because of a wider carbon and nitrogen ratio and lower biodegradability due to high lignin content, coir pith is still not considered as a good carbon source for use in agriculture. Coir pith is composted to reduce the wider C: N ratio, reduce the lignin and cellulose content and also to increase the manorial value of pith (Table 2). Composting of coir pith reduces its bulkiness and converts plant nutrients to the available form. Every part of the coconut plant (Plate 5) has its value to humans; there are 20 more uses of coconut. These include as (1) tableware- the shell of a coconut can be used to make tableware such as bowls, serving tray, spoon or ladle; (2) fashion accessories-the coconut shell can also be carved out to make fashion accessories like necklaces, bangles, pendants, earrings and so on. In fact, the Hawaiians use coconut shell to make the buttons for their Hawaiian shirts; (3) furniture-the trunk of the coconut palm can be turned into furniture; (4) fuel source- the coconut shell and coconut husks are good sources of fuel; (5) brooms- The midribs of the coconut leaves are usually bundled and tied up with strings to become brooms; (6) barbecue skewers-the midribs or the coconut leaves can also be used as barbecue skewers; (7) ropes- the fiber material of coconut husk, which is also known as coir, can be used to make ropes; (8) brushes- the fiber material of coconut husk can also be used to make brushes which are usually used to scrub floors; (9) woven products- the coconut leaves can be woven to create products like baskets, bags and mats; (10) musical instruments-halved coconut shells are knocked together to make up rhythms and beats. The coconut shells are also used to make the base of musical instruments such as the Chinese "ban-hu" and "yea-hu"; (11) roof- the dried coconut leaves are used to make roofing materials for tropical huts in the ancient times. However, nowadays roof made up of coconut leaves is still being used for chalets at tourist attractions; (12) hair oil- coconut oil is regularly used as natural hair oil by people in some parts of the world like in India and in Southeast Asia. It is believed that coconut oil can promote healthy hair growth and also to treat hair infected by lice; (13) bridge-the palm wood of the coconut trunk is also used as a bridge to cross river in the olden days; (14) canoe- the hollowed trunk of coconut palm is used as small canoe by the Hawaiians; (15) soap- coconut oil is used as a basic ingredient in some of the cosmetic soap products; (16) toothpaste- coconut oil is also used as a basic ingredient to make toothpaste for sensitive teeth; (17) relieve minor skin irritations- coconut oil can also be applied to skin to treat minor irritations like insect bites and sunburn; (18) compost- the fiber of coconut husk is also used as compost in horticulture; (19) medical benefits of virgin coconut oil- virgin or pure coconut oil is believed to be

Plate 6. Date palm Tree in the field, NIFOR Sub-Station, Dutse.

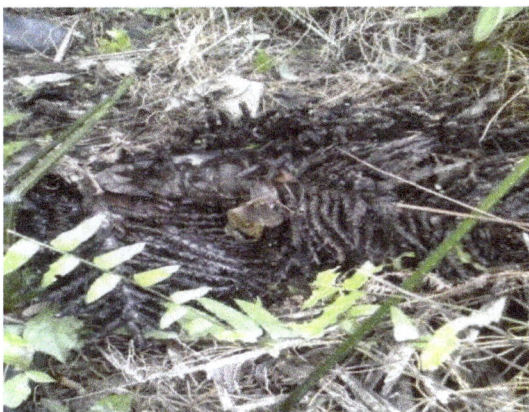

Plate 7. Empty raphia palm fruit bunches.

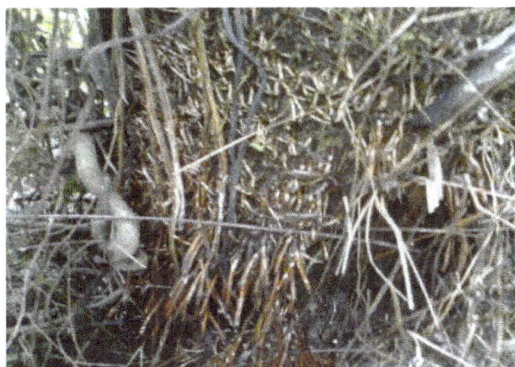

Plate 8. Raphia roots.

beneficial to human. More and more research efforts are being put forward to prove that it has many medicinal values such as an important dietary supplement to boost up immune system, assists constipation or digestive problems, assists weight control, anti-cancer agents, anti- aging properties and many more; (20) dye- the roots of the coconut plant are used as a dye.

Date palm (*P. dactylifera*)

The date palm is a very important and popular tree crop found growing mostly in the arid regions of the world (Plate 6). It is useful because it produces fruits which are a source of high energy supplying food and raw materials for building and construction (Morton, 1987). The biomass produced by date palms includes the following: Leaves/fronds, fruit clusters (Plate 15), inflorescence (Plate 16) and tree trunk. With the exception of the trunk, the other biomass can be used as plant material in the production of organic fertilizer. These biomasses also have alternate uses. Young leaves are cooked and eaten among people of the near East and North Africa while mature leaves are used for religious purposes (Palm Sunday). Mature leaves are also woven into mats, screens, baskets, crates, fans, hats etc. The stripped fruit clusters are used for brooms while the wood from the trunk is used for construction and left over parts burned as fuel.

Date palm biomass can be composted with other nitrogenous sources such as poultry droppings and cow dung (Plate 17). Currently, the composting of date palm biomass (Table 3) is on-going at the Nigerian Institute for Oil palm Research (NIFOR) Sub- Station Dutse, Jigawa State (Plate 18).

Raphia palm (*R. hookeri*)

The origin of Raphia palm is traced to tropical West Africa based on the reports of several explorers, investigators and scientist (Daziel, 1937; Russel, 1965; Otedoh, 1974). Raphia palms are usually associated with swampy coastal areas of West Africa and tapped for wine. Over 28 species of Raphia palm has been identified and 8 of them found in Nigeria. *R. hookeri* is our primary focus here. The components of *R. hookeri* biomass include: The roots (Plate 8), the trunks, the leaves, the inflorescence and empty fruit bunch (Plate 7).

The roots can be used alternatively for medicinal purposes by the local farmers and traditional healers. The trunks are used for building materials, pulp for paper making, fire wood for cooking and warming of houses, breeding grounding useful larvae. The leaves are used as piassava, bamboo, raffia, brooms, roofing mats and black fiber while raphia inflorescence is tapped for palm wine, production of local gin and gaso oil. The fruits are used as growth regulators, fish poison (though no longer encouraged), human food, raphia oil and raphia seed.

Shea tree (*V. paradoxa*)

Available Shea tree (Plate 9) biomass includes the fruits, flowers, husks, foliage, roots and trunk. The husks make a good mulch and fertilizer (FAO, 1988a), and are also used as fuels on three stone fires (Plate 10). Shea tree

Plate 9. A typical Shea tree.

Plate12. Shea nuts.

Plate 10. Shea fruits, seeds, nuts and leaves.

Plate 11. Shea fruits.

The following are some alternative uses of shea tree biomass: - The flowers are made into edible fritters. The fruit pulp, being a valuable food source, is also taken for its slightly laxative properties. Shea nut (Plate 12) cake is used for cattle feed (Salunkhe and Desai, 1986), and also eaten raw by children (Faegri, 1966) while the residual meal is used as a waterproof agent to repair and mend cracks in the exterior walls of mud huts, windows, doors and traditional beehives. The sticky black residue, which remains after the clarification of the butter, is used for filling cracks hut walls (Greenwood, 1929) and as a substitute for kerosene when lighting firewood (Wallace-Bruce, 1995).

The husks make a good mulch and fertilizer (FAO, 1988a), and are also used as fuels on three stone fires. Shea leaves (Plate 10) are used as medicine to treat stomach ache in children (Millee, 1984). A decoction of young leaves is used as a vapor bath for headaches. The leaves in water form a frothy opalescent liquid, with which the patients head is bathed. A leaf decoction is also used as an eye bath (Abbiw, 1990; Loupe, 1994). The leaves (Plate 11) are sources of saponin, which lathers in water and can be used for washing (Abbiw, 1990). Branches may be hung in the door way when woman goes into labour to protect the new born baby. Branches may also be used to cover the dead prior to their burial (Agbahungba and Depommier, 1989). The roots are used as chewing sticks in Nigeria, most commonly in savannah areas (Isawumi, 1978). Roots and bark are grounded to paste and taken orally to cure jaundice (Ampofo, 1983). These are used for the treatment of diarrhea and stomach ache (Millee, 1984). Mixed with tobacco the roots are used as poison by the Jukun of northern Nigeria. In horses, chronic sores are treated using boiled and pounded root bark of this plant (Dalziel, 1937). Infusions of the bark have shown to have selective anti-microbial properties, as being effective against *Sarcina lutha* and *Staphylococcus mureas* but not mycobacterium. Macerated with the bark of *Ceiba pentandra*, and salt, bark infusion have been used to treat cattle with worms in the tundra region of Senegal

husks have a capacity to remove considerable amounts of heavy metal ions from aqueous solutions for example with waste water. These were found to be more effective than the melon seed husks for absorption of lead, Pb (11) ions (Eromosele and Otitolaye, 1994). There is need to conduct further research on the potentials of Shea husk.

and Guinea (Ferry et al., 1974). The infusions have been used to treat leprosy in Guinea Bissau (Daziel, 1937) and for gastric problems (Booth and Wickens, 1988) as well for diarrhoea or dysentery. A bark decoction is used in Cote d'ivoire in baths and therapeutic sit-baths to facilitate delivery of women in labour, and is drunk to encourage lactation after delivery (Abbiw, 1990; Loupe, 1994). However, in northern Nigeria such a concoction is said to be lethal (Daziel, 1937). A bark infusion is used as an eye wash to neutralize the venom of spitting cobra and also, in Ghana, as a footbath to help extract jiggers. Greenwood (1929) noted that the stripping of the bark for medical purposes may have a severe impact on the health of Shea trees and may even be fatal. The wood is used only when the individual Shea tree is not valued for butter production.

This is because of the long gestation period of the tree crop as such farmers are reluctant to cut down any tree unless it is proven to be of inferior value for butter production. The latex is heated and mixed with the palm oil to make glue. It is chewed as a gum and made into balls for children to play with (Louppe, 1994). In Burkinafaso, musicians use it to repair cracked drums and punctured drumheads (Millee, 1984). It contains only 15 to 25% of carotene and, therefore, is not suitable for the manufacture of rubber.

COMPOSTING

Composting is the conversion of refuse into stable humus like substance under aerobic conditions. Composting offers an opportunity to recover and reuse a portion of the nutrients and organic fraction in agricultural wastes. Important factors in the process include intimate mixing of wastes, small particle size and oxygen for the microbial degradation of wastes, time to accomplish the composting and moisture. There are 2 main methods of composting organic materials. These include: - Pit method and windrow method. In the pit system, the sizes of the dug pit should be manageable in size $(2 \times 2 \times 1)$ m^3 according to Gordon (1982). The wastes are piled up in layers inside the pit and in alternate manner. Nitrogenous source of fertilizer is placed between the pilled layers e.g. poultry droppings. The pit is left for 2 weeks after which it is transferred to another pit and then stirred. The windrow method is similar to the pit system except that composting is carried out on the floor surface rather than in a pit. Poles are inserted between the piles while preparing the compost to ensure better air circulation and then removed later (Chude et al., 2001).

Stages in compost preparation

There are 3 stages in compost preparation and each stage is associated with peculiar groups of microorganisms. The 3 stages include: The initial, active and curing stage.

Maturity of compost is determined when pH reading ranges from 4.5 to 6.0 and temperature is stabilized at 30°C. Compost Preparation follows the concept of:

Mesophilic ⟶ Thermophilic ⟶ Mesophilic ⟶ Curing

In the initial stage, Mesophiles (fungi) predominate. Other organisms are exposed to the compost before the composting process starts. Mesophiles release heat. The active stage is the second phase. Here, most organic matter is converted to CO_2 and humus. Many microbes remain to be discovered and described due to limitations with isolation techniques. Few genera of bacteria isolated from thermophilic stage include *Bacillus*, *Clostridium* and *Thermus*. Properly ventilated composting pile maintains an appropriate temperature of 131 to 155°F. High temperatures ensure rapid organic matter processing while simultaneously providing optimal conditions for the destruction of human and plant pathogens as well as weed seeds. Mixing prevents temperatures from exceeding 160°F which effectively stops all microbial activity. Air pores created also serves as a passage for oxygen required by microbes to efficiently breakdown organic matter. Overheating occurs at 170°F upwards. Most microbes die and microbial activity ceases. Surviving microbes are usually in spores and return/germinate when the temperature becomes favourable. If the compost pile is too low in readily utilizable organic substrates, pile may not be able to support the microbial activity needed to return to thermophilic conditions. It may then be necessary to supplement the composting pile with additional feedstock to ensure maximum degradation and pathogen removal. At the curing stage, a properly functioning pile will deplete itself of a majority of easily degradable substrates leaving behind some cellulose, mostly lignin and humic materials for mesophiles that is, fungi and actinimycetes to act upon. This is the final stage of compost curing and is characterized by inability to identify the plant or other organic parts. Humification is characterized by increase in concentration of humic acids from approximately 4 to 12% and a decrease in C/N ratio from 30 in original material to 10 in the final product.

Fungi and actinomycetes use extracellular enzymes to degrade chitin, cellulose and lignin which are insoluble in water. Fungi, though they grow and respond more slowly than bacteria are well suited for exploiting an environment rich in complex recalcitrant organic compounds like those found in compost at curing stages.

Quality of raw materials for organic fertilizer production

Studies on the utilization of oil palm by-products such as mulching with pruned fronds, empty fruit bunches (EFB) and palm oil mill effluent (POME) as replacement for

Table 1. Nutrient contents in by-products obtained from a hectare of oil palm (kg/ha/year).

Part of palm	N	P	K	Mg	Ca
Annual pruning	107.9	10.0	139.4	17.2	25.6
Empty bunches	5.4	0.4	35.3	2.7	2.3
Fibre	5.2	1.3	7.6	2.0	1.8
Shell	3.0	0.1	0.8	0.2	0.2
Effluent (raw)	12.9	2.1	26.6	4.7	5.4

Source: Tarmizi (2000).

Table 2. Nutritive value of raw and composted coir pith compost.

S/N	Parameter	Raw coir pith (%)	Composted coir pith (%)
1	Lignin	30.00	4.80
2	Cellulose	26.52	10.10
3	Carbon	26.00	24.00
4	Nitrogen	0.26	1.24
5	Phosphorous	0.01	0.06
6	Potassium	0.78	1.20
7	Calcium	0.40	0.50
8	Magnesium	0.36	0.48
9	Iron(ppm)	0.07	0.09
10	Manganese(ppm)	12.50	25.00
11	Zinc(ppm)	7.50	15.80
12	Copper(ppm)	3.10	6.20
13	C:N ratio	112.1	24:1

Source: Department of Environmental science; Centre for Soil and Crop Management Studies (2008) Tamil Nadu Agricultural University.

Table 3. Nutritive value of different parts of the date palm biomass.

S/N	Sample description	P (%)	Ca (%)	Mg (%)	K (%)	Na (%)	N (%)
A	DTPBHS	0.058	0.275	0.160	0.969	0.047	0.228
B	DTPB	0.032	0.281	0.190	0.958	0.043	0.082
C1	DTPB	0.043	0.257	0.112	0.922	0.045	0.179
C2	DTPB	0.088	0.160	0.065	0.935	0.030	0.077

A = DTPBHS = Date palm bunch stem; B = DTPB = Sticks of bunches; C = DTPB = Flowers. Source: Aisueni et al., Date palm task execution report, NIFOR (2009).

chemical fertilizers have been done extensively with encouraging outcomes. Table 1 shows the various nutrient contents that can be obtained from a hectare of oil palm. Composition analysis of EFB reveals that 1 tonne of EFB (fresh weight) has the fertilizer equivalent of 7 kg urea, 2.8 kg rock phosphate, 19.3 kg muriate of potash and 4.4 kg kieserite (Singh et al., 1999).

The interaction between EFB application and chemical fertilizer on inland soil improves oil palm growth and increases yield by up to 75% (Singh et al., 1999), while on coastal soil, the response depends on the type of alluvium. After taking into account the transport costs for

the EFB, the savings in fertilizer can reach 28% less than the prior cost (Nasir, 2001).

In the coir pith composting technology, coir pith (pulverized and sieved coconut husk) is collected from the coir industry without any fiber. If fibrous materials are present, it is removed by sieving at the source itself. Otherwise, it has to be removed at the end of composting at the compost yard. These fibrous materials will not get composted and it will hinder with composting process. It is advisable to bring fiber free coir pith for composting. In choosing a site for composting, a separate area should be earmarked. It is better to have

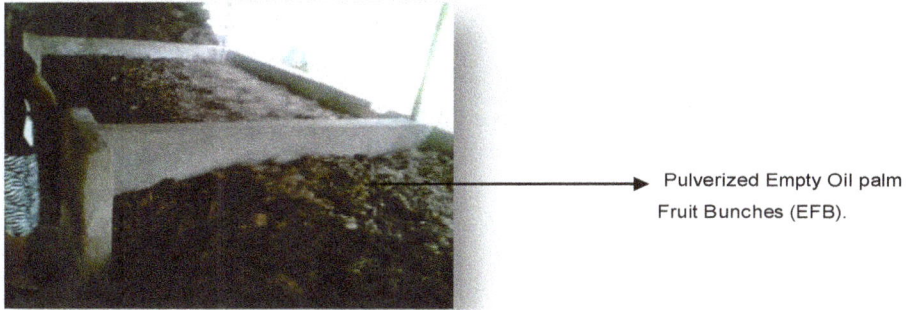

Pulverized Empty Oil palm Fruit Bunches (EFB).

Plate 13. Typical cells used in composting, NIFOR main station, Benin City, Edo State.

Plate 14. Coir pith heap.

Plate 17. Cow dung.

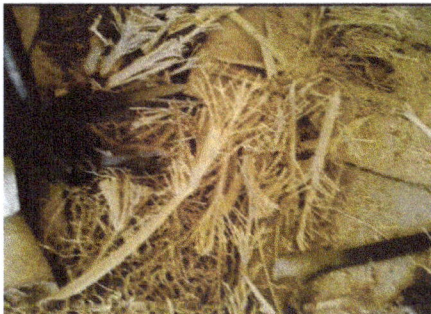

Plate 15. Sticks of bunches.

Plate 18. Compost cells under construction.

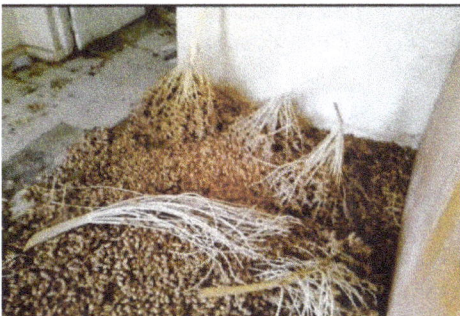

Plate 16. Flowers.

an elevated place for composting. In between coconut trees, shade under any tree is good for composting. The shady area conserves the moisture in the composting material. The floor of the compost making area should be leveled. If earthen floor is available the floor can be made hard by hard pressing and also by applying cow dung slurry. Presence of roof over the composting material is advantageous, since it protects the material from rain and severe sunshine (Plate 13).

Initially coir pith should be put up for 3 inch height and thoroughly moistened. After moistening, nitrogenous

Table 4. Tolerable contents of some micronutrients and heavy metals in soils with regard to their plant compatibility.

Nutrients	Tolerable limits (proposal) (Mg kg)	Percentage
Zn	300	0.03
Cu	100	0.01
Be	100	0.01
F	500	0.05
Cr	100	0.01
Ni	100	0.01
Co	50	0.005
As	50	0.005
Se	10	0.001
Mo	10	0.001
Cd	5	0.0005
Hg	5	0.0005
Pb	100	0.1

source material should be added. The nitrogenous source may be in the form of urea or fresh poultry litter. If urea is applied, it is recommended that 5 kg urea is required for one ton of coir pith. This 5 kg equally divided into five portions and in alternative layer of coir pith one kg of urea should be applied. If fresh poultry litter is applied, it is recommended at 200 kg for one ton of coir pith (1:5, poultry droppings to coir pith). The compost heap should be turned once in 10 days to allow the stale air trapped inside the compost material to go out and fresh air will get in. The composting process is an aerobic one; the organisms decomposing the materials require oxygen for their metabolic activity.

Composting of date palm biomass can be done using pits or cells/ compost compartments. The material being composted (empty date palm bunches) are cut to smaller sizes following the addition of cow dung sub-samples to the various parts of the date palm biomass pieces. The cow dung and date palm biomass are mixed together and turned from one cell to another to allow even distribution and decomposition of the organic materials. For Raphia palm and Shea tree, research needs to be carried out to explore the possibilities and determine the most suitable parts of the trees for compost making.

Compost as a soil conditioner impacts the following properties when applied to soils: Lightens heavy soil, improves the texture of light sandy soil, increases water holding capacity, base saturation, cation exchange capacity and buffering capacity of soil, contributes to the enlargement of the root systems of plants and bioremediation of oil spills or derelict lands. Characterization of various sources of nutrient from organic sources reveal the following limitations: Low levels of plant nutrients, non-available forms of plant nutrients, sub-optimal contents of trace elements, excessively high content of obnoxious and toxic metals,

high C:N ratio and objectionable odour (Table 4).

Specification/standards for organic fertilizer

The under listed criteria must be satisfied before any material of organic origin is accepted as organic fertilizer (FAO, 1994):

1. The minimum content of N: 1 to 4%
2. The minimum content of P: 1.50 to 3%
3. The minimum content of K: 1 to 1.5%
4. Moisture content should not exceed: 15 to 25%
5. Total organic matter content should at least: 20% or more
6. Carbon: Nitrogen (C/N Ratio)10: 1 to 15: 1
7. pH: 6.5 to 7.5
8. Odour: Free
9. Non –biodegradable materials such as glass, metal
10. Splinters: Free
11. Colour: Variable
12. Texture: Variable
13. Pathogens: Free

Acceptability, standardization and certification of organic fertilizer

Acceptability of the fertilizer depends on its availability. These fertilizers are not as readily available as inorganic fertilizers. This tends to increase the cost of the product. The cost of these organic fertilizers is more expensive than the conventional inorganic but with the transfer of the knowledge of composting to farmers, organic fertilizers can be produced at little or no extra cost. Handling- The locally produced composted organic fertilizers are bulky as a result farmers may prefer to go for the inorganic fertilizers still. It is however established that for mature palms, EFB 37 tonnes/ha/year or 258.7 kg/palm/year (Loong et al., 1987). For nursery seedlings, 150 g compost / nursery bag is sufficient (Aisueni et al., 2000). The standardization of any organic fertilizer follows or should conform with the rules and standards specific to that country. In the case of acceptability, the exportation of such a product will have to conform to the standard of the importing country. The International Federation of Organic Movement (IFOAM) is a body responsible for the standardization of production and processing of organic products. The IFOAM accreditation criteria for bodies certifying organic production and processing are called IFOM Norms. In Nigeria the Nigerian Organic Agriculture Network (NOAN) with its head quarters in the Faculty of Agriculture, University of Ibadan Oyo State is a non-governmental organization created to coordinate the activities of all stakeholders in organization production. More information on the criteria, acceptability levels and conditions required for the preparation and utilization of organic fertilizers from composted materials can be found in the IFOAM Norms (www.ifoam.org).

BENEFITS OF ORGANIC FERTILIZERS TO THE NATIONAL ECONOMY

The use of organic fertilizers will provide an environmentally friendly, naturally sustainable, safe and affordable means for maintaining soil fertility and increasing crop production. It will also lower the cost on the importation of inorganic fertilizers. This will ensure food security and improve the standard of living of Nigerians.

ACKNOWLEDGEMENTS

The authors wish to express their thanks to the Former Executive Director of NIFOR, Dr. D. A. Okiy, the former Director/Head Biology and Crop Production Department, Dr. C. O. Okwuagwu and the former head, Date Palm Programme Dr. J. O. Odewale.

REFERENCES

Abbiw D (1990). The useful plants of Ghana. Intermediate Technology publications/Royal Botanic Gardens, London/kew. P. 337.

Agbahungba G, Depommier D (1989). Aspects du parc a karites-neres (vitellaria paradoxa Gaertn F. parkia biglosa jacq. Benth.) dans le sud du Borgou (Benin). Bois et forets des Triopiques. 222:41-54.

Aisueni NO, Oviasogie PO, Brown G, Eguagie EI (2009). Evaluation of Cow dung based compost for commercial production of Bio-fertilizer. Date palm task execution report, NIFOR sub –station Dutse, Jigawa State.

Aisueni NO, Omoti U, Ekhator F, Oviasogie P (2000). Effects of compost on soils supporting Nursery seedling Production of oil palms. Nigerian J. tree crops Res. 4(2):43-51.

Ampofo O (1983). First aid plant medicine. Ghana rural reconstruction movement, Yentsi Centre, Mampong.

Asian Development Bank, ADB (2006). Technology overview–palm oil waste management- Appendix VII.

Basri MW, Maizura I, Siti Nor Akmar A, Norman K (2003). Oil Handbook of Industrial Crops". (Eds. V Haworth Press, New York. (In press).

Booth FE, Wickens GE (1988). Non-timber uses of selected arid zone trees and shrubs in Africa. FAO conservation guide, 19:1-176.

Chude VO, Osno AO, Uzoigwe GA (2001). Proceedings of an In – House Workshop On organic fertilizer development.

Daziel JM (1937). Useful plant of West Africa. Crown Agents, London. P. 612.

Department of Environmental science Centre for Soil and Crop Management Studies (2008). Composting technology and organic waste utilization in Agriculture. Tamil Nadu Agricultural University, Coimbatore-641 003, 2008.

Eromosele IC, Otitolaye OO (1994). Binding of iron, zinc and lead ions from aqeous solution by shea butter (Butyrospermum pakii) seed husks.bulletin of Environ. Contaminat. Toxicol. 52 530-537.

Faegri k (1966). Some problems representativity in pollen analysis. Palaeobotanist,15:135-140.

FAO (1988a). Appendix 5, forest generic resource priorities. 10. Africa. Report of sixth session of the FAO panel of experts on Forests Gene Resources, held in Rome, Italy, December 8-11, 1985, pp 86-89. FAO, Rome. p.79.

FAO (1994). Standards for organic fertilizers. In: organic recycling in Asia and the pacific. RAPA Bulletin 10:86.

Ferry MP, Gessain M, Geeain R (1974). Ethnobotanique tenda. Documents du centre de researches Anthropologiques du muse de l"Homme, paris.

Greenwood M (1929). Shea nuts and shea butter. Bulletin of the Agricultural Department, Nigeria. pp.59-100.

Hartley KT, Greenwood M (1933). The effect of small applications of farmyard manure on the yield of cereals in Nigeria. Emp. J. Exp. Agric.1:119-121.

Isawumi MA (1978). Nigerian chewing sticks. Nigerian field. 43:50-58.
Karen C (2008). Palm oil In: Environmental Management for Palm Oil Mill by Kittikun, A.H. et al., http://www.hi.sierraclub.org/maui/palmoil.html.

Loong SG, Nazeeb M, Letchumanan A (1987). Optimising the use of EFB mulch on oil palms on two different soils.1987 Kuala Lumpur, Malaysia.

Loupe D (1994). Le karate en cote d"i international en Researche Agronomique pour le Development, Montpellier. P. 28.

Mille JK (1984). Secondary products of species native to the Dinderesso Forest Reserve, Forest Educational and Development project USAID, Ouagadougou.

Morton J (1987). Date: In Fruits of Warm Climates. F. J. Morton. P 5-11.

Miami FI, Odeyemi O (1991). Development of Brady rhizobium inoculant (Bio fertilizer) for legume Inoculant. In proceedings of the first National seminar on organic fertilizer in Nigerian Agriculture; Present and Future, kaduna, 5-8 March, pp. 86-98.

Otedoh MO (1974). Raphia oil, its extraction,proparties and utilization. J.Nig.Institut. Oil palm Res. 5(19):45-49.

Rafiu-Adio OT, Fadare TA, Osonubi O, Idowu EO, Amure A, Awoyemi AO, Olaniyan A (2000). Evaluationof the effects of micorrhiza on crop yield. A paper Pesented at the 13th South West Zonal Refils Workshop, I.A.R. & T., Moor Plantation Ibadan., Feb. 14-18th.

Rahman SHA, Choudhury JP, Ahmad AL (2006). Production of xylose from oil palm empty fruit bunch fiber using sulfuric. Biochem. Eng. J. 30:97-103.

Russel TA (1965). The Raphia palms of West Africa. Kew Bull. 19(2):173-196.

Salunkhe DK, Dessai BB (1986). Postharvest biotechnology of oilseeds. CRC, Boca Raton. P. 264.

Singh CP, Amberger AA (1991). Solubilization and availability of palm. „In L Chopra and K V Peter) The and urine. Biol. Agric. Hort. 7:261-269.

Toyota K, Kuninaga S, (2006). Comparison of soil microbial community between soils amended with or without farmyard manure. Appl. Soil Ecol. 33:39–48.

Yusoff S (2006). Renewable energy from palm oil-innovation effective utilization waste. J. Cleaner Prod. 14:87-93.

Wallace-bruce Y (1995). Shea butter extraction in ghana, In: Do it Herself. Women and technical innovation. (Ed. By H. Appleton), pp.157-161.

20

Effect of sowing dates, fertility levels and cutting managements on growth, yield and quality of oats (*Avena sativa* L.)

Intikhab Aalum Jehangir[1], H. U. Khan[1], M. H. Khan[2*], F. Ur-Rasool[1], R. A. Bhat[1], T. Mubarak[1], M. A. Bhat[1] and S. Rasool[1]

[1]Division of Agronomy, SKUAST-Kashmir, Shalimar-191 121, India.
[2]Central Institute of Temperate Horticulture, ICAR, Srinagar (J&K) - 190 007, India.

A two year study was conducted during *rabi* seasons of 2009-10 and 2010-11 at Research Farm of Sher-e-Kashmir University of Agricultural Sciences and Technology of Kashmir to find out the influence of sowing dates, fertility levels and cutting management on growth, yield and quality of oats. The results revealed that September 30 sowing recorded significant improvement in green fodder yield over October 10 sowing. Crude protein content was highest in October 10 sown crop, whereas crude fibre was highest in September 20 sown crop. The fertility level of 150:70:40 ($N:P_2O_5:K_2O$ kg ha^{-1}) significantly increased both green and dry fodder yield as well as crude protein content over 125:60:30 and 100:50:20 ($N:P_2O_5:K_2O$ kg ha^{-1}), however crude fibre content significantly decreased with increase in fertility level. Double cut crop recorded 14.75 and 16.24% increase in green fodder yield and 3.70 and 1.36% in dry fodder yield over single cut crop during 2009-10 and 2010-11, respectively. Moreover, double cut crop recorded higher crude protein content but lower crude fibre content.

Key words: Sowing dates, fertility levels, cutting management, green and dry fodder yield, crude protein, crude fibre, oats.

INTRODUCTION

Oats (*Avena sativa* L.) rank fifth in terms of world production of cereals and is widely used as a companion crop for under-seeding of forage legumes (Dost, 1997). It is the most important winter cereal fodder which is rich source of energy, protein, vitamin B_1, phosphorus, iron and other minerals and is mainly grown in temperate and cool sub-tropical environments. A chronic fodder shortage, most serious in winter, is a major limiting factor for livestock production. There are two traditional fodder deficit periods especially in temperate regions including, December to March, when traditional winter fodder crops like berseem (*Trifolium alexandrinum*), oats (*A. sativa*)

and lucerne (*Medicago sativa*) are dormant and May to June (when the main summer season fodder crops such as maize (*Zea mays*), pearl millet (*Pennisetum glaucum*) and sorghum (*Sorghum bicolor*) have just begun growth and the winter fodder season is over. As a result of deficit periods, fodder becomes available for livestock feeding in late April thereby resulting in drastic reduction in milk and meat production. Sharma and Bhunia (2000) reported that higher fodder yield was recorded with increasing levels of nitrogen and when cutting was taken at 85 days after sowing. Similarly, Demetrio et al. (2012) obtained higher fodder yield by using up to two cuts in the vegetative stage, or a single one in the flowering stage. In view of this, an effort was made to adjust sowing date of oats in such a way that some green fodder becomes available to the livestock just at onset of winter without

*Corresponding author. E-mail: drmhkhan8@gmail.com

Table 1. Growth characters of oats as affected by sowing dates, fertility levels and cutting management.

Treatments	Plant height (cm)				Tillers m^{-2}				Leaf area index			
	2009-10		2010-11		2009-10		2010-11		2009-10		2010-11	
	1st cut	2nd cut	1st cut	2nd cut	1st cut	2nd cut	1st cut	2nd cut	1st cut	2nd cut	1st cut	2nd cut
Sowing dates												
September, 20	77.05	83.56	75.50	74.23	377.11	324.28	355.20	316.77	2.50	4.02	2.48	3.93
September, 30	68.01	108.48	67.14	106.78	333.83	331.14	330.98	326.46	2.33	5.47	2.32	5.39
October, 10	20.17	110.56	20.00	108.68	259.30	334.10	254.44	336.04	0.48	5.33	0.45	5.11
SE(m)±	1.42	2.00	1.74	1.82	3.88	1.94	3.75	2.25	0.10	0.16	0.10	0.13
CD (p=0.05)	4.10	5.78	4.25	5.26	11.22	5.60	10.82	10.82	0.28	0.47	0.29	0.39
Fertility levels (N:P$_2$O$_5$:K$_2$O kg ha^{-1})												
150:70:40	59.96	108.01	59.38	103.32	353.55	363.27	347.35	359.01	2.08	5.60	2.04	5.28
125:60:30	55.21	102.10	54.43	97.16	332.22	331.87	322.47	329.26	1.84	5.18	1.80	4.98
100:50:20	50.40	92.50	49.40	89.15	284.46	294.96	270.74	290.99	1.42	4.14	1.40	4.17
SE(m)±	1.42	2.00	1.74	1.82	3.88	1.94	3.75	2.25	0.10	0.16	0.10	0.13
CD (p=0.05)	4.10	5.78	4.25	5.26	11.22	5.60	10.82	6.51	0.28	0.47	0.29	0.39
Cutting levels												
Single cut (Cut at 50% flowering)	58.04	113.29	57.00	108.44	328.48	390.75	320.30	384.28	1.83	5.66	1.83	5.51
Double cut (Cut on 15th Dec. and 50% flowering)	58.01	88.45	57.03	84.64	330.34	269.25	318.74	268.56	1.83	4.28	1.79	4.44
SE(m)±	0.94	1.63	0.96	1.40	3.17	1.29	3.06	1.49	0.06	0.10	0.06	0.09
CD (p=0.05)	NS	4.71	NS	4.29	NS	4.57	NS	5.30	NS	0.38	NS	0.32

reduction in the total yield.

MATERIALS AND METHODS

A field experiment was undertaken during *rabi* seasons of 2009-10 and 2010-11 at Research Farm of Sher-e-Kashmir University of Agricultural Sciences and Technology of Kashmir on silty clay loam soil low in available nitrogen (261.48 kg ha^{-1}), medium in available phosphorus (20.83 kg ha^{-1}) and potassium (165.0 kg ha^{-1}) with neutral pH (6.8). The treatments consisting of three sowing dates (September 20, September 30 and October 10), three fertility levels (150:70:40, 125:60:30,100:50:20 kg N:P$_2$O$_5$:K$_2$O ha^{-1}) and two cuttings managements (Single cut - cut at 50% flowering and double cut - cut on 15th December and 50% flowering) were laid out in randomized block design replicated thrice. Oat variety "*Sabzar*" was sown as per treatment in rows 23 cm apart with a seed rate of 100 kg ha^{-1} in the plots of 13.8 m^2 area with 15 rows per plot. The fertilizers were applied as per treatment with half dose of nitrogen and full dose of phosphorus and potassium in the form of urea, diamonuum phosphate (DAP) and muriate of potash (MOP) as basal, and the remaining half of nitrogen was top dressed in two equal splits one each at 30 DAS and 1st week of March. All other operations were carried as per recommended package and practices. The observations on the plant height (cm), number of tillers m^{-2}, leaf area index, green fodder yield and dry fodder yield (sundried) in q ha^{-1} were recorded both at 1st and 2nd cut. Plant samples from green fodder yield of each treatment were sun dried followed by oven drying at 60 to 65°C to a constant weight and were finely ground for analysing of nitrogen content by microKjeldal method (Jackson, 1967), which was multiplied with 6.25 to represent protein content and calculate protein productivity. Crude fibre was determined by the method given by AOAC (1995). The data were analysed by the methods given by Cochran and Cox (1963).

RESULTS AND DISCUSSION

Sowing dates

Data (Table 1) revealed that crop sown on September 20th recorded significantly higher plant height, tillers m^{-2} and leaf area index at 1st cutting, whereas at 2nd cutting (September 30th and October 10th) they were statistically similar but significantly higher than September 20th during both the years of experimentation. The data presented in Table 2 revealed that September 30th and September 20th sown crops, at par with one

Table 2. Green fodder and dry matter yield of oats (q ha^{-1}) as affected by sowing dates, fertility levels and cutting management.

Treatments	Green fodder yield				Dry matter yield			
	2009-10		2010-11		2009-10		2010-11	
	1st cut	2nd cut	1st cut	2nd cut	1st cut	2nd cut	1st cut	2nd cut
Sowing dates								
September, 20	150.29	229.18	148.09	211.31	27.26	70.05	25.31	65.22
September, 30	146.49	236.31	146.03	218.27	25.50	71.99	24.80	67.10
October, 10	25.61	244.02	25.09	228.91	5.10	74.06	5.06	69.95
SE(m)±	-	3.86	-	3.71	-	1.04	-	1.16
CD (p=0.05)	-	11.45	-	10.73	-	3.01	-	3.35
Fertility levels (N:P$_2$O$_5$:K$_2$O kg ha^{-1})								
150:70:40	122.23	248.31	118.48	236.10	21.60	75.22	20.32	71.92
125:60:30	107.59	236.84	110.56	223.15	19.18	72.12	19.00	68.42
100:50:20	92.47	224.46	90.37	199.10	16.66	68.76	15.85	61.92
SE(m)±	-	3.86	-	3.71		1.04	-	
CD (p=0.05)	-	11.45	-	10.73		3.01	-	
Cutting levels								
Single cut (Cut at 50% flowering)	-	320.32	106.47	301.44	-	81.17	-	76.06
Double cut (Cut on 15th Dec. and 50% flowering)	107.44	260.15		243.94	19.25	62.89	18.32	58.78
SE(m)±	-	3.24		3.03		0.69	-	0.77
CD (p=0.05)	-	9.34		8.76	0.43	2.45	-	2.73

another recorded increase in the total green fodder yield to the tune of 41.97 and 40.73%, respectively, in 2009-10 and 43.56 and 41.55% in 2010-11 and dry matter yield by 23.45 and 22.92% in 2009-10 and 22.51 and 20.69% in 2010-11, respectively over October 10th sown crop. The higher temperatures available to the early sown crop resulted in the better growth of crop in terms of plant height and tiller production thereby producing more tonnage at 1st cut on December 15th. Sood et al. (1992) also reported higher yield in early sown oat crop compared to delayed sowing. Khalil et al. (2011) reported that forage dry matter production in wheat cut at 90 DAS was significantly higher than cut at 75 DAS. October 10th sown crop recorded higher crude protein but lower fibre content than September 20th and September 30th sown crop (Table 3). Higher protein content in October 10th sown crop could be attributed to higher nitrogen content in the plant at 1st cut and 2nd cutting. Protein content is inversely proportional to fibre content, hence lower fibre content in October 10th sown crop (Dost, 2004).

Fertility levels

The plant height and number of tillers m^{-2} showed significant and consistent increase with increase in fertility level from 100:50:20 to 150:70:40 kg N:P$_2$O$_5$:K$_2$O ha^{-1}, however, leaf area index recorded with fertility level 125:60:30 and 150:70:40 kg N:P$_2$O$_5$:K$_2$O ha^{-1} remained at par but significantly higher than 100:50:20 kg N:P$_2$O$_5$:K$_2$O ha^{-1} during both the years of experimentation (Table 1). The highest fertility level 150+70+40 kg N:P$_2$O$_5$:K$_2$O ha^{-1}

significantly improved the oats yield with superiority of 7.58 and 16.% and 6.25 and 22.48% in total green fodder yield and 5.63 and 13.38% and 5.51 and 18.60% in total dry matter yield over 125+60+30 and 100+50+20 kg N:P$_2$O$_5$:K$_2$O ha^{-1} during 2009-10 and 2010-11, respectively (Table 2). The abundant supply of nitrogen may have increased protoplasmic constituents and accelerated the process of cell division and elongation which has resulted in luxuriant vegetative growth in terms of plant height there by higher biomass and dry matter yield. Besides, phosphorus is involved in energy transfer and phosphorus dose at 70 kg P$_2$O$_5$ ha^{-1} may have significantly increased the tiller number especially at early crop growth stage thereby resulting in higher tonnage. These results corroborate the findings of Singh et al. (1997), Bali et al. (2003) and Malik and Paynter (2010).

The crude protein content showed significant improvement with increasing levels of fertility but crude fibre content remained unaffected (Table 3). Application of higher doses of nitrogen may have increased the nitrogen concentration in the plant and hence the crude protein content. Similar findings have also been made by Pandey et al. (1998).

Cutting management

The double cut crop recorded higher fodder yield with an increase of 14.75 and 16.24% in total green fodder yield and 3.70 and 1.26% in total dry matter yield over single cut crop during 2009-10 and 20110-11, respectively (Table 2). Double cut crop harvested on December 15th

Table 3. Crude protein and crude fibre content (%) of oats as affected by sowing dates, fertility levels and cutting management.

Treatments	Crude protein				Crude fibre			
	2009-10		2010-11		2009-10		2010-11	
	1st cut	2nd cut	1st cut	2nd cut	1st cut	2nd cut	1st cut	2nd cut
Sowing dates								
September, 20	17.37	8.56	17.31	8.43	19.33	22.77	19.10	22.66
September, 30	19.06	8.62	19.00	8.50	19.26	22.59	19.02	22.51
October, 10	19.62	8.87	19.56	8.75	18.86	22.05	18.74	21.95
SE(m)±	-	0.07	-	0.05	-	0.18	-	0.17
CD (p=0.05)	-	0.21	-	0.14	-	0.53	-	0.48
Fertility levels (N:P_2O_5:K_2O kg ha^{-1})								
150:70:40	19.56	8.93	19.43	8.87	19.01	22.38	18.89	22.24
125:60:30	19.12	8.68	19.06	8.50	19.19	22.45	19.92	22.35
100:50:20	17.43	8.43	17.43	8.31	19.26	22.58	19.04	22.52
SE(m)±	-	0.07	-	0.05	-	0.18	-	0.17
CD (p=0.05)	-	0.21	-	0.14	-	NS	-	NS
Cutting levels								
Single cut	-	8.50	-	8.43	-	23.10	-	22.94
Double cut	18.68	8.87	18.56	8.68	19.15	21.84	18.95	21.80
SE(m)±	-	0.06	-	0.04	-	0.15	-	0.13
CD (p=0.05)	-	0.17	-	0.11	-	0.43	-	0.39

for 1st cut before arrest of growth due to chilling temperatures and snowfall damage provide good quantity of green fodder especially early sown crop thereby total fodder yield in two cuts was significantly higher than in single cut crop. Previously, Shah and Hasan (1999) and Singh and Dubey (2007) also recorded higher green fodder and dry matter yield in double cut compared to single cut crop. It was found that double cut crop recorded significantly higher crude protein and lower crude fibre content than single cut crop (Table 3). Higher nitrogen concentration in double cut crop might have resulted in higher crude protein content. Lower lignin concentration in the oat stems of double cut crop due to more softness of stem restricted the crop to become more fibrous.

REFERENCES

AOAC (1995). Official Methods of Analysis. 16th Edn., Association of Official Analytical Chemists, Washington, DC., USA.

Bali AS, Wani MA, Shah MH (2003). Growth and grain yield of Oat (Avena sativa L.) as influenced by varying row spacing and fertility levels. SKUAST J. Res. 5:217-221.

Cochran GC, Cox MM (1963). Experimental Designs. Asia Publishing House, Bobmay, pp. 293-316.

Dost M (1997). End of Assignment report on fodder component. PAK/86/027. FAO/UNDP, Gilgit, Pakistan.

Dost M (2004) Fodder oats in Pakistan. Fodder Oats: a world overview. Food and Agriculture Organization of the United Nations. Plant Production and Protection Series p.33

Jackson ML (1967). Soil Chemical Analysis. Prentice Hall Inc., England, Chliffs, N.J. p. 498.

Khalil SK, Khan F, Rehman A, Muhmmad F, Amanullah, Khan AZ, Wahab F, Akhtar S, Zubair M, Khalil JH, Shah MK, Khan H (2011). Dual purpose wheat for forage and grain yield in response to cutting, seed rate and nitrogen. Pak. J. Bot. 43(2):937-947.

Malik RK, Paynter B (2010). Influence of N fertilization on yield and quality of oats hey and grain in Western Australia. Worked on grass of Soil Science, Soil Solution for acquiring World, 1-6 August, pp. 187-189.

Pandey TD, Namdeha KL, Saxena RR (1998). Crop compatibility and fertility levels in maize forage crop under bastar agro-climatic conditions. Forage Res. 24(1):57-59.

Shah WA, Hasan B (1999). Grain and fodder yield of Oats (Avena sativa L.) as influences by nitrogen levels and cutting schedules. Forage Res. 24(4):185-190.

Singh SD, Dubey SN (2007). Soil properties and yield of fodder oat (Avena sativa L.) as influenced by sources of plant nutrient and cutting management. Forage Res. 33(2):101-103.

Singh J, Rana DS, Joon RK (1997). Effect of sowing time, cutting management and phosphorus levels on growth, fodder and grain yield of oats. Forage Research 23(2): 115-117.

Sood BR, Singh R, Sharma VK (1992). Effect of sowing dates and cutting management on forage yield and quality of oat (Avena sativa). Forage Res. 18(2):130-134.

Demetrio JV, Costa ACTda, Oliveira PSRde (2012). Biomass yield of oat cultivars under different cutting management systems. Pesquisa Agropecuaria Tropical. 42(2):198-205.

Sharma SK, Bhunia SR (2000). Response of oat (Avena sativa) to cutting management, method of sowing and nitrogen. Indian J. Agron. 46(3):563-567.

21

An assessment of andrographolide production in *Andrographispaniculata* grown in different agro-climatic locations

Phurailatpam Arunkumar[1] , Bishoyi Ashok[2] and Maiti Satyabrata[2]

[1]Central Agricultural University, Pasighat, Arunachal Pradesh, India.
[2]Directorate of Medicinal and Aromatic Plants Research, ICAR, Anand, Gujarat, India.

An interlocational study was carried out to assess the effect of different agro-climatic locations on the secondary metabolite production in the medicinal plant *Andrographis paniculata*. Three accessions of *A. paniculata* were grown in three different locations of India differing in climatic conditions in RBD experimental plots. The andrographolide content in the plant samples of three accessions of *A. paniculata* was estimated from plants collected from these three locations, namely Anand, Kalyani and Solan. The three accessions with no morphological differences were characterised using RAPD and ISSR profiles. As the plants were grown in different agro-climatic conditions, as expected, there was variation in the andrographolide content in the three accessions. Samples collected from Kalyani were found to contain maximum andrographolide content as compared to the other two locations.

Key words: *Andrographispaniculata*, andrographolide, locations, high-performance liquid chromatography (HPLC), random amplified polymorphic DNA (RAPD).

INTRODUCTION

Plant growth and development are complex biological phenomena that depend upon genetic and environmental variables (Waller and Nowacki, 1978). There is indication by many workers that gene × environment interaction plays a crucial role for production of active ingredients in medicinal plants. Both growing location and seasonality play important roles in the production of active metabolites in the plant. Plant growth development in relation to season has distinct impact on accumulation of secondary metabolites (Teiz and Zeiger, 1998). *Andrographispaniculata* is an important medicinal plant. The leaves contain andrographolide, an important medicinal chemical. It is used both in indigenous and modern systems of medicine for treating several

diseases. The plant was reported to respond to application of organic and fertilizer nutrients (Rajeswara et al., 2004). In the present experiment, three accessions of *Andrographispaniculata* were grown in three locations to study the influence of agro-climatic conditions of these locations and plant age on the secondary metabolite (andrographolide) production.

MATERIALS AND METHODS

Three accessions of *Andrographispaniculata* were planted at three locations namely Kalyani, West Bengal (tropical type), Anand, Gujarat (arid type) and Solan, Himachal Pradesh (semi-tropics type) differing widely in agro-climatic conditions in an RBD and the

Table 1. Experimental materials and locations.

S/N	Accessions	Place of collection
1	1	Trichy, Tamil Nadu
2	2	Dangs, Gujarat
3	3	Anand, Gujarat

Experimental locations

S/N	Location of study	State	Temp. °C (max, min)	MSL (m)	Average rainfall (mm)	Co-ordinates
1	Kalyani	West Bengal	4, 45	13.5	1496	22.99°N; 88.45°E
2	Anand	Gujarat	12.5, 45	44	730	22.60°N; 72.93°E
3	Solan	Himachal Pradesh	-10, 22.8	1600	910	30.91°N; 77.09°E

Figure 1. plant habit of (a) Accession 1, (b) accession 2, (c) accession 3 at 80 DAT; different growth phases of A. paniculata, (d) seedling at transplanting stage, (e) 40 DAT, (f) 60 DAT, (g) 80 DAT, (h) 100 DAT, (i) 120 DAT.

samples were collected at 20 days interval (Table 1) (Figure 1). Herbage samples were collected at four regular intervals that is, starting from 60 Days After Transplanting (DAT) (DoS -2nd June and DoT - 18th July) till 120 DAT at 20-day intervals from the three

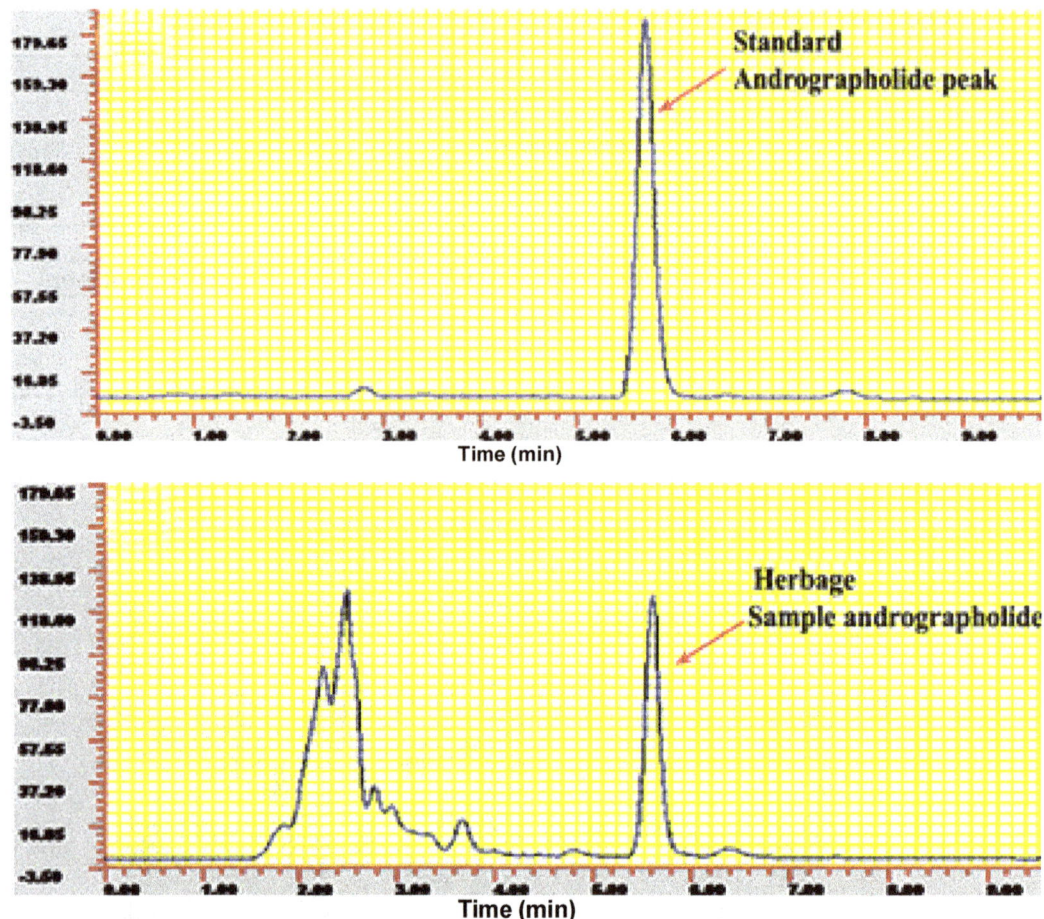

Figure 2. HPLC graph showing the concentration of andrographolide in herbage.

locations. The HPLC analysis was done for the samples for estimating the andrographolide content. The Shimadzu HPLC system used for the estimation of andrographolide consists of LC-10AD *VP* pump, Rheodyne sample injector, SPD 10A UV-VIS detector along with Aimil chromatograph data station for data collection and analysis. With the help of this system, selected phytochemicals (standardization, screening and quantification) were carried out at 229 nm wavelength (Chauhan et al., 2000) (Figure 2). The experiment was conducted to investigate whether agro-climate has any influence on the accessions for the andrographolide accumulation. The data obtained from the three locations on the three accessions were analyzed statistically using MSTATC software in the analysis of variance (two factors randomized complete block design with split plot combined over locations) as given by Gomez and Gomez (1976) and the analysis of variance showed differential influences of the locations and accessions on the andrographolide content.

Since the accessions studied in the present investigation were devoid of distinct phenotypic characters, molecular characterization through RAPD and ISSR were carried out for establishing any differences among the accessions at genetic level. For the RAPD analysis, polymerase chain reaction (PCR) was performed based on the protocol of Williams et al. (1990) with minor modifications. Amplification reactions were performed with 2.5 µl of 10X PCR buffer (Bangalore Genei, dNTPs (Fermentas, USA), 5 pmole of the

primer, 1U of Taq DNA polymerase (Bangalore Genei, India), 30 ng of genomic DNA. DNA amplification was performed in a thermal cycler (EppendorfAG, Hamberg, Germany) programmed for 43 cycles. ISSR analysis was performed using ISSR primers obtained from BangloreGenei, India. Polymerase chain reaction (PCR) was performed based on the protocol of Zietwiecki et al. (1994) with some modifications.

RESULTS AND DISCUSSION

Differentiation of the three accessions at genetic level

Since the accessions studied in the present investigation were morphologically indistinguishable molecular characterization through RAPD and ISSR were carried out for establishing differences among the accessions at genetic level (Figures 3 and 4). Molecular analysis using Jaccard's coefficient, cophenetic correlation, principal coordinate analysis (PCA) and dendrogram (Figures 5 and 6) of the three accessions also clearly showed the differences and relationship amongst them. Molecular

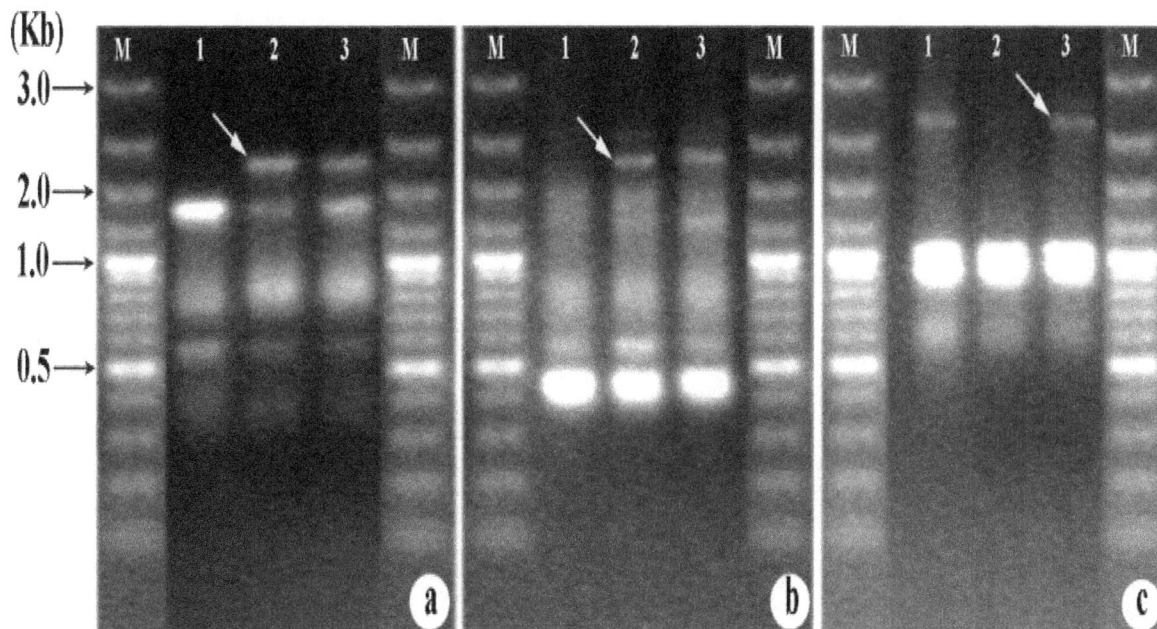

Figure 3. RAPD banding pattern of *A. paniculata* with a) OPA08, B) OPA18 and c) OPN02 [M-100 bp DNA ladder plus, 1- accession 1, 2- accession 2 and 3- accession 3]. *Arrow indicate variation in banding pattern in the three accessions.

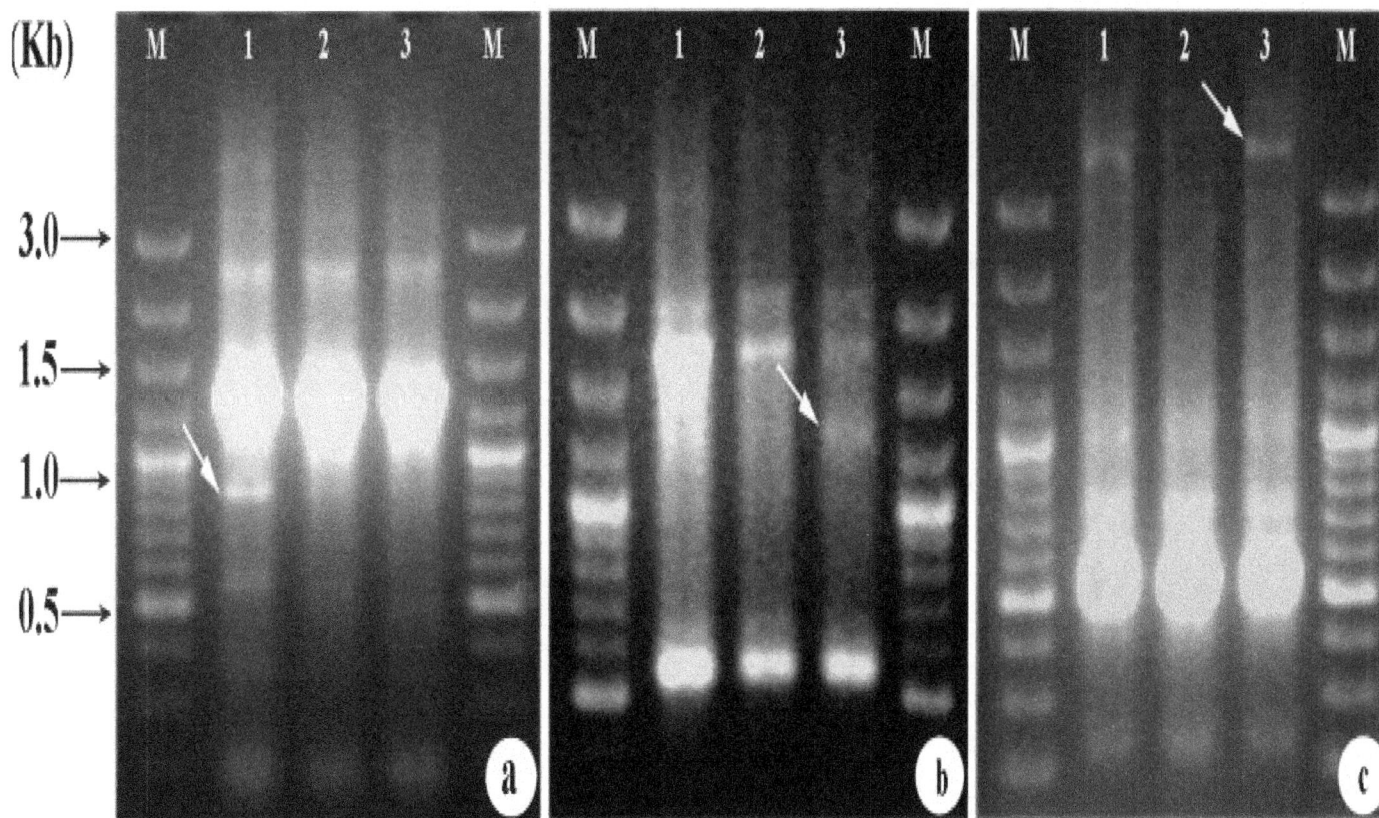

Figure 4. ISSR banding pattern of *A. paniculata* with a) (CT) 9G, B) b) (CT) 8RC and c) (CA) 8AT [M-100 bp DNA ladder plus, 1- accession 1, 2- accession 2 and 3- accession 3]. *Arrow indicates variation in banding pattern in the three accessions.

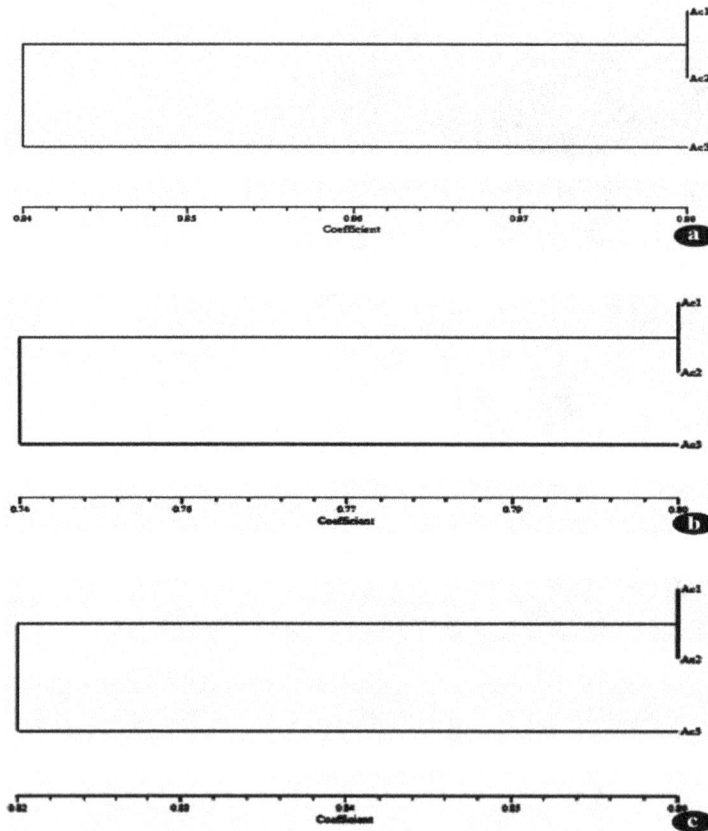

Figure 5. Dendrogram (a) RAPD, (b) ISSR and (c) combined RAPD-ISSR.

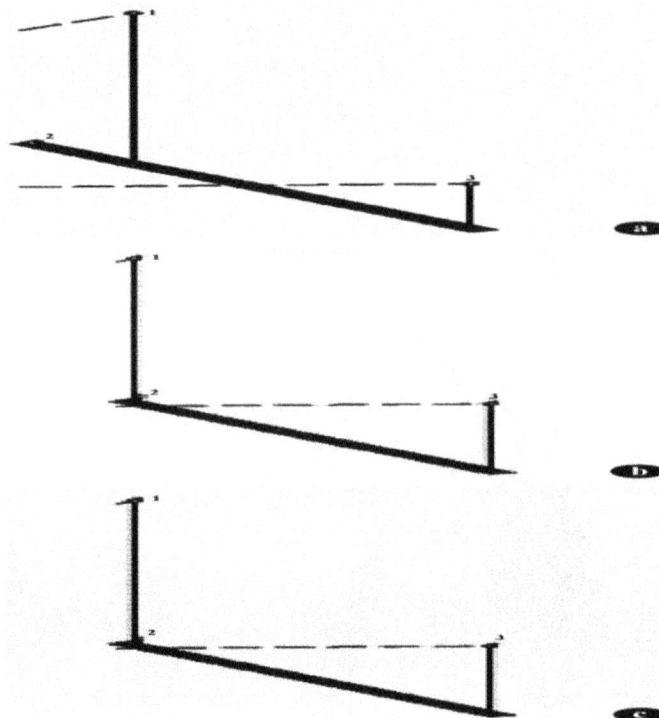

Figure 6. PCA (a), RAPD, (b) ISSR and (c) combined RAPD-ISSR.

Table 2. Effect of locations and accessions and their interaction on herbage andrographolide content (%) at 120 DAT.

Accessions	Locations			Mean
	L₁ (Anand)	L₂ (Kalyani)	L₃ (Solan)	
1	0.909 (1.174)	1.777 (1.483)	0.592 (1.033)	1.09 (1.23)
2	1.492 (1.388)	1.463 (1.384)	0.308 (0.895)	1.09 (1.22)
3	1.261 (1.313)	1.659 (1.446)	0.605 (1.036)	1.17 (1.27)
Mean	1.221 (1.292)	1.633 (1.438)	0.502 (0.988)	

Lsd$_{Ac}$ (p = 0.05): 0.034; CV$_{Ac}$ (%): 5.71; Lsd$_L$ (p = 0.05): 0.047; CV$_L$ (%): 7.66; Lsd$_{AcxL}$ (p = 0.05): 0.058; CV$_{AcxL}$ (%): 5.71. Values in parentheses denote transformed values.

Table 3. Interaction between different locations and stage of plant growth on herbage andrographolide content (%) andrographolide content (%).

Stages of plant age	Locations			Mean
	L₁ (Anand)	L₂ (Kalyani)	L₃ (Solan)	
S1 (60 DAT)	0.62 (1.05)	**2.33 (1.68)**	0.62 (1.06)	1.19 (1.26)
S2 (80 DAT)	1.97 (1.57)	2.20 (1.64)	0.80 (1.13)	1.66 (1.44)
S3 (100 DAT)	1.35 (1.36)	1.31 (1.34)	0.40 (0.95)	1.02 (1.21)
S4 (120 DAT)	0.94 (1.19)	0.69 (1.09)	0.18 (0.82)	0.60 (1.03)

Lsd$_{SxL}$ (p = 0.05): 0.067; CV$_{SxL}$(%): 7.66; Lsd$_S$(p = 0.05): 0.054; CV$_S$ (%): 7.66. *Figures in parentheses denote transformed values.

markers have demonstrated its usefulness to find out genetic similarities and differences between accessions even when a classical morphological description is severely limited. Despite the importance of the crop, very little research has been done to assess the genetic variation of this species using molecular markers (Padmesh et al., 1999).

Effect of locations, stages of plant and accessions on andrographolide content

The samples were collected at different growth stages, where the vegetative stage is found up to 60 to 65 DAT and after that the reproductive stage starts. The study showed that the locations significantly influenced the andrographolide accumulation in the herbage irrespective of accessions or age of the plant. Maximum andrographolide content was recorded in Kalyani (1.63%) with moderate mean temperature (< 28°C) followed by Anand (1.221%) which has dry climate with high mean temperature(> 30°C) and content was lowest at Solan (0.502%) a high altitude (1350 m above msl) which is having temperate climate (20 to 25°C) (Table 2). Chatterjee and Raychaudhuri (1992) expressed that although the biosynthesis of secondary metabolites is genetically controlled, there is a considerable environmental effect on their accumulation. Altitude, photoperiod, developmental stage, NPK application,

mineral nutrients and physiological and biochemical factors greatly influence the accumulation of these chemicals. Significant influence of accessions was also noted on andrographolide content irrespective of the locations and sampling times. Andrographolide content was highest in accession 3 (1.17%) which was followed by accession 1 (1.09%) which was at par with that of accession 2 (1.09%) (Table 2). Stages of plant age also influenced the andrographolide content irrespective of the locations and accessions. Significantly highest andrographolide content was recorded at plant age 80 DAT (1.66%), thereafter it was reduced to 1.02% at plant age 100 DAT and 0.60% at plant age 120 DAT (Table 3).

Interaction of locations and accessions significantly influenced the herbage andrographolide content irrespective of stages of plant growth. Maximum andrographolide content was recorded in accession 1 at location 2 (1.777%) which was at par with accession 3 at location 2 (1.659%). However, the minimum andrographolide content was observed in accession 2 at location 3 (0.308%) (Table 2). Location and different stages of plant age had significant influence on the herbage andrographolide content irrespective of the accessions. Maximum content of andrographolide was at location 2 at plant age 60 DAT (2.33%) which was at par with location 2 at plant age 80 DAT (2.20%). The minimum herbage andrographolide content (%) was observed in location 3 sampled at plant age 120 DAT (0.18%) (Table 3). The herbage andrographolide content

Table 4. Interaction between accessions and stages of plant growth on herbage andrographolide content (%).

Stages of plant age	Accessions		
	1	2	3
S1 (60 DAT)	1.18 (1.26)	1.03 (1.20)	1.36 (1.33)
S2 (80 DAT)	1.73 (1.47)	1.59 (1.40)	1.66 (1.46)
S3 (100 DAT)	0.99 (1.21)	0.98 (1.19)	1.10 (1.25)
S4 (120 DAT)	0.47 (0.98)	0.76 (1.10)	0.58 (1.02)

Lsd$_{AcxS}$ (p = 0.05): 0.067; CV (%)$_{AcxS}$: 5.71. *Figures in parentheses denote transformed values.

Table 5. Interaction of accessions, locations and sampling dates on herbage andrographolide content (%).

Stages of plant age	L$_1$ (Anand)			L$_2$ (Kalyani)			L$_3$ (Solan)		
	Accessions 1	Accession s2	Accession s3	Accessions 1	Accessions 2	Accession s3	Accession s1	Accessions 2	Accessions 3
S1 (60 DAT)	0.51 (1.00)	0.58 (1.04)	0.76 (1.12)	2.33 (1.68)	2.07 (1.60)	2.59 (1.76)	0.71 (1.10)	0.44 (0.97)	0.73 (1.11)
S2 (80 DAT)	1.59 (1.44)	2.49 (1.73)	1.84 (1.53)	2.58 (1.75)	1.91 (1.55)	2.11 (1.61)	1.01 (1.23)	0.36 (0.92)	1.04 (1.24)
S3 (100 DAT)	0.94 (1.20)	1.54 (1.43)	1.59 (1.44)	1.58 (1.44)	1.19 (1.30)	1.17 (1.29)	0.46 (0.98)	0.21 (0.84)	0.54 (1.02)
S4 (120 DAT)	0.61 (1.05)	1.36 (1.36)	0.85 (1.16)	0.62 (1.06)	0.69 (1.09)	0.77 (1.13)	0.19 (0.83)	0.22 (0.85)	0.11 (0.78)

Lsd$_{AcxLxs}$ (p = 0.05): 0.116; CV (%)$_{AcxLxs}$: 5.71. *Figures in parentheses denote transformed values.

was significantly influenced by the interaction of stages of plant growth and the accessions. Maximum herbage andrographolide content was found in accession 1 at plant age 80 DAT (1.73%) which was at par with the content of andrographolide in accession 3 at plant age 80 DAT (1.66%). The minimum herbage andrographolide content (%) was observed in accession 1 at plant age 120 DAT (0.47%) and was at par with the andrographolide content of accession 3 at plant age 120 DAT (0.58%) (Table 4). The locations, accessions and stages of plant growth had significant interaction effect in

influencing the andrographolide content of herbage. Maximum herbage andrographolide content was recorded at location 2, in accession 3 at plant age 60 DAT (2.59%) which was at par with accession 1 at plant age 80 DAT at the same location (2.58%), location 1 in accession 2 at plant age 80 DAT (2.49%) and location 2 in accession 1 at plant age 60 DAT (2.33%). Minimum herbage andrographolide content was noted at location 3, in accession 3 at plant age 120 DAT (0.11%) (Table 5) (Figure 7).

Accumulation pattern of andrographolide content in three accessions in three distinctly

diverse locations clearly showed that location 2 (Kalyani) was superior to the other two locations that is, Anand and Solan. Location 3 (Solan) has temperate climate (<25°C), low relative humidity (~60%) and high altitude (1600.0 m) and is not suitable for this crop since this species is a native of subtropical region and well adapted to areas of high rainfall (1000 to 2500 mm) and high humidity (> 70%) with moderate temperature (24 to 37°C). Andrographolide content was 3 times higher in Kalyani as compared to Solan. Similarly, it did not perform better in Anand because Anand is a dry area (mean RH ~65%) having moderate rainfall

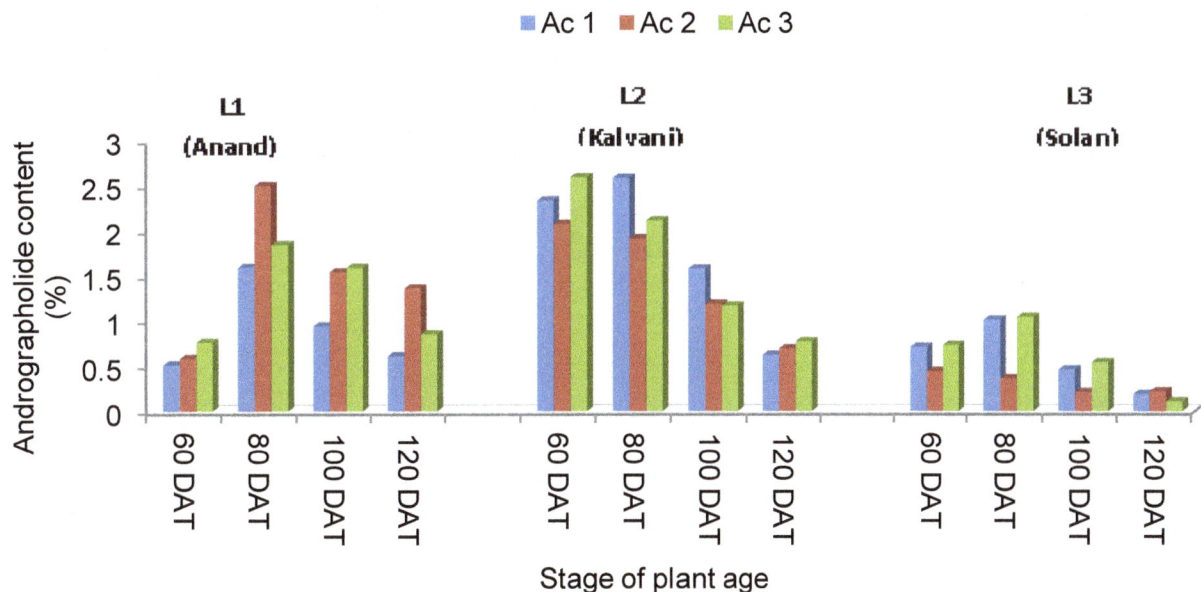

Figure 7. Accumulation pattern of herbage andrographolide content in three accessions across locations.

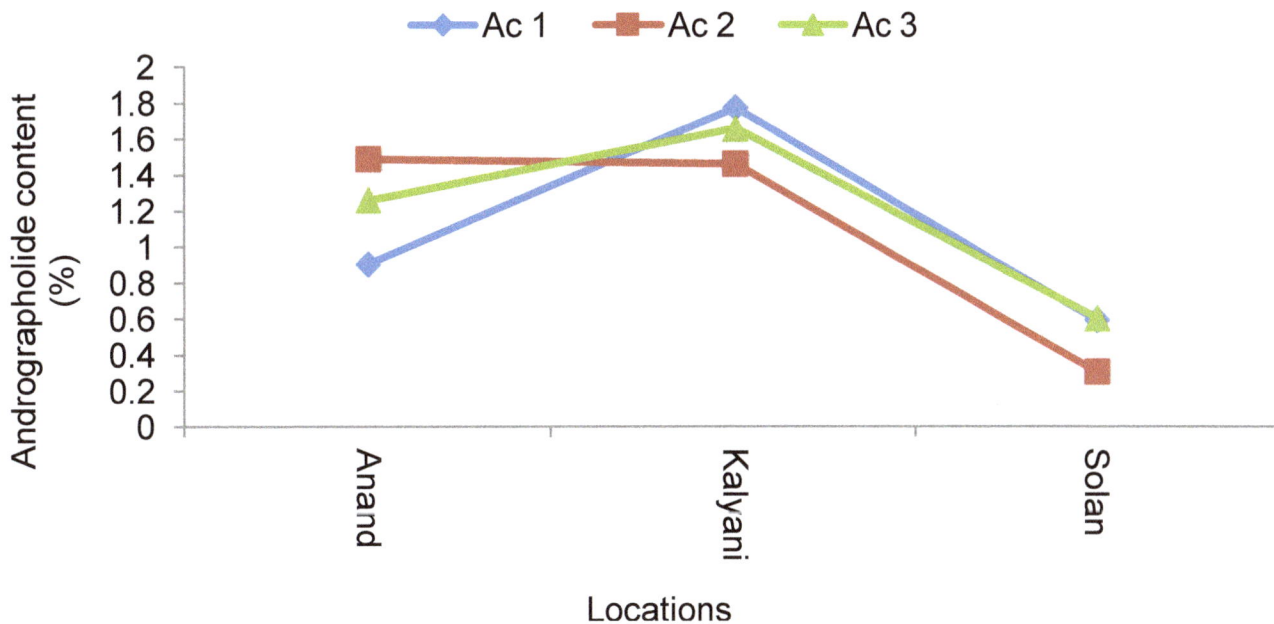

Figure 8. Influence of locations on andrographolide accumulation in three accessions of *A. Paniculata*.

(700 to 800 mm) with high day temperature (> 38.5°C). Performances of all the accessions were superior in Kalyani, although accession 1 was the best because of its adaptability to the weather conditions of Kalyani (average temperature < 28°C, annual rainfall - 2030 mm, mean humidity -84%) which is similar to the place from where the accessions was collected that is, Trichy in Tamil Nadu (Table 1) (Figure 8).

REFERENCES

Chatterjee SK, Raychaudhuri SP (1992). Cultivation and quality improvement of medicinal plants in West Bengal. In Recent advances in medicinal aromatic and spice crops, 2:397-407:New Delhi, India.

Chauhan SK, Singh B, Agrawal S, Singh B, Agrawal S (2000). A HPLC determination of andrographolide in*Andrographispaniculata*. Indian J. Nat. Prod. 16(2):10-13.

Gomez KA, Gomez TA (1976). Statistical Procedures for Agricultural Research. John Wiley & Sons Publication. Singapore.

Padmesh P, Sabu KK, Seeni S, Pushpangadan P (1999). The use of RAPD in assessing genetic variability in *Andrographis paniculata* Nees, a hepatoprotectivedrug. Curr. Sci. 76:833-835.

Rajeswara RBR, Rajput DK, Sastry KP, Kothari SK, Singh K, Singh CP, Bhattacharya AK (2004). Integrated nutrient management in king of bitters.Proceedings of National Symposium on Organic Farming-Prospects and Challenges in the New Millennium. Society of Agronomists, Acharya N.G. Ranga Agricultural University, Hyderabad, pp. 149-152.

Teiz L, Zeiger E (1998). Plant defences: surface protectants and secondary metabolites.In: Plant Physiology.Sinauer Associates Inc. Publishers, Sundland, Massachusetts. 2nd ed. pp. 671-757.

Waller GR, Nowacki EK (1978). Alkaloid Biology and Metabolism in Plants. Plenum Press, New York.

Williams JGK, Kubelic AR, Livak KJ, Rafalsky JA, Tingey SV (1990). DNA polymorphisms amplified by arbitrary primer are useful as genetic markers. Nucleic Acid Res.18:6531-6535.

Zietwiecki E, Rafalski A, Labuda D (1994). Genome fingerprinting by simple sequence repeat (SSR)–anchored polymerase chain reaction amplification. Genomics 20:176-183.

22

Genetic variability, heritability and genetic advance for yield and its components snake gourd (*Trichosanthes anguina* L.)

N. Deepa Devi and S. Mariappan

Department of Horticulture, Agricultural College and Research Institute, Tamil Nadu Agricultural University, Madurai- 625104, India.

Genetic variability and heritability were studied in 50 genotypes of snake gourd (*Trichosanthes anguina* L.) to determine the magnitude of variability in the population and to identify genotypically diverse and economically desirable genotypes for utilization in crop improvement. The phenotypic coefficient of variation was found to be slightly higher than the genotypic coefficient of variation for all the characters under consideration, indicating that the apparent variation is not only genetic, but also due to influence of the growing environment in expression of the genotypes. The estimates of genotypic variance also showed considerable variation for majority of the characters. Higher values of phenotypic coefficient of variation and genotypic coefficient of variation were recorded for fruit length, fruit weight and number of fruits per plant while, moderate phenotypic and genotypic coefficient of variation were observed for yield, fruit girth and number of seeds per fruit indicating the extent of available genetic variability for these traits. High heritability along with high genetic advance observed for fruit length, fruit yield, fruit girth and number of fruits per plant is indicative of additive gene action in control of these traits and phenotypic selection based on these traits in the segregating population is likely to yield desired individuals.

Key words: Snake gourd, genotypic coefficient of variation (GCV), phenotypic coefficient of variation (PCV), heritability, genetic advance.

INTRODUCTION

Snake gourd (*Trichosanthes anguina* L.) belonging to the family Cucurbitaceae is an important summer vegetable rich in minerals and fiber making the food wholesome and nutritious (Ahmed et al., 2000). It is also considered to be of medicinal importance and is one of the few vegetables capable of giving more yield per unit area. However the average productivity of the crop is low as large numbers of local types are in cultivation. No serious attempts have so far been made to upgrade the productivity of this development of superior high yielding genotypes depends on the improvement of yield

components in snake gourd. Yield being complex character and associated with many other contributing traits, which are simply inherited (Rao et al., 1990). The assessment of variability present in any crop species is the essential pre-requite for formulating an effective breeding programme as the existing variability can be used to enhance the yield level of the cultivars following appropriate breeding strategies (Patil et al., 2012). The information on heritability alone may not help in identifying characters for enforcing selection and heritability estimates in conjunction with predicted genetic

Table 1. Estimates of genetic parameters for some economic characters in snake gourd.

Characters	GCV (%)	PCV (%)	Heritability (%)	GA (% of mean)
Vine length	15.40	15.41	99.82	31.70
Internode length	12.64	13.38	89.19	24.58
First female flower appear on node	16.47	18.87	76.18	29.62
Days to first female flower appearance	5.08	5.82	76.18	9.14
Days to first male flower appearance	6.89	10.20	45.64	9.59
Number of seeds per fruit	22.77	22.77	99.97	46.89
Fruit length	47.15	47.15	99.99	97.13
Fruit girth	24.66	24.67	99.94	50.78
Fruit weight	46.23	121.59	14.46	36.21
Number of fruits per plant	31.81	31.86	99.69	65.42
Yield	29.46	29.46	99.99	60.69

advance is more reliable (Johnson et al., 1955). Heritability gives the information on the magnitude of inheritance of characters from parent to off spring, while genetic advance will be helpful in finding the actual gain expected under selection. The present investigation thus aims at to assessing the variability by estimating phenotypic and genotypic coefficient of variation as well as heritability and genetic advance in snake gourd for future breeding programme.

MATERIALS AND METHODS

The present investigation was carried out at the department of Horticulture, Agricultural College and Research Institute, Madurai during 2011 to 2012. Totally, 50 genotypes were assembled from different geographical location and utilized for the study. The 50 genotypes comprised of 40 genotypes from NBPGR, New Delhi, three varieties from Tamil Nadu Agricultural University viz., PKM1, MDU1 and Co 2 and seven local types from the following locations in Tamil Nadu viz., Kulithalai, Kumbakonam, Palayajeyamkondam, Nagappattinam, Jeyamkondam, Madurai and Coimbatore. The experiment was laid out in a Randomized Block Design with three replications. The seeds were sown at a spacing of 2 × 2 m. The plants were supported by trellis and other intercultural operations such as weeding, irrigation and plant protection measures were performed as and when needed. Observations on yield and ten yield contributing characters viz., vine length, inter nodal length, days to first female flower appearance, days to first male flower appearance, fruit length, fruit girth, fruit weight, number of seeds per fruit, number of fruit per plant and fruit yield were recorded. Standard statistical procedures were used for the analysis of variance, coefficient of variation (Burton and De Vane, 1952) and heritability (Lush, 1940).

RESULT AND DISCUSSION

Genetic variability

The estimates on genotypic coefficient of variation, phenotypic coefficient of variation, heritability and genetic advance as percent of mean for the traits under study are

furnished in Table 1. The phenotypic coefficient of variation was found higher than the genotypic coefficient of variation for most traits studied. The genotypic coefficients of variation obtained for various yield, yield attributing ranged from 5.08 to 47.15. The highest genotypic coefficient of variation was observed for fruit length (47.15) followed by fruit weight (46.23) and number of fruits per plant (31.81), while the genotypic coefficients of variation observed for yield (29.46), fruit girth (24.66) and number of seeds per fruit (22.77) were moderate. The genotypic coefficient of variation was low for days to first male flower appearance (6.89) and days to first female flower appearance (5.08). The phenotypic coefficient of variation observed was high for fruit weight (121.59) followed by fruit length (47.15) and number of fruits per plant (31.86), and moderate phenotypic coefficients of variation was observed for yield (29.46), fruit girth (24.66) and number of seeds per fruit (22.77). The phenotypic coefficient of variation was low for days to first male flower appearance (10.20) and days to first female flower appearance (5.82).

Study of genotypic and phenotypic coefficient of variation indicated the extent of variability for different traits in snake gourd and those results are in conformity to the findings of Rahman et al. (2002). Higher phenotypic and genotypic coefficient of variation recorded for fruit length, fruit weight and number of fruits per plant indicates that, these genotypes exhibit much variation among themselves with respect to these characters offering more scope for selection. Phenotypic coefficient of variation (PCV) and genotypic coefficient of variation (GCV) recorded for days to first male flower appearance and days to first female flower appearance were low and was in agreement with the findings of Miah et al. (2000) and Rahman et al. (2002).

Heritability and genetic advance

High heritability coupled with low genetic advance, low

heritability with high genetic advance or low heritability and low genetic advance offers less scope for selection, as they indicates the role of non additive genetic effects. High heritability coupled with high genetic advance is indicative of greater proportion of additive genetic variance and consequence a high genetic gain expected from selection (Singh and Rai, 1981). The characters having heritability with low genetic advances as percent of mean appeared to be controlled by non-additive gene action and selection for such characters may not be effective (Singh and Singh, 2007).

The genotypes recorded high heritability values for all the characters under study. Fruit length (99.99%), fruit yield (99.99%) fruit girth (99.94%), number of fruits per plant (99.69%) and vine length (99.82%) had recorded high heritability value. Genetic advance as percent of mean ranged from 9.13% for days to first female flower appearance to 97.14% for fruit length. High genetic advance was also recorded for number of fruits per plant (65.42%), fruit yield (60.69%), fruit girth (50.78%) and number of seeds per fruit (46.89%). While moderate genetic advance was recorded for vine length (31.69%), fruit weight (29.61%) and internodal length (24.58%). High genetic advance indicated that, additive genes govern these characters and selection will be rewarding for improvement of these traits. The above finding supports the results of Rahman et al. (2002). The genetic advance recorded was low for days to first female flower appearance (9.13%) and days to first male flower appearance (9.59%).

Conclusion

The genetic improvement in snake gourd is possible through selection exercised for fruit length, fruit weight, number of fruits per plant, fruit yield and number of seeds per fruit which, showed high values of phenotypic coefficient of variation (PCV) and genotypic coefficient of variation (GCV) coupled with high heritability and genetic advance. This will provide an opportunity to select better recombinants for various characters and thereby creating large variability for these characters in the future generations. However, characters predominantly controlled by additive gene action would be amenable for to conventional breeding methods.

REFERENCES

Ahmed MS, Rasul MG, Bashar MK, Mian ASM (2000). Variability and heterosis in snake gourd (Trichosanthes anguina L.). Bangladesh J. Plant Breed. Genet. 13:27-32.

Burton GW, De Vane (1952). Quantitative inheritance in grasses. Proc. 6th Int. Grassland Congress 1:277-283.

Johnson HW, Robinson HF, Comstock RE (1955). Estimation of Genetic variability and environmental variability in soybean. J. Agron. 47:314-318.

Lush JL (1940). Intra – sire correlation and regression of offspring on dams as a methods of estimating heritability of characters. Proc. Am. Soc. Anim. Produces 33:293-301.

Miah MA, Rahman MM, Uddin MS, Rahman AKMM, Ullah MH (2000). Genetic association in bitter gourd (Momordica charantia L.). Bangla. J. Sci. Techol. 2:21-25.

Patil PR, Surve VH, Mehta HD (2012). Line x Tester analysis in Rice (Oryza sativa L.). Madras Agric. J. 99:210-213.

Rahman MA, Hossain MD, Islam MS, Biswas DK, Ahiduzzaman M, (2002). Genetic variability, heritability and path analysis in snake gourd (Trichosanthes anguina L.). Pak. J. Biol. Sci. 5(3):284-286.

Rao DSRM, Singh H, Singh B, Khola OPS, Faroda AS (1990). Correlation and path coefficient analysis of seed yield and its components in sesame (Sesamum indicum L.). Haryana Agric. Univ. J. Res. 20:25-30.

Singh AK, Singh N (2007). Studies on genetic variability and heritability in balsam (Impaties balsamina). J. Ornamental Hort. 10:128-130.

Singh RP, Rai JN (1981). Note on the heritability and genetic advance in chilli. (Capsicum annuum L.). Prog. Hortic. 13(1):89-92.

23

Peroxidase isozyme characterization of elite genotypes of Pearl millet (*Pennisetum glaucum* (L.) R. Br)

Manivannan, A.[1,2,3], **Somveer Nimbal**[3] and **Chhabra, A. K.**[3]

[1]Department of Plant Breeding and Genetics, AC&RI, Madurai, -625104, India.
[2]Directorate of Maize Research, Pusa campus, New Delhi, -110012, India.
[3]Department of Plant Breeding and Genetics, College of Agriculture, CCS HAU, Hisar, -125004, India.

Isozyme markers are the oldest among molecular markers. Isozyme markers have been successfully used in several crop improvement programmes. Peroxidase (POX) isozyme has proven to be reliable genetic marker in breeding and genetic studies of Pearl millet. The present study was conducted to characterize 21 genotypes (7 hybrids, 6 female parents, 5 male parents and 3 open pollinated varieties (OPV) of Pearl millet by using POX isozyme. Five bands were found with Rm/Rf value ranges from 0.5 to 0.64. All bands were found to be polymorphic in nature except band with Rm value of 0.55 which was present in all genotypes. Intensity of bands varied with each genotype. Only 3 genotypes (H 77/29-2, HMS 7A and HHB 94) out of 21 genotypes were differentiated from other genotypes. Similarity indices based on POX banding pattern revealed that hybrid HHB 50, HHB 60, HHB 67 and HHB 146 completely resembled (SI 1.000) with their female parents MS 81A, MS 843A and ICMA 95222A. This showed the maximum contribution of female parent in comparison with male parent towards the development of hybrid.

Key words: Peroxidase, characterization, banding pattern.

INTRODUCTION

Pearl millet (*Pennisetum glauccum* (L.) R. Br.) provides stable food for millions of people of African countries and Indian sub-continent. It is the sixth important cereal, primarily gown for grain and fodder production. Pearl millet growing in environments in these areas are characterized by low and erratic rainfall, high temperature and poor soil fertility. In these environments, Pearl millet is the only successful cereal and a major source of energy for the poor farming community. With ability to adopt diverse agro-ecological conditions, it plays a unique position in world agriculture. Pearl millet is a summer annual grass originating from Africa, from where it was introduced into other regions of the world with diverse agro-climatic conditions, that is, from the hot area of Africa to the hot area of temperate zones. Therefore, a

large number of diversity is found within and among pearl millet cultivars. Due to its highly out-crossing breeding behaviour, Pearl millet was originated from several independent domestication events and wide range of stressful environmental conditions, in which it had been traditionally cultivated. Pearl millet exhibits a tremendous amount of diversity at both phenotypic and genotypic levels (Poncet et al., 1998; Liu et al., 1994).

Estimation of genetic diversity and identification of superior genotypes are some of the prime objectives of any crop improvement programmes. Highly diverse genotypes or accessions can be utilized as parents in hybridization programmes to produce superior varieties/hybrids. Therefore, there is a need to evaluate available genotypes for their genetic diversity. In the early days, crop breeders used morphological markers for the assessment of genetic diversity and choosing parents for developing new cultivars. Morphological markers data are affected by the interaction of the genotype with the environment in which it is expressed. Moreover, due to the high out-crossing breeding nature and structure of genetic diversity in pearl millet species, the morphological data/markers are inadequate in providing reliable information for the calculation of genetic distance and pedigree studies.

Isozyme markers are the oldest among the molecular markers. Isozyme markers have been successfully used in several crop improvement programmes (Glaszmann et al., 1989; Baes and Custsem, 1993). Isozymes have proven to be reliable genetic markers in breeding and genetic studies of plant species (Heinz, 1987), due to consistency in their expression, irrespective of environmental factors. Isozymes provide useful evidences in the study of variation between cultivars in terms of intensity of common bands and presence or absences of other bands (William and Mujeeb, 1992). Study of the isozyme pattern is considered as an important tool for understanding the genetic relationship between individuals and also for the identification of hybrids. Genetically, the production of isozyme of multiple forms or molecular weight is accounted to the allelic variation of the organism. Therefore, isozymes of a particular molecular weight can be considered as a direct manifestation of the blue print of the specific gene loci (Abiden and Vijayakumar, 2002). The utility of isozymes as genetic marker (Cheniany, 2007) is generally attributed to their polymorphism, codominence, simple inheritance, simple assay and obliquity in plant tissues or organs (Simpson and Withers, 1986). Moreover, isozymes study may be useful to diversity analysis in plants (Philomina and Surendran, 2003). Peroxidase (POX) isozyme has been widely used for characterization of plant germplasm (Li and Li, 1996; Ju-Zheng et al., 1997; Gupta et al., 2008). An attempt was made to study the isozyme diversity in some of elite Pearl millet varieties, hybrids and their parental lines.

MATERIALS AND METHODS

Seven (7) hybrids viz., HHB 50[H1] (MS 81A × H 90/4-5), HHB 60[H2] (MS 81A × H 77/833-2), HHB 67[H3] (MS 843A × H 77/833-2), HHB 68[H4] (MS 842A × H 77/833-2), HHB 94[H5] (ICMA 89111A × G 73-107), HHB 117[H6] (HMS 7A × H 77/29-2) and HHB 146[H7] (ICMA 95222A × HTP 94/54); six (6) male sterile lines viz., MS 81A[F1/F2], MS 843A[F3], MS 842A[F4], ICMA 89111A[F5], HMS 7A[F6] and ICMA 95222A[F7]; five (5) restorer lines viz., H 90/4-5[M1], H 77/833-2[M2/M3/M4], G 73-107[M5], H 77/29-2[M6] and HTP 94/54[M7] and three (3) open pollinated varieties (OPV) viz., HC 4[OPV1], HC 10[OPV2] and HC 20[OPV3]. These 21 genotypes comprised the experimental materials for the present study (Table 1).

To record the electrophoregram of POX, the method followed was that of Mitra et al. (1970). The POX was displayed as brown bands. Based on polyacrylamide gel, bands were scored as present

{1} and absent {0} in data sheet to form a {1, 0} matrix. Then data were analyzed and similarity matrix was constructed from binary data with Jaccard's coefficient (Jaccard, 1908) and dendrogram were generated with Unweighted Pair Group Method Arithmetic Average (UPGMA) algorithm using NTSYSPC - version 2.01 software (Rohlf, 2000).

RESULTS AND DISCUSSION

Five bands were found with Rm/Rf value ranges from 0.5 to 0.64 (Table 2). All bands were found to be polymorphic in nature except band with Rm value of 0.55 which was present in all genotypes. Intensity of bands varied with each genotype (Table 3). Band of Rm value 0.5 showed dark band for MS 842A(F4), HHB68(H4) and HHB94(H5), medium intensity for HHB 60(H2), very light for MS81A(F1/F2), HMS 7A(F6) and HHB 50(H1). Band at Rm value 0.52 showed medium intensity for H 90/4-5(M1), MS 843(F3), ICMA 95222A(F7), HHB 67(H3), HHB 146(H7), dark for H 77/833-2 (M2/M3/M4), G 73/107(M5), H 77/29-2 (M6), MS 81A(F1/F2), MS 842 A(F4), ICMA 89111A(F5), HHB 50(H1), HHB 68(H4), HHB 117 (H6) and light for HTP 94/54(M7) and HHB 60(H2). Band at Rm value of 0.55 showed dark intensity for G 73-107(M5), H 77/29-2(M6), MS81A(F1/F2), ICMA 89111A(F5), HMS 7A(F6), ICMA 95222A(F7), HHB 50(H1), HHB 94(H5), HHB 117(H6), medium intensity for H 90/4-5(M1), MS 842A(F4), HHB 68(H4), and light intensity for H77/833-2 (M2/M3/M4), ICMA 95222A(F7), MS843A(F3), HHB 60(H2), HHB 67(H3). Band at Rm value of 0.61 showed dark for ICMA 89111A(F5), ICMA 95222A(F7), HHB94(H5), HHB146(H7), medium intensity for H77/833-2(M2/M3/M4), G73-107(M5), HHB68 (H4), HHB117(H6), light for HTP94/54(M7), MS81A(F1/F2), MS843A(F3), HHB50(H1), HHB60(H2), HHB67(H3) and very light intensity for H90/4-5(M1). Band at Rm value of 0.64 showed dark intensity for HMS 7A(F6), HHB 117(H6), medium intensity for ICMA 89111A(F5), HHB 94(H5), light intensity for H77/833-2(M2/M3/M4), G73-107(M5), H77/29-2(M6), HTP 94/54(M7), MS 81A(F1/F2), MS 843A(F3), ICMA 95222A(F7), HHB 50(H1), HHB 60(H2), HHB 67(H3) and very light intensity for H 90/4-5(M1)and HHB 146(H7). Out of 21 genotypes, only 3 genotypes [H 77/29-2(M6), HMS 7A(F6) and HHB 94(H5)] were differentiated from other genotypes. H77/29-2(M6) showed the presence of band with Rm value of 0.52 but absence of other polymorphic bands (Rm = 0.50, 0.61). HMS 7A(F6) had shown only one of these three bands (Rm = 0.50), whereas a band of Rm = 0.52 value was absent in HHB 94 and having other two bands (Rm = 0.50 and 0.61). Based on zymogram, HHB 60(H2) showed light band at RM value of 0.5. H90/4-5(M1) showed very light band at Rm value of 0.61. Band of 0.52 RM value showed light band for HTP 94/54(M7) and HHB 60(H2). Very light band was observed at RM value of 0.64 for H 90/4-5(M1) and HHB 146(H7).

Table 1. List of Pearl millet genotypes and their pedigree.

S/N	Genotype	Status	Pedigree	Year of release	Origin
1	HHB 50	Hybrid	MS 81A × H 90/4-5	1987	CCS HAU, Hisar
2	HHB 60	Hybrid	MS 81A × H 77/833-2	1988	CCS HAU, Hisar
3	HHB 67	Hybrid	MS 843A × H 77/833-2	1990	CCS HAU, Hisar
4	HHB 68	Hybrid	MS 842A × H 77/833-2	1993	CCS HAU, Hisar
5	HHB 94	Hybrid	ICMA 89111A × G 73-107	1999	CCS HAU, Hisar
6	HHB 117	Hybrid	HMS 7A × H 77/29-2	2002	CCS HAU, Hisar
7	HHB 146	Hybrid	ICMA 95222A × HTP 94/54	2002	CCS HAU, Hisar
8	MS 81A	CMS	Derived from Tift 23D$_2$ after irradiation	1981	ICRISAT, Hyderabad
9	MS 843A	CMS	Selected from AKM 2068 for Downy mildew resistance	1984	ICRISAT, Hyderabad
10	MS 842A	CMS	Re Selected from AKM 2068 for Downy mildew resistance	1984	ICRISAT, Hyderabad
11	ICMA 89111A	CMS	881A cytoplasm source(B$_1$) backcrossed to ICMB 89111	1989	ICRISAT, Hyderabad
12	HMS 7A	CMS	Developed by backcrossing from the cross 81 A × 35(81B × 69B)	1991	ICRISAT, Hyderabad
13	ICMA 95222A	CMS	81A cytoplasm(A$_1$) source back crossed to ICMB 95222	1995	ICRISAT, Hyderabad
14	H 90/4-5	Restorer	Developed by selecting selfed progenies form synthetic HSI	1976	CCS HAU, Hisar
15	H 77/833-2	Restorer	Developed by selfing a Haryana land race population	1976	CCS HAU, Hisar
16	G 73-107	Restorer	Developed by selecting selfed progenies of GAM 73	1976	CCS HAU, Hisar
17	H 77/29-2	Restorer	Developed by selecting selfed plants from Rajasthan landrace	1976	CCS HAU, Hisar
18	HTP 94/54	Restorer	Developed by selecting selfed progenies of high tillering of Tago population	1992	CCS HAU, Hisar
19	HC 4	OPV	Developed by intermating seven inbred lines	1985	CCS HAU, Hisar
20	HC 10	OPV	Bred by random mating 15 S1 progenies of NELC population	1999	CCS HAU, Hisar
21	HC 20	OPV	Bred by random mating S1 progenies from gene pool selected for good yield and drought stress	2000	CCS HAU, Hisar

Table 2. Banding pattern of POX isozyme in 21 genotypes of pearl millet.

Band	Rf/Rm	H 90/4-5	H 77/833-2	H 77/833-2	H 77/833-2	G 73-107	H 77/29-2	HTP 4/54	MS 81A	MS 81A	MS 843A	MS 842A	ICMA 89111A	HMS 7A	ICMA 95222A	HHB 50	HHB 60	HHB 67	HHB 68	HHB 94	HHB 117	HHB 146
		M1	M2	M3	M4	M5	M6	M7	F1	F2	F3	F4	F5	F6	F7	H1	H2	H3	H4	H5	H6	H7
1	0.5	-	-	+	+	-	+	+	+	+	-	+	+	+	-	+	+	-	+	+	-	-
2	0.52	+	+	+	+	+	+	+	+	+	+	+	+	-	+	+	+	+	+	-	+	+
3	0.55	+	+	+	+	+	+	+	+	+	+	+	+	+	+	+	+	+	+	+	+	+
4	0.61	+	+	+	+	+	-	+	+	+	+	-	+	-	+	+	+	+	+	+	+	+
5	0.64	+	+	+	+	+	+	+	+	+	+	-	+	+	+	+	+	+	+	+	+	+

+, Present of band; -, absent of band. **Note:** HHB 50[H1] (MS 81A × H 90/4-5), HHB 60[H2] (MS 81A × H 77/833-2), HHB 67[H3] (MS 843A × H 77/833-2), HHB 68[H4] (MS 842A × H 77/833-2), HHB 94[H5] (ICMA 89111A × G 73-107), HHB 117[H6] (HMS 7A × H 77/29-2) and HHB 146[H7] (ICMA 95222A × HTP 94/54); Six male sterile lines viz., MS 81A[F1/F2] MS 843A[F3] ,ICMA 89111A[F5], MS 842A[F4], HMS 7A[F6] and ICMA 95222A[F7]; Five restorer lines viz., H 90/4-5[M1], H 77/833-2[M2/M3/M4], G 73-107[M5], H 77/29-2[M6] and HTP 94/54[M7]. H1- Hybrid (HHB50), F1- MS 81A(Female parent of H1), M1- H 90/4-5 (Male parent of H1).

Table 3. Schematic zymogram of POX isozyme in 21 genotypes of pearl millet.

Band	Rf/Rm	H 90/4-5 M1	H 77/833-2 M2	H 77/833-2 M3	H 77/833-2 M4	G 73-107 M5	H 77/29-2 M6	HTP 94/54 M7	MS 81A F1	MS 81A F2	MS 843A F3	MS 842A F4	ICMA 89111A F5	HMS 7A F6	ICMA 95222A F7	HHB 50 H1	HHB 60 H2	HHB 67 H3	HHB 68 H4	HHB 94 H5	HHB 117 H6	HHB 146 H7
1	0.5	-	-	-	-	++++	-	-	+	+	-	++++	-	+	-	+	++	-	++++	++++	-	-
2	0.52	+++	+++	+++	+++	++++	++++	++	++++	++++	+++	++++	+++	-	+++	++++	+++	+++	++++	-	+++	++
3	0.55	+++	++	++	++	+++	+++	++	++	+++	++	+++	+++	+++	+++	+++	++	++	++	+++	+++	+++
4	0.61	+	+++	+++	+++	+++	-	++	++	++	++	-	+++	-	++	++	++	++	++	++	+++	+++
5	0.64	+	+++	++	++	+++	++	++	++	++	++	-	+++	++++	++	++	++	++	++	++	++++	+

+, Very light; ++, light; +++, medium; ++++, dark; -, absent of bands.

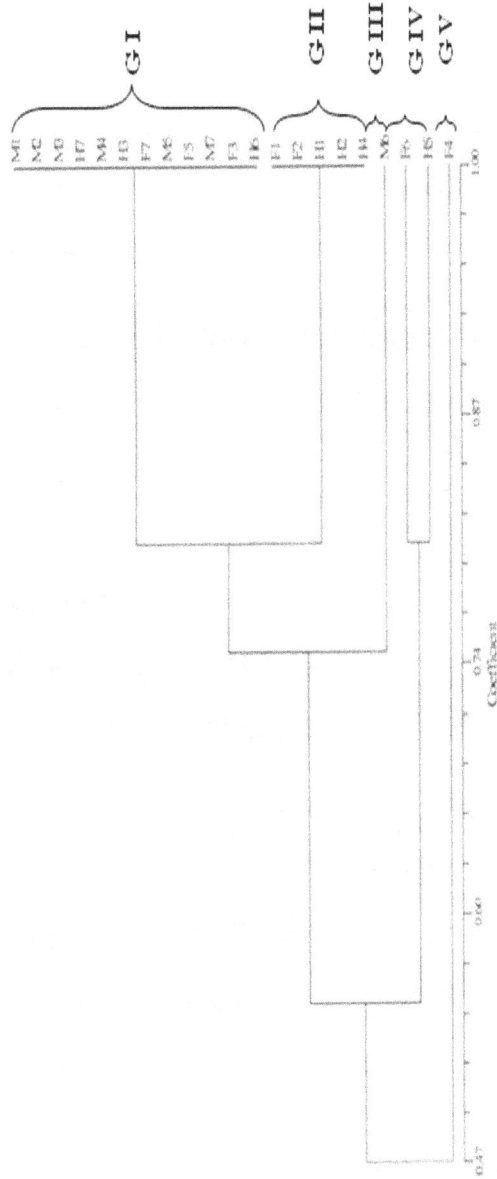

Figure 1. Dendrogram of 21 genotypes of Pearl millet based on POX banding pattern. H1-HHB50, F1-MS 81A, M1-H90/4-5; H2-HHB60, F2- MS81A, M2- H77/833-2, M3- H77/833-2; H3-HHB67, H4-HHB68, F4- MS842A, M4- H77 /833-2; H5-HHB94, F5- ICMA89111A, M5- G73-107; H6-HHB117, F6- HMS7 A, M6- H77/29-2; H7-HHB146, F7- ICMA95222A, M7- HTP94/54.

Clustering

All the experimental materials could be grouped into as much as five clusters based on less than 50% Jaccard's similarity coefficient. MS 842A(F4) was only component of a distinct cluster. HHB94(H5) and HMS 7A(F6) of one group; H77/29-2(M6) alone formed another group, MS81A(F1/F2), HHB50 (H1), HHB 60(H2), HHB 68(H4) of another group and H90/4-5(M1), H77/833-2(M2/M3/M4), HHB 67(H3), ICMA 89111A(F7), G73-107(M5), ICMA 89111A(F5), HTP94/54(M7), MS 843A(F3), HHB 117(H6) were the four group of four different distinct clusters (Figure 1). It was also found that

Table 4. Similarity indices between hybrid and their parents.

Hybrid	Female parent	Male parent
HHB 50(H1)	1.0000 MS 81A(F1)	0.8000 H 90/4-5 (M1)
HHB 60(H2)	1.0000 MS 81A(F2)	0.8000 H 77/833-2 (M2)
HHB 67(H3)	0.6000 MS 843A(F3)	1.0000 H 77/833-2 (M3)
HHB 68(H4)	0.6000 MS 842A(F4)	0.8000 H 77/833-2 (M4)
HHB 94(H5)	0.4000 ICMA 89111A(F5)	0.6000 G73-107 (M5)
HHB 117(H6)	0.4000 HMS 7A(F6)	0.8000 H77/29-2 (M6)
HHB 146(H7)	1.0000 ICMA 95222A(F7)	1.0000 HTP94/54(M7)

in each cluster, except the cluster having MS843A (F4) as sole member, the members were 100% similar to each other. Besides, the similarity coefficient ranged from 0.4000 to 0.8000 in all other combinations. Similarity indices based on POX banding pattern (Table 4) revealed that hybrid HHB 50(H1), HHB60 (H2), HHB 67(H3)and HHB 146(H7) completely resembled (SI 1.000) with their female parents MS 81A(F1/F2), MS 843A(F3) and ICMA 95222A(F7). Hybrids HHB 68(H4) and HHB 117(H6) closely resembled (SI 0.8000) towards their male parent than female parents.

The genotypic variation in respect of band numbers indicated differences among the genotypes. Similar type of POX polymorphism among some hexaploid wheat was observed by Gupta et al. (2008). This finding also corroborates the previous findings of Pushpam and Rangasamy (2006), Khandelwal et al. (2004) in rice, Manjunatha et al. (2003) in sugarcane, Philomina and Surendran (2003) in neem, Abideen and Vijayakumar (2002) in *Acasia* species and Roy et al. (2001) in grass pea. Commonness in band numbers as well as Rm values found in the present experiment are indicative of their genetic closeness, whereas band number and their relative mobility values when found different in three genotypes, indicated their genetic distinctness in the molecular level. Difference in band intensity as well as band width is indicative of differences in POX activities. POX activity is increased in plant tissues as defensive response to water stress (Badiani et al., 1990). As there is a significant correlation between root characteristics and water stress condition, the information gathered from POX diversity may help in breeding of Pearl millet varieties with higher POX activity. This can identify cultivar variation which can be used for identifying diverse lines for use as parents in further studies. They can also be used towards a better understanding of phylogentic relationships of different genotypes. However, it is necessary to use molecular markers like random-amplified polymorphic DNAs (RAPDs), restriction fragment length polymorphisms (RFLPs), amplified fragment length polymorphisms (AFLPs) and expressed sequence tags (ESTs) to map the entire Pearl millet genome, which could be used to generate novel cultivars through marker-assisted selection, map based cloning and transgenic works.

REFERENCES

Abiden ZM, Vijayakumar NK (2002). Isozyme variation in four *Acacia* species. Indian J. Genet. 62(4):373-374.

Badiani M, Biasi MD, Colognola M, Artemi, Biasi MGD (1990). Catalase, peroxidase and superoxide dismutase activities in seedling submitted increasing water deficit. Agrochimica 34(1-2):90-102.

Baes P, Van Custsem (1993). Electrophoretic analysis of eleven isozyme systems and their possible use as biochemical markers in breeding chicory (*Cychorium intybus* L.). Plant Breed. 110:16-23.

Cheniany M, Ebrahimzadeh H, Salimi A, Niknam V (2007). Isozyme variation in some populations of wild diploid wheats in Iran. Biochem. Syst. Ecol. 35(6):63-371.

Glaszmann JC, Fautret A, Noyer JL, Feldmann P, Lanaud C (1989). Biochemical genetic markers in sugarcane. Theor. Appl. Genet. 78(4):537-543.

Gupta S, Ali MN, Bhattacharyay S, Sarkar HK (2008). Peroxidase diversity in hexaploid wheat [*Triticum aestivum*]. Indian J. Crop Sci. 3(1):129-131.

Heinz DJ (1987). Sugarcane Improvement through Breeding, Elsevier, Amsterdam. pp. 273-311.

Jaccard P (1908). Nouvelles recherches sur la distribution forale. Bulletin de la Société vaudoise des sciences naturelles 44:223-270.

Ju-Zheng C, Sun Lan Z, Ju-Zc, Sun LZ (1997). A comparative study on peroxidase isozyme of K-type and V-type male sterile lines of common wheat, and their maintainer lines. Acta Agric. Boreali-Sinica 12(2):7-11.

Khandelwal V, Sharma V, Singh D (2004). Stability for grain yield in sorghum [*Sorghum bicolor* (L.) Moench]. Indian J. Genet. 65(1):53-54.

Li SH, Li SH (1996). Zymogram analysis of peroxidase isoenzyme of spring wheat varieties in Ningxia. Ningxia J. Agric. Forest. Sci. Technol. 2:10-13.

Liu CJ, Witcombe JR, Hash CT, Busso CS, Pittaway TS, Nash M, Gale MD (1994). Witcombe J R, Duncan R R, (eds): Use of molecular marker in sorghum and pearl millet breeding for developing countries. Overseas Development Administration: London U.K. pp. 57-69.

Manjunatha BR, Virupakshi S, Naik GR (2003). Peroxidase isozyme polymorphism in popular Sugarcane cultivars. Curr. Sci. 85(9):1347-1349.

Mitra R, Jagannath DR, Bhatia CR (1970). Discelectrophoresis of analogous enzymes in Hordeum. Phytochemistry 9:1843-1850.

Philomina D, Surendran C (2003). The application of isozymes to diversity studies in neem (*Azadirachta indica*. A Juss). Indian J. Genet. 63(1):93-94.

Poncet V, Lamy F, Enjalbert J, Joly H, Sarr A, Robert T (1998). Genetic analysis of the domestication syndrome in pearl millet (*Pennisetum glaucum* L, Poaceae): inheritance of the major characters. Heredity 81:648-658.

Pushpam R, Rangaswamy SR (2006). Varietal differences in peroxidase isozyme and protein profiles in rice seedling growing under salinity stress. Adv. Plant Sci. 19(1):313-314.

Rohlf PJ (2000). NTSYSpc. Numerical taxonomy and multivariate analysis system, version 2.01. Applied Biostatistics, New York.

Roy M, Mondal N, Das PK (2001). Seed protein characterization and isozyme diversity for cultivar identification in grass pea. Indian J. Genet. 61(3):246-249.

Simpson MJ, Withers LA (1986). Characterization of plant genetic resources using isozyme electrophoresis. A guide to the literature, IBPGR, Rome, P. 258.

William DHM, Mujeeb KA (1992). Isozyme and cytological markers of some Psathyrostachys juncea accessions. Theor. Appl. Genet. 84(5-8):53.

24

Effect of chemical thinning on yield and quality of peach cv. Flordasun

S. B. Meitei[1], R. K. Patel[2], Bidyut C. Deka[2], N. A. Deshmukh[2] and Akath Singh[3]

[1]College of Post Graduate Studies (CAU), Barapani-793 103, Meghalaya, India.
[2]Division of Horticulture, ICAR Research Complex for NEH Region, Umiam-793 103, Meghalaya, India.
[3]CAZRI, Jodhpur, India.

A field trial was conducted at experimental farm of the Division of Horticulture, ICAR Research Complex for North Eastern Hill Region, Umiam, Meghalaya, India during 2009 to 2010 to evaluate the effect of chemical fruit thinning and their effect on yield and quality of peach cv. Flordasun. Four chemical thinners *viz.* Thiourea at 2.5 and 5%, GA_3 at 75 and 100 ppm, Urea at 4 and 6% and Ethrel at 100 and 150 ppm after fruit set were applied. All the thinning treatments increased fruit drop and fruit colour but reduced the fruit firmness. All the chemical treatments except (GA_3 75 ppm and Urea 4%) after fruit set advanced the time of harvesting. Ethrel at 150 ppm and thiourea at 5% reduced the crop load and improved the physico-chemical characteristics of fruits and fetch premium price in the market due to their attractive colour appearance than control.

Key words: Peach cv. Flordasun, chemical thinning, fruit quality.

INTRODUCTION

Peach (*Prunus persica*. Batsch L.) belonging to the family Rosaceae and sub- family Prunoideae is the third most widely distributed temperate fruit in the world. It is basically a temperate zone plant and its commercial production area is confined between the latitude of 30 and 40° N and S. In India, however, its cultivation is confined to mid hill zone of Himalayas extending from Jammu and Kashmir, Himachal Pradesh, Punjab, Haryana and parts of Uttar Pradesh, Tamil Nadu to North eastern Hill region. Low chilling peaches are grown in sub mountainous region and western Uttar Pradesh. In sub tropical climate of NEH region, peach is being grown in limited scale and has a great potential for its cultivation due to its diverse topography, altitude and climatic condition. In Punjab, introduction of low chilling, high yielding and early ripening cultivars of superior quality traits has brought about miraculous change in peach cultivation (Nijjar, 1977). Due to early access to the

commercial markets, peach cultivation has become a highly economic proposition and area under this crop has increased at a faster rate since it is being planted in solid blocks as well as a filler tree in orchards of mango, litchi, pear, etc.

It is well established that heavy bearing of peach trees adversely affects the size and quality of fruits resulting in poor returns to the growers. In addition, breakage of limbs under heavy crop load and increasing susceptibility to late winter frost particularly in the temperate zones are the other adverse effects of heavy bearing. Fruit thinning is essential commercial practices to optimize fruit size, maximize crop value, improve fruit colour, shape, and quality, promote return bloom and to maintain tree growth and structure (Byers et al., 2003). The practice of fruit thinning often leads to improvement in the quality and size of the fruits. Thinning of the peach fruit, advanced the fruit maturity by 4 to 7 days increase the fruit size by

20 to 35% in different cultivars as compared to unthinned fruits (Chanana et al., 2002).

To accomplish the job of thinning, various methods such as hand thinning, mechanical thinning and chemical thinning are used. Hand thinning fruitlets at 45 to 50 days after bloom is the standard commercial practice in most peach producing areas. Hand thinning of peach is the single most expensive management practice of growing peach (Stover et al., 2004). The horticulturists all over the world have been trying to evolve some chemical treatments to thin out the excessive crop load so that the quality of the remaining fruits are improved. Continuous efforts have established suitability of number of chemicals which could be applied to thin out the fruits economically and without deleterious effect in the tree or fruit quality. However, such chemicals had been observed to be specific as regard to their efficacy in different agro-climatic conditions and also differential response of different cultivars. In recent years, use of chemical thinners in various fruit crops like apple, peach, plum and apricot etc. have been advocated. Out of various chemicals employed to thin out the excessive fruit load in different fruit plants, gibberallic acid, NAA, ethephon (ethrel), CCC, 3- chlorophenoxy propionic acid, MH, DNOC, 2,4,5- Trichloro - phenoxy acetic acid, thiourea, savin etc. have somepromising results in different fruit plants. Considering the above facts in view, the present investigation was undertaken to evaluate the effect of chemical fruit thinning on fruit yield and quality of Flordasun peach grown under mid hill sub tropics of Meghalaya

MATERIALS AND METHODS

Study site

The present investigation was carried out at Horticultural Research Farm of ICAR Research complex for North Eastern Hills Region Umiam, Meghalaya, India during 2009 to 2010. The experimental site was situated at 25° 41'-21" North latitude and 91° 55'-25"East longitude and at an elevation of 1010 m above mean sea level. The climate of the site can be characterized as subtemperate with minimum and maximum temperatures ranging from 6 to 29° C and with average annual rainfall of 2841 mm.

Selection of plant

Eight year old peach cv. Flordasun planted at a spacing of 5×5 m was selected from the peach orchard of the Experimental Farm. The selected trees were marked with metal tag for recording observation.

Experimental design

The experiment was laid out in Randomized Block Design with eight treatments viz., T_0 (Control -water sprayed), T_1 (Urea 4%), T_2 (Urea 6%), T_3 (Thiourea 2.5%), T_4 (Thiourea 5%), T_5 (GA$_3$ 75 ppm), T_6 (GA$_3$ 100 ppm), T_7 (Ethrel 100 ppm), T_8 (Ethrel 150 ppm). Three trees was the unit of treatment and each treatment was replica-

ted three times and the 5 m distance was maintained between treatments.

Preparation and application of solution

Fresh solutions of the chemicals were prepared in the field by dissolving the required quantity of Urea, Thiourea, GA$_3$ and Ethrel in water. Three liters of respective solutions were applied per tree in the form of fine spray at the time of fruit setting through Knapsack hand sprayer during evening. All other cultural operations were followed as suggested by Patel et al. (2008).

Statistical analysis

The data were subjected to statistical analysis as per the method of Gomez and Gomez (1984). Least significant of difference (LSD) at 5% level was used for finding the significance differences if any, among the treatment means.

Fruit drop

Four shoots which were evenly distributed on all the directions were randomly selected on each tree and tagged. Number of fruit on each shoots was counted at the time of fruit set, before application of chemical thinners. Fruit drop was calculated by making subsequent count of fruit dropped after application of thinning treatments.

Days taken from fruit setting to fruit maturity

Number of days from fruit set to the date of first harvest was recorded. The maturity was adjudged when the shoulder near the suture line of the fruit lost firmness and showed highest total soluble solid content (TSS).

Fruit yield

Fruit yield per tree was calculated by multiplying the number of fruits with mean fruit weight and expressed in kg/tree.

Fruit weight

Weight of 10 fruits was recorded with the help of electronic pan balance and weight of individual fruit was calculated and expressed as mean fruit weight in gram.

Fruit size (length and breadth)

Length and breadth of 10 fruits was measured with the help of Digital Vernier calliper and their mean fruit size (length and breadth) was expressed in millimeters.

Fruit colour

An arbitory four-point system was followed to evaluate the fruit colour as 1 to 2 (poorly coloured), 2 to 3 (moderately coloured) and 3 to 4 (highly coloured). An average score of ten fruits was calculated in each replication keeping in mind the characteristic colour. The fruits were subjected to a panel of judges for colour rating.

Fruit firmness

Firmness of fresh five fruits was measured using a Stable Micro System TA-XT-plus texture analyzer (Texture Technologies Corp., UK) fitted with needle. Firmness value was considered as mean peak compression force and expressed in kg. The studies were conducted at a pre test speed of 1 mm/s, test speed of 2 mm/s, distance of 5 mm and load cell of 50 kg.

Pulp weight and stone weight

Weight of pulp and weight of stones of fresh five fruits was recorded with the help of an electronic pan balance and expressed as mean weight in gram.

Pulp to stone ratio

Weight of pulp and stone of fresh five fruits was recorded with the help of electronic pan balance and pulp: stone ratio was determined by dividing the pulp weight by stone weight.

Total soluble solids

Total soluble solids of the fresh fruits were recorded by using a digital refractometer at room temperature and expressed in ^{0}Brix. The refractometer was cleaned with distilled water after each observation.

Acidity

Total titratable acidity of fresh fruits was estimated by taking 10 ml of juice which was diluted with distilled water to make the final volume 100 ml. 10 ml of this sample was titrated against N/10 NaOH using phenolphthalein as the indicator (Ranganna, 1997).

Sugars

25 ml of the fresh fruit solution used for the estimation of reducing sugars was taken and 2.5 ml of concentrated HCl was added and kept overnight. Next day, the solution was neutralized with 1N NaOH and the volume was made up to 75 ml. It was then titrated against Fehling's solution A and B (5 ml each) as done in case of reducing sugars. From the following formula, total sugar was calculated (Ranganna, 1997).

$$\text{Total sugar as invert sugar (\%)} = \frac{\text{mg of invert sugar} \times \text{dilution} \times 100}{\text{Titre value} \times \text{Weight or volume of sample}}$$

Ascorbic acid

Ascorbic acid content was determined by using 2, 6-Dichlorophenol-indophenol dye method of Freed (1966). 5 g of the fresh fruit sample was grounded with about 25 ml of 4% oxalic acid and filter through Whatman no. 4 filter paper. The filtrate was collected in a 50 ml volumetric flask and the volume was made up with 4% oxalic acid and titrated against the standard dye to a pink point. The amount of ascorbic acid was calculated using the following formula and expressed as mg/ 100 g.

$$\text{Ascorbic acid (mg/ 100 g)} = \frac{\text{Titre value} \times \text{Dye factor} \times \text{Volume make up} \times 100}{\text{Aliquot} \times \text{Weight of the sample}}$$

Total minerals

Total minerals were estimated as per the method suggested by Srivastava and Kumar (2002). 2 g of the samples was weighed in a previously weighed silica crucible. The crucibles were then heated over a low flame to volatize as much of the organic matter as possible and then heated in a muffle furnace at 600°C for 3 to 4 h. The samples were then cooled in a desiccator and weighed. To ensure complete ashing, the crucible is again heated in the furnace for 30 min, cooled and weighed. Total mineral was determined by using the following formula.

$$\text{Total mineral (\%)} = \frac{\text{Weight of ash}}{\text{Weight of sample}} \times 100$$

Phenol content

Total phenol content was determined using the Folin-Ciocalteu's reagent (Singleton and Rossi, 1965). 0.5 g of fresh sample was crushed with 80% alcohol in crucible and put in tubes. Then the material was centrifuged at 10000 rpm for 20 min and supernatant was transferred to Petridish. Residue was centrifuged again by putting little amount of alcohol in the tubes. Supernatant was transferred to Petridish again and evaporated to dryness. Next day 5 ml of water was added in the petridish. 1 ml of supernatant, 9 ml water and 5 ml folin reagent was transferred to 25 ml volumetric flask. After 3 min sodium carbonate was added and the OD was recorded at 650 nm and total phenol content was calculated by using standard graph.

Anthocyanin content

Total anthocyanin content was estimated by taking 10 g of fresh fruit pulp which was blended with ethanolic HCL. Then transfer to 100 ml volumetric flask and made up the volume up to 100 ml with ethanolic HCL and kept overnight at 4°C. Next day filter through whatman No. 1 filter paper, then the filtrate was taken and its O.D value at 535 nm was measured and recorded (Ranganna, 1997).

$$\text{Total anthocyanins (mg/100 g)} = \frac{\text{O.D} \times \text{volume made up} \times 100}{\text{Weight or volume of samples}}$$

Total carotenoids

Total carotenoids were estimated as per the method described by AOAC (1980). 3 g fresh sample was grounded with 50 ml of acetone-LR. To the separating funnel, 50 ml of petroleum ether was added followed by the coloured acetone extract. Then 300 ml of distilled water was added slowly along the wall of the separating funnel. On a 50 ml volumetric flask, a small funnel was placed and a small quantity of Na_2SO_4 was placed on the cotton and the upper layer was filtered into the volumetric flask. Volume was made upto 50 ml with petroleum ether and optical density was recorded at 450 nm using petroleum ether as blank and expressed in mg/100 g. Total carotenoids were estimated by the following formula.

Table 1. Effect of chemical thinning on fruit drop, days to maturity and fruit yield.

Treatments	Fruit drop (%)	Days to maturity	Fruit yield (kg/ tree)
T_1: Thiourea (2.5%)	60.86	73	36.69
T_2: Thiourea (5%)	66.14	73	31.73
T_3: GA$_3$ 75 ppm	40.77	73	39.57
T_4: GA$_3$ 100 ppm	45.73	72	33.64
T_5: Urea (4%)	39.78	78	38.67
T_6: Urea (6%)	47.36	74	35.48
T_7: Ethrel 100 ppm	63.58	70	29.62
T_8: Ethrel 150 ppm	68.32	70	28.96
T_0: Control	23.08	75	44.27
LSD $p_{(0.05)}$	2.85	-	1.45

$$\text{Total carotenoids (mg/100 g)} = \frac{A \times \text{Volume} \times 10^4}{A_m^x \times \text{Weight of sample}}$$

Where, A = Absorbance at 450 nm, A_m^x = Absorption coefficient of β- carotene in petroleum ether (2592).

RESULTS AND DISCUSSION

Effect of treatment on fruit drop, days to maturity and fruit yield

It was observed from the findings that all the thinning treatments significantly increased the percent of fruit abscission (Table 1). Amongst the chemical thinners, ethrel at 150 ppm (68.32%) sprayed after fruit set (AFS) was found to be the most effective in promoting fruit abscission as compared to other treatments. However, the rate was higher with increase in concentration of chemicals. The increase in percentage of fruit drop could be attributed due to the reduction of the translocation of ^{14}C sucrose from the leaves to the developing fruits after application of these chemicals and also due to the reduction in auxin transport before fruit drop which led to increased ethylene production thereby reducing auxin biosynthesis directly or indirectly by changing the mobilizing ability of the fruits resulting in the abscission of such fruits. Results supporting to these finding were also reported by Retamales et al. (1990) with 3-CPA and CGA-15281(an ethylene releasing compound); Abdel Hamid (1999) with ethrel and Babu and Yadav (2002) with thiourea. From the present investigation it was found that all the treatments reduced the number of days from fruit setting to fruit maturity in comparison to control sprayed after fruit set except urea at 4%. Among chemical thinners, ethrel at 100 ppm and 150 ppm advanced the fruit maturity by 5 days. The advancement in fruit maturity might be due to the increase in ethylene production in the fruit of treated plants during the final growth period which led to increased physiological activity

such as climacteria like respiration. On the other hand, fruit maturity delayed by 3 days with the application of urea (4%) after fruit set (AFS) and it could be attributed due to severe phytotoxicity resulting in a large reduction in leaf area and subsequent new growth, which in turn slowed down the rate of fruit development (Durner et al., 1990). Similar observation with respect to enhancement of fruit maturity had been reported by Chahill et al. (1980) with ethephon; Sandhu and Singh (1983) with ethephon; Allan et al. (1992) with ethephon and Sharma et al. (2001) with ethrel. All the chemical thinning treatments reduced the fruit yield per tree. The control (T_0) tree gave the maximum yield (44.27 kg) followed by 39.57 kg in T_3 (GA$_3$ at 75 ppm), whereas the minimum yield (28.96 kg) was produced by T_8 (Ethrel at 150 ppm). But, lower rate of urea and GA$_3$ did not affect the fruit yield and it was comparable to control. Reduction in yield could be attributed to decrease in number of fruits per tree under these treatments as a result of higher percentage of fruit drop. The present findings are in agreement with those of Gianfagna (1990) who observed reduction in yield with ethephon treatments.

Effect of treatment on physical characteristics of peach fruit

All the chemical thinning treatments significantly increased the fruit weight in comparison to control (T_0), where it was minimum (43.21 g). The fruit weight was increased with the increase in concentration of chemicals. The fruit weight was maximum (47.48 g) in T_8 (Ethrel at 150 ppm) which was statistically at par with T_2 (46.59 g), T_7 (45.82 g) and T_1 (45.61 g). Increase in fruit weight might be due to the reduction in the number of fruits per tree thereby increasing the leaf to fruit ratio which resulted in increased availability of photosynthates and lesser nutritional competition among the developing fruits, thus improving the fruit weight. These results get support from the findings of Chahill et al. (1980), Vitagliano et al. (1985), Khalil and Stino (1987), Zhang

Table 2. Effect of chemical thinning on fruit weight, fruit size (length and breadth), fruit colour, fruit firmness, pulp weight, stone weight and pulp stone ratio.

Treatments	Fruit weight (g)	Fruit length (mm)	Fruit breadth (mm)	Fruit color (4 point basis)	Fruit firmness (kg/cm^2)	Pulp weight (g)	Stone weight (g)	Pulp to stone ratio
T_1: Thiourea (2.5%)	45.61	44.86	44.26	2.83	0.1323	41.72	3.92	10.64
T_2: Thiourea (5%)	46.59	45.93	44.36	2.91	0.1316	42.64	4.01	10.63
T_3: GA$_3$ 75 ppm	43.76	43.61	42.32	2.48	0.1413	39.86	3.79	10.52
T_4: GA$_3$ 100 ppm	44.56	44.23	43.72	2.74	0.1358	40.65	3.83	10.62
T_5: Urea (4%)	43.78	43.21	42.82	2.41	0.1382	39.92	3.80	10.51
T_6: Urea (6%)	44.91	44.62	43.21	2.62	0.1346	40.72	3.84	10.60
T_7: Ethrel 100 ppm	45.82	45.56	44.68	2.96	0.1321	41.92	3.95	10.61
T_8: Ethrel 150 ppm	47.48	46.82	45.64	3.06	0.1278	43.69	4.04	10.81
T_0: Control	43.21	43.12	42.81	2.36	0.1432	38.54	3.71	10.39
LSD p$_{(0.05)}$	2.11	1.28	1.45	0.14	0.004	2.03	0.18	0.21

(1990) and Abdel Hamid (1999).

All the chemical thinning treatments significantly increased fruit length and breadth as compared to control. There was progressive increase in fruit length and breadth with the increase in concentration of chemicals. Maximum fruit length (46.82 mm) and breadth (45.64 mm) was found in T_8 (Ethrel at 150 ppm) which was at par with T_2 (Thiourea at 5 %) and T_7 (Ethrel at 100 ppm), whereas minimum fruit length (43.12 mm) and breadth (42.81 mm) were observed in control (T_0). Increase in fruit size could be attributed due to increase in leaf to fruit ratio as a result of thinning, thus increasing the availability of photosynthates and nutrients to the remaining fruits thereby increasing the size of individual fruits. Increase in fruit size was also observed by Chahill et al. (1980) with ethephon, Li et al. (1991) with GA$_3$ and Sharma et al. (2001) with ethrel. The fruit colour was improved by all chemical thinning treatments. There was a progressive increase in fruit colour with the increase in concentrations of all the chemical thinners (Table 2). Highest concentrations of ethrel (150 ppm) when applied after fruit set produced maximum fruit colour (3.06) compared to other treatments. Since higher concentrations of these chemicals suppressed the vegetative growth, the fruits on the trees remained better exposed to sunlight and aeration, thereby resulting in better colour development. Change in fruit colour depends upon the degradation of chlorophyll and accumulation of colouring pigments like anthocyanins and carotenoids. There was a positive correlation between fruit colour and photo synthetically active radiation (Correlli-Grappadelli and Coston, 1991). These findings are in line with the findings of Sandhu and Singh (1983), Sinha et al. (1983) and Modic (1989). All the chemical thinning treatments reduced the fruit firmness but significant reduction was observed only with ethrel at 150 ppm (0.1278 kg/cm^2). Firmness had been reported to decrease with increase in fruit size (Von Mollendorg et

al., 1992). Generally, the decrease in fruit firmness under chemical thinning treatments might be due to higher accumulation of nitrogen in the fruit resulting in fruit softening via activation of cell wall softening enzymes. These findings are in line with Trevisan et al. (2000) who observed the fruit firmness with increasing NAA concentration and Saini et al. (2003) also observed that fruit firmness was reduced by chemical thinning.

Pulp weight is the important quality characters for assessing the effect of different treatments. The data (Table 2) revealed that all the chemical thinning treatments had increased the pulp weight in comparison to control. Maximum pulp weight (43.69 g) was recorded in T_8 (Ethrel at 150 ppm), whereas minimum pulp weight (38.54 g) was found in control (T_0). The data presented in Table 2 showed that all the treatments significantly increased the stone weight of the fruit. Maximum stone weight (4.04 g) and pulp to stone ratio (10.81%) was found in T_8 (Ethrel at 150 ppm), whereas minimum stone weight (3.71 g) pulp to stone ratio (10.39 %) was recorded in control (T_0). The increase in pulp weight, stone weight and pulp to stone ratio could be attributed to the fact that fruit thinning increased fruit size which resulted in higher proportionate pulp weight and marginal increase in stone weight. These results get support from the findings of Chahill et al. (1980) and Khalil and Stino (1987) with ethephon (100 ppm).

Effect of treatment on chemical characteristics of peach fruit

It was found that the total soluble solids, reducing sugars, non-reducing sugars and total sugars were increased by increasing concentrations of all the chemical thinners when applied after fruit set (Table 3). Amongst all the chemical thinners the maximum increase in total soluble solids (13.22°Brix), reducing sugar (1.86%), non-reducing

Table 2. Effect of chemical thinning on fruit weight, fruit size (length and breadth), fruit colour, fruit firmness, pulp weight, stone weight and pulp stone ratio.

Treatments	Fruit weight (g)	Fruit length (mm)	Fruit breadth (mm)	Fruit color (4 point basis)	Fruit firmness (kg/cm^2)	Pulp weight (g)	Stone weight (g)	Pulp to stone ratio
T_1: Thiourea (2.5%)	45.61	44.86	44.26	2.83	0.1323	41.72	3.92	10.64
T_2: Thiourea (5%)	46.59	45.93	44.36	2.91	0.1316	42.64	4.01	10.63
T_3: GA$_3$ 75 ppm	43.76	43.61	42.32	2.48	0.1413	39.86	3.79	10.52
T_4: GA$_3$ 100 ppm	44.56	44.23	43.72	2.74	0.1358	40.65	3.83	10.62
T_5: Urea (4%)	43.78	43.21	42.82	2.41	0.1382	39.92	3.80	10.51
T_6: Urea (6%)	44.91	44.62	43.21	2.62	0.1346	40.72	3.84	10.60
T_7: Ethrel 100 ppm	45.82	45.56	44.68	2.96	0.1321	41.92	3.95	10.61
T_8: Ethrel 150 ppm	47.48	46.82	45.64	3.06	0.1278	43.69	4.04	10.81
T_0: Control	43.21	43.12	42.81	2.36	0.1432	38.54	3.71	10.39
LSD p$_{(0.05)}$	2.11	1.28	1.45	0.14	0.004	2.03	0.18	0.21

Table 3. Effect of chemical thinning on TSS (^{0}Brix), acidity, reducing sugars, non-reducing and total sugars of peach.

Treatments	TSS (^{0}Brix)	Acidity (%)	Total sugars (%)	Reducing sugars (%)	Non – reducing sugars (%)
T_1: Thiourea (2.5%)	12.97	0.68	6.19	1.81	4.38
T_2: Thiourea (5%)	13.18	0.66	6.28	1.83	4.45
T_3: GA$_3$ 75 ppm	12.52	0.71	6.03	1.73	4.30
T_4: GA$_3$ 100 ppm	12.61	0.70	6.11	1.76	4.35
T_5: Urea (4%)	12.47	0.73	6.05	1.71	4.34
T_6: Urea (6%)	12.49	0.77	6.10	1.74	4.40
T_7: Ethrel 100 ppm	13.15	0.67	6.24	1.84	4.41
T_8: Ethrel 150 ppm	13.22	0.64	6.42	1.86	4.56
T_0: Control	12.26	0.71	5.71	1.69	4.02
LSD p$_{(0.05)}$	0.58	0.03	0.20	0.08	0.19

sugar (4.56%) and total sugar (6.42%) was recorded in ethrel at 150 ppm when applied after fruit set. Improvement in total soluble solids, reducing sugar, non-reducing sugars and total sugars might be attributed to reduced crop load due to thinning, consequently increasing the leaf to fruit ratio, which resulted in more synthesis, transport and accumulation of sugars in the remaining fruits, thus improving the total soluble solids and sugars. The increase in total soluble solids and reducing sugars had been reported by Sandhu and Singh (1983) with ethephon; Abdel Hamid (1999) with ethrel and Saini et al. (2003), whereas increase in total sugars had also been reported by Bhullar et al. (1981) with pre-harvest calcium nitrate application; Sandhu and Singh (1983) with ethephon; Zhang (1990) with paclobutrazole; Gupta and Kaur (2004) with ethrel in plum.

All the chemical thinning treatments decreased the acidity of the fruit except urea (4 and 6%); however, thinning with ethrel at 150 ppm was most effective in reducing the acidity as compared to other treatments. Maximum increase in acidity (0.77%) was recorded in

urea at 6% (T_6) treatment, whereas minimum reduction in acidity (0.64%) was recorded in ethrel at 150 ppm when applied after fruit set (Table 3). Reduction in acidity under chemical thinning treatments might be due to conversion of organic acids into sugar. The present findings are in close conformity with the findings of Chahill et al. (1980) who had also reported decrease in acidity with 100 ppm ethephon treatment, whereas increase in acidity under chemical thinning might be due to the increase in biosynthesis of organic acid with slower rate of nitrogen application.

The ascorbic acid, total minerals, phenols, anthocyanin and carotenoids content were increased by increasing concentrations of all the chemical thinners when applied after fruit set (Table 4). Amongst all the chemical thinners the maximum ascorbic acid (52.87 mg/100 g), total minerals (1.62%), phenols (150.43 mg/100 g), anthocyanin (4.81 mg/100 g) and carotenoids (10.87 mg/100 g) was recorded in ethrel at 150 ppm when applied after fruit set. Increase in anthocyanin and carotenoid content might be due to the fact that fruit

Table 4. Effect of chemical thinning on ascorbic acid, minerals, phenols, anthocyanin and carotenoid of peach.

Treatments	Ascorbic acid (mg/100 g)	Minerals (%)	Phenol content (mg/100 g)	Anthocyanin content (mg/100 g)	Carotenoid (mg/100 g)
T_1: Thiourea (2.5%)	52.34	1.52	149.85	4.58	10.64
T_2: Thiourea (5%)	52.56	1.59	150.22	4.67	10.73
T_3: GA$_3$ 75 ppm	50.87	1.29	148.86	4.22	10.39
T_4: GA$_3$ 100 ppm	51.43	1.38	149.32	4.36	10.50
T_5: Urea (4%)	50.95	1.27	148.87	4.23	10.41
T_6: Urea (6%)	51.59	1.35	149.46	4.34	10.52
T_7: Ethrel 100 ppm	52.49	1.55	149.91	4.65	10.69
T_8: Ethrel 150 ppm	52.87	1.62	150.43	4.81	10.87
T_0: Control	50.69	1.23	148.50	4.15	10.25
LSD $p_{(0.05)}$	1.25	0.08	0.95	0.17	0.32

thinning resulted in reducing the inter-fruit competition for minerals, metabolites and precursors, which increased the faster accumulation of colouring pigments, thus anthocyanin and total carotenoids content of the fruit were increased, whereas increase in ascorbic acid might be due to the lower rate of conversion of ascorbic acid to dehydro-ascorbic acid. These results are in line with the findings of Abdel Hamid (1999) who reported increase in anthocyanins content with ethrel 100 or 200 ppm treatment when applied after fruit set, on the other hand increase in ascorbic acid had also been reported by Babu and Yadav (2002) with GA$_3$ and thiourea treatments when applied at full bloom stage in peach.

Conclusion

From the study it could be inferred that ethrel at 150 ppm and thiourea 5% found most effective chemical thinner to reduced the crop load and improved the physico-chemical characteristics of peach cv. Flordasun. Hence, ethrel at 150 ppm and thiourea at5% may be applied to thin out the fruits without deleterious effect in the fruit quality and to fetch the premium price in the market due to their attractive colour appearance than control.

REFERENCES

Abdel Hamid N (1999). Effect of chemical thinning and thinning pattern on yield and quality of Flordaprince peach. Arab Universities J. Agric. Sci. 7:159-173.

Allan P, George AP, Rasmussen T, Nissen RJ (1992). Effect of different methods of thinning low chilling Flordaprince peaches. J. South Afr. Soc. Hort. 2:24-27.

AOAC (1980). Officials Methods of Analysis. Association of official Agricultural Chemists, 13th Edn., Wasington, D.C.

Babu KD, Yadav DS (2002). Chemical thinning in peach (*Prunus persica* Batsch.). Res. Crops 3:573-578.

Bhullar GS, Dhillon BS, Randhawa JS (1981). Effect of pre and post-harvest calcium nitrate sprays on the ambient and cold storage of Flordasun peach. J. Res. Punjab Agric. Univ. 18:282-286.

Byers RE, Costa G, Vizzotto G (2003). Flower and fruit thinning of peach and other Prunus. Hort. Rev. 28:285-292.

Chahill BS, Grewal SS, Dhatt AS (1980). Effect of thinning on fruit retention and some physio-chemical characteristics of peach. Punjab Hort. J. 20:70-73.

Chanana YR, Kaundal, GS, Kanwar, JS, Arora NK, Saini RS (2002). Flower and fruit thinning of peach and other Prunus. Horti. Rev. 28:351-392.

Correlli-Grappadelli L, Coston DC (1991). Thinning pattern and light environment in peach tree canopies influence fruit quality. Hort. Sci. 26:1464-66.

Durner EF, Gianfiana TJ, Rooney FX, Teiger GS, Seiler MJ, Cantarella MJ (1990). Harvest date and size distribution of peach fruit are altered with fall application of ethephon. Hort. Sci. 25:911-913.

Freed M (1966). Method of vitamin assay. Interscience Publication Inc., New York.

Gianfagna TJ (1990). The effect of Lab 173711 and ethephon on flowering and cold hardiness of peach flower buds. In: Proceedings of Plant Growth Regulators Society of America, 17th Ann. Meet, 5-9 Aug., Ithaca.

Gupta M, Kaur H (2004). Effect of flower and fruit thinning on maturity and quality of Plum (*Prunus salicina* Lindl) cv. Satluj purple. Indian J. Hort. 6:32-34.

Khalil FA, Stino GR (1987). Effect of hand thinning on yield and quality of Sunred nectarines. Assiut. J. Ag. Sci. 18:71-82.

Li SH, Bussi C, Defrance H (1991). Improvement of chemical thinning in peach using gibberellic acid. Studies over five years on different cultivars. Revue Suisse de Viticulture, d'Arboriculture et d'Horticulture 23:147-153.

Modic D (1989). The effect of manual and chemical thinning on productivity and quality of peach cvs. Collins and Jerseyland. Jugoslovenko Vocarstvo 23:459-466.

Nijjar GS (1977). New peach introductions in Punjab. In: Fruit Breeding in India. Oxford and IBH Publishing Co., New Delhi.

Patel RK, Singh Akath, Deka Bidyut C, Ngachan SV (2008). Handbook of fruit production. Director, ICAR Research Complex for NEH Region, Umiam-Meghalaya. P. 127.

Ranganna S (1997). Manual of analysis of fruits and vegetable products. Tata McGraw Hill Publishing Company Limited, New Delhi.

Retamales J, Cooper T, Bangerth F (1990). Effects of CGA-15281 and 3-CPA as thinning agents in nectarines. J. Hort. Sci. 65:639-647.

Saini RS, Singh G, Dhaliwal GS, Chanana YR (2003). Effect of crop regulation in peaches with urea and ammonium thiosulphate on yield and physico-chemical characteristics of fruits. Haryana J. Hort. Sci. 32:187-191.

Sandhu AS, Singh Z (1983). Effect of ethephon on maturity and fruit quality of peach (*Prunus persica* Batsch.). Punjab Hort. J. 23:172-174.

Sharma N, Bir Singh, Singh RP, Singh B (2001). Influence of chemical and hand thinning on maturity, quality and colour of fruits in Redheaven peaches. Hortic. J. 14:6-10.

Singleton VL, Rossi JA (1965). Colorimetry of total phenolics with phosphomolybdic-phosphotungstic acid reagents. Am. J. Eno. Viticult. 16:144-158.

Sinha MM, Tripathi SP, Teweri JP, Misra RS (1983). Effect of Alar and CCC on flowering and fruiting of peach cv. Alexander. Punjab Hort. J. 23:43-46.

Srivastava RP, Kumar S (2002). Fruits and vegetables preservation-Principles and Practices, International book Distributing Co. pp. 353-363.

Stover E, Davis D, Wirth F (2004). Economics of fruit thinning: A review focusing on apple and citrus. Hort. Tech. 14:282-289.

Trevisan R, Faria JLC, Fachinelo JC, Silva JGC (2000). Chemical thinning of peach (*Prunus persica* L. Batsch) cv. BR-2 trees. Revista Cient. Rural. 5:49-55.

Vitagliano C, Testolin R, Vishi M (1985). The effect of gibberallic acid on pattern on abscission and growth in fruits and shoots of peach cv. Glohaven. Rivista della ortoflorofrutticoltua Italiana 69:225-233.

Von Mollendorg LJ, Jacob G, De Villiers OT (1992). The effect on storage temperature and fruit size on firmness, extractable juice, wolliness and browning on two nectarines cultivars. J. Hort. Sci. 67:647-654.

Zhang ZH (1990). The effects of paclobutrazole on peach tree on early fruiting stages. Jiangsu Nongye Kexue 1:49-50.

25

Effect of job characteristics on satisfaction and performance: A test in Egyptian agricultural extension system

Hazem S. Kassem[1] **and Ahmed M. Sarhan**[2]

[1]Department of Agricultural Extension and Rural Society, Faculty of Agriculture, Mansoura University, Egypt.
[2]Department of Agricultural Extension, Faculty of Agriculture, South-Valley University, Egypt.

The current study tested core dimensions of the job characteristics model (JCM) among extension agents in Egyptian agricultural extension system. Agricultural extension system was chosen due to its importance in achieving sustainable agricultural strategy 2030. The paper examines the effect of core job dimensions on both affective responses represented by satisfaction, and behavioral responses represented by performance. Core job dimensions are skill variety, task identity, task significance, autonomy, and feedback. 230 extension agent were selected by formula. Data were elicited from extension agents who attended the weekly meeting which had been held in the sub-directorates in administrative districts during the period from September to October 2008 in Dakahalia and Qena governorates. Frequencies, percentages, arithmetic mean, reliability coefficient, multiple correlation, and multiple regression were used to analyze data statistically. Regression analysis revealed that performance was not related to the core job dimensions while satisfaction was. The findings of this study offer several implications for the JCM as a theory especially, in agreement with most research, due to ability of job characteristics to predict levels of job satisfaction. The managers of Egyptian agricultural extension system should put job characteristics into consideration for job redesign to enhance satisfaction and performance of extension agents.

Key words: Job characteristics, satisfaction, performance, agricultural extension.

INTRODUCTION

Much of the history of management and motivation theory is rooted in the desire to understand the factors that contribute to increased levels of job performance and workplace productivity. Not surprisingly, ratings of job satisfaction have consistently served as one of the highest correlates of job performance and productivity (Gardner and Pierce, 1998; Judge et al., 2001b). Accordingly, job satisfaction has been the most widely studied construct in the history of industrial/organizational psychology (Judge et al., 2001a).

Critical organizational outcomes have been associated with work design elements. However, debate among researchers is active in terms of what outcomes are really determined by work design. More specifically, it seems to be accepted by researchers that the various job dimensions have their most significant effects on intrinsic motivation and satisfaction, while the effects on actual work behaviors such as performance and turnover are not well established (Ambrose and Kulik, 1999).

One of the most popular models outlining the central antecedents of job satisfaction is known as the job characteristics model (JCM). Hackman and Oldham's

(1976) job characteristics model describes the relationship between job characteristics and individual response to work. The model identified five "core job characteristics". These are:

(i) Skill variety: The degree to which a job requires a worker to use different skills, abilities, or talents;
(ii) Task identity: The degree to which a job involves performing a whole piece of work from start to finish;
(iii) Task significance: The degree to which a job has an impact on the lives or work of other individuals;
(iv) Autonomy: The degree to which a job allows a worker the freedom and independence to schedule work and decide how to carry it out;
(v) Feed back: The degree to which performing a job provides a worker with clear information about his or her effectiveness.

The model goes on to specify the above five core job characteristics as determinants of three "critical psychological states". These are experienced meaningfulness, experienced responsibility, and knowledge of results. In turn, the specified critical physiological states will lead to higher internal work satisfaction, high quality performance, high satisfaction with the work, and lower absenteeism and turnover. Hackman and Oldham (1975) developed the Job Diagnostic Survey (JDS) to measure these five core job characteristics. According to Boonzaaier et al. (2001), the JDS can be used to:

(i) Diagnose jobs considered for redesign in order to establish the current potential of a job for enhancing motivation and satisfaction;
(ii) Identify those specific characteristics that are most in need of enrichment;
(iii) Assess the 'readiness' of employers to respond positively to improved jobs.

In Egypt the agricultural extension service is still largely the responsibility of the government through ministry of agriculture. Over the last decade, extension service started experiencing some challenges due to socio-economic changes and agricultural sector reforms taking place in the country. Extension agents are personnel who are responsible for meeting the goals of extension system.

Accordingly, the current study aims to further address the above concern. Specifically, this paper will test the impact of core job dimensions on satisfaction (affective response) and performance (behavioral response) of extension agents in Egypt. Despite the wide research interest, it seems that the agricultural extension environment, especially in the local level, did not receive adequate attention from work design research. So, another key objective of this study is to fill this knowledge gap. In this regards, the study is designed to assess the effects of the five core job dimensions according to

Hackman and Oldham (1980) on extension agents' satisfaction and self-perceived performance.

Hypotheses

Based on the review of the literature and the general discussion, the following two hypotheses are advanced:

H_1: There is a significant positive relationship between the five core dimensions and extension agents' job satisfaction.
H_2: There is no significant relationship between the five core dimensions and extension agents' self-perceived performance.

METHODS

Population and sample

The population for this study was all extension agents employed by the extension service in Dakahalia and Qena governorates. 230 extension agents were selected for this study by Krejcie and Morgan (1970) formula. Data were elicited from extension agents who attended the weekly meeting which had been held in the sub-directorates in administrative districts during the period from September to October 2008.

Instruments

Extension agents' perceptions of the five job characteristics and their level of job satisfaction were obtained utilizing a modified version of the job diagnostic survey developed by Hackman and Oldham (1980). The job diagnostic survey consists of seven different sections, the first five of which were used in this study. An additional section containing 8 questions created by the researcher was added to the end of the questionnaire to collect selected demographic characteristics of the respondents.

The JDS and job satisfaction consists of 27 items. Items were rated on a 5-point scale ranging from strongly agree to strongly disagree. The self-assessed performance scale comprised of 16 items on a 5-point scale ranging from strongly agree to strongly disagree.

RESULTS

Scale reliabilities

As a first step, scale reliability coefficients (cronbach alphas) for all measures adopted in this study were computed. Nunnally (1978) maintains that reliabilities which are less than 0.6 are considered poor, while those above are acceptable, while those above 0.8 are good. Results showed that reliability for JDS, satisfaction, and performance was 0.77, 0.74 and 0.72 respectively.

Descriptive statistics

Majority of the participants (81.3%) in this study were

Table 1. Descriptive statistics for the JDS.

Job characteristics	Mean statistic	Std. deviation Statistic	Skewness		Kurtosis	
			Statistic	Std. Error	Statistic	Std. Error
Task variety	4.1761	0.52914	-0.453	0.160	0.772	0.320
Task significance	3.9054	0.72202	-0.325	0.160	-0.521	0.320
Task identity	3.5696	0.80858	-0.202	0.160	-0.333	0.320
Task autonomy	2.8826	1.26047	0.118	0.160	-1.118	0.320
Feed back	3.0725	0.93275	0.329	0.160	-1.121	0.320

Table 2. Correlation matrix of all variables.

Variables	1	2	3	4	5	6	7
1-Task variety	1						
2-Task significance	0.495(**)	1					
3-task identity	0.474(**)	0.588(**)	1				
4-Task autonomy	0.208(**)	0.222(**)	0.359(**)	1			
5-Feed back	0.166(*)	0.394(**)	0.353(**)	0.151(*)	1		
6- Satisfaction	0.433(**)	0.468(**)	0.438(**)	0.113	0.362(**)	1	
7-Performance	0.236(**)	0.184(**)	0.227(**)	0.199(**)	0.115	0.073	1

*Correlation is significant at the 0.05 level ** Correlation is significant at the 0.01 level.

male, having an average of 44.3 years. This was a well educated sample; 18% of respondents held masters or doctoral degrees, the remainder holding either bachelors or associate degrees. Participants had been with ministry of agriculture an average of 18.3 years, serving in extension service for an average of 12.7 years. The descriptive statistics for the JDS scales for extension agents are set out in Table 1. The variability of the means, standard deviation, skewness and kurtosis reflects how the respondents responded to the different scales. The variability indicates that the data which were collected and analyzed were normally distributed.

Correlations

The correlation matrix among all variables in this study is summarized in Table 2. There is no correlation between the dependent variables (r =0.073). Most of the correlation coefficients between satisfaction and job dimensions were statistically significant and moderately correlated except for task autonomy (r =0.113). Meanwhile, self-perceived performance is significantly and low correlated with job dimensions except for feed back (r =0.115).

Hypotheses testing: Multiple regression

Two model hierarchical linear regression analyses were performed to test the two hypotheses of this study. Table 2 shows results of the multiple regression with

satisfaction as dependent variable and the five core dimensions as independent variables. The first hypothesis was that job characteristics factors would predict levels of job satisfaction. To test this hypothesis, the five job characteristics factors of task variety, task significance, task identity, task autonomy and feed back were entered into the first regression model as it shown in Table 3. All five variables except task autonomy were found to be significant, positive predicators of job satisfaction levels. Combined, the five job characteristics accounted for 33% of the variance in job satisfaction. These findings provide partial support for the first hypothesis, with the job characteristics of autonomy failing to demonstrate a clear factor predictor. Results of the second model are shown in Table 4. The dependent variable was self-perceived performance, and the five core dimensions as independent variables. The second hypothesis was that job characteristics factors would not predict levels of self-perceived performance to test this hypothesis. Unlike the first model all five variables except for task variety were found to be non significant, positive predicators of self-perceived performance levels. Combined, the five job characteristics accounted only 8.8% of the variance in self-perceived performance. These findings provide partial support for the second hypothesis except for task variety which succeeded to demonstrate a clear factor predictor.

DISCUSSION

The first regression model's finding that all job

Table 3. Results of multiple regression between job satisfaction as dependent variable and core job dimensions.

Analysis of variance					
Model	Sum of squares	Df	Mean square	F	Sig.
1 Regression	42.479	5	8.496	22.078	0.000
Residual	86.198	224	0.385		
Total	128.677	229			

Variables in the equation					
Model	Unstandardized coefficients		Standardized coefficients	T	Sig.
	B	Std. Error	Beta	B	Std. Error
1 (Constant)	-0.024	0.340		-0.071	0.944
Task variety	0.344	0.093	0.243	30.704	0.000
Task Significance	0.193	0.076	0.186	20.529	0.012
Task identity	0.156	0.068	0.168	20.278	0.024
Task Autonomy	-0.041	0.035	-0.069	-10.181	0.239
Feedback	0.161	0.049	0.200	30.301	0.001

Multiple R: 0.575, R Square: 0.330, Adjusted R Square: 0.315, Std. Error: 0.62033.

Table 4. Results of multiple regression between self-perceived performance as dependent variable and core job dimensions.

Analysis of variance					
Model	Sum of squares	Df	Mean square	F	Sig.
2 Regression	7.431	5	1.486	4.347	0.001(a)
Residual	76.591	224	0.342		
Total	84.022	229			

Variables in the equation					
Model	Unstandardized coefficients		Standardized coefficients	T	Sig.
	B	Std. Error	Beta	B	Std. Error
2 (Constant)	0.796	0.320		20.485	0.014
Task variety	0.179	0.087	0.156	20.047	0.042
Task significance	0.011	0.072	0.014	0.158	0.875
Task identity	0.065	0.064	0.087	10.016	0.311
Task autonomy	0.061	0.033	0.127	10.853	0.065
Feedback	0.022	0.046	0.034	0.482	0.630

Multiple R: 0.297, R Square: 0.088, Adjusted R Square: 0.068, Std. Error: 0.58474.

characteristics except task autonomy significantly and positively predicted levels of job satisfaction provides support for the first hypothesis , as well as the applicability of the JCM in agricultural extension work context. In the workplace, regardless of title, position or skill set, employees seem to prefer and respond positively to environments characterized by the four factors of task significance, task variety, task identity and feed back. Employees express higher levels of job satisfaction in jobs where they also believe that their tasks are important for the welfare of others, where opportunity is given to perform a variety of tasks, where involvement in projects is from inception to completion so

as to facilitate understanding, and where regular feedback is provided concerning the quality of work performance. Efforts to create workplaces characterized by high levels of job satisfaction and workplace productivity, therefore, should design jobs that maximize these job characteristics.

The job characteristics of autonomy did not load cleanly on a latent factor. Although, everyone needs a degree of individual autonomy, but to measure individual autonomy in team setting, it may me important to frame individual autonomy in the context of team involvement. The failure of autonomy to load on its own factor in this study is at least partly due to the difference in meaning between individually based and team based autonomy. The second regression model showed the what we predicted regarding self-perceived performance. In this study, all of the core job dimensions, except for task variety came out as non-significant related to performance. Performance in this case is related to skill variety, but not other core job dimensions. This is another interesting result. It seems that extension agents perceive task variety as driver for performance. Variety of extension services is incentive for extension agents to use and acquire different skills and abilities which reflect on the performance, specifically they see task variety is a source of satisfaction.The last conclusion about satisfaction-performance relationship. The findings showed no correlation between them. It seems that satisfaction not always follows performance. This result ensure that satisfaction in such a heavy expatriate environment could be related more to extrinsic factors such salaries, benefits, contract renewals, etc.

Ideas for future research

This study has helped fill a gap in the research literature for the applicability of the JCM to extension work, however, much more remains to be studied in this area. Future studies looking at the JCM would benefit by being longitudinal in nature, to assess the stability of perceptions. In addition, this study used self-perceived performance which is a limitation and it would be of value to try to independently measure performance. Also, worthy of scholarly attention is the assessment of effects that experience, level of skills, career aspirations have on satisfaction and performance. In addition, role of growth needs strength as a moderator between job characteristics and satisfaction could be examined to know how to motivate extension agents to recognize their need of growth, and how to create jobs that fulfill this basic human need.

Conclusion

This study has provided support for the applicability of the JCM to agricultural extension work. By broadening the viability of the job characteristics of task significance, task variety, task identity, and feed back, it gives credence to theories espousing their universal importance across work setting. So far as the evidence at this early stage suggests, Egyptian agricultural extension system will benefit by looking into the impact of job design by training their managers to acquire redesign skills. There might be added value in terms of satisfaction and performance of extension agents if extension system refines the process by which they design tasks and jobs.

REFERENCES

Ambrose ML, Kulik CT (1999). Old friends,New faces: Motivation research in the 1990s, J. Manage. 25(3):231-292.
Boonzaaier B, Ficker, F, Rust B (2001). A review of research on the job characteristics model and the attendant job diagnostic survey, South Afri. J. Bus. Manage. 32(1):11-29
Gardner DG, Pierce JL (1998). Self-esteem and self efficacy within the organizational context. Group and Organization Manage. 23:48-70.
Hackman JR, Oldham GR (1975). development of the job diagnostic survey. J. Appl. Psycol. 60:159-170.
Hackman JR, Oldham GR (1976). Motivation through the design of work : Test of a theory. Organizational behavior &human performance 16:250-279.
Hackman JR, Oldham GR (1980). Work redesign, Reading, MA:Addison-Wesley.
Judge TA, Parker SK, Colbert AE, Heller D, Ilies R (2001a). Job satisfaction: A cross-cultural review. n: N.Anderson DS. Ones, HK. Sinangil C, Viswesvaran (Eds.), Handbook of industrial,Work &Organizational Psychol. London: Sage 2:25-52.
Judge TA, Thoresen CJ, Bono JE, Patton GK (2001b). The job satisfaction–job performance relationship: A qualitative and quantitative review. Psychol. Bulltin 127(3):376-407.
Krejcie RV, Morgan DW (1970). Determining sample size for research activities. Educational and Psychological Measurement, 30:607-610.
Nunnally J (1978). Psychometric theory. New York: McGrow-Hill.

26

Effect of foliar application of nutrients and biostimulant on nut quality and leaf nutrient status of pecan nut cv. "Western Schley"

Naira Ashraf[1]*, Moieza Ashraf[2], Gh Hassan[3], Munib-U-Rehman[3], N. A. Dar[4], Inayat. M. Khan[4], Umar Iqbal[3] and S. A. Banday[3]

[1]Department of Fruit Science, Dr Y S Parmar University of Horticulture and Forestry, Nauni, Solan, Himachal Pradesh 173230, India.
[2]P. G. Department of Environmental Science, Kashmir University, J&K, India.
[3]Division of Fruit Science, S.K. University of Agricultural Sciences and Technology, Kashmir, India.
[4]S.K. Universities of Agricultural Sciences and Technology, Kashmir, India.

A field study on pecan nut (cv. Western Schley) was conducted in the experimental orchard of the Department of Fruit science, Dr. Y. S. Parmar University of Horticulture and Forestry, Nauni, Solan during the year 2008 to 2009. The study comprised of one experiment in which pecan trees under investigation were subjected to foliar spray of 0.5% urea, 0.1% boric acid, 0.5%, zinc sulphate, 5 ml/L supramino and their combinations. The results revealed that the foliar application of 0.5% urea, 0.1% boric acid, 0.5%, zinc sulphate and 5 ml/L supramino resulted in better nut quality in comparison to control. It was found that leaf nutrient contents (N, P, K, Ca and Mg) were also recorded maximum in trees treated with 0.5% urea+0.1% boric acid+0.5% zinc sulphate+5 ml/L supramino. Maximum leaf iron content was recorded in trees treated with 0.5% urea and 5 ml/L supramino whereas, trees sprayed with 0.5% urea, 0.1% boric acid, 0.5% zinc sulphate and 5 ml/L supramino was found to have highest leaf Zn, Mn and Cu contents.

Key words: Pecan nut, foliar sprays, leaf nutrient, nut quality.

INTRODUCTION

Out of the 20 species of the genus *Carya*, *Carya illinoensis* W. is extensively cultivated on commercial scale in united states of America, Australia, Canada and Western Europe. Pecan is considered as the "queen of nuts" in U.S.A because of its value both as a wild and cultivated nut (Woodroof, 1979). Its nuts have high nutritional and calorific value, so, pecan is most acceptable in comparison to other nuts. Pecan nut contains high content of proteins (12.5%), fats (71.42%), P_2O_5 (0.46%), K_2O (0.23%) and is rich in oil content and some varieties have shown as high as 76% oil. Besides having large export potential as nut, its timber is also expensive and used in gun-stock, carving, cabinet manufacture of high class and many other uses. Pecan nut can be grown in areas having 450 to 1550 m elevation which are free from severe spring frost and excessive heat in summer and receive annual rainfall ranging from 750 to 2000 mm. It requires warm temperate climate. It requires 240 to 280 days growing under warm climate with a mean temperature of above 26.7°C. Although, it can be grown in a wide range of soils but deep friable soils rich in organic matter and having 5.5 to 6.0 pH are the most suitable for its cultivation. Pecan nut has a long tap root and good soil depth of 6 m is desirable for its successful cultivation (Hanna, 1987). Major problem faced with its cultivation is the pecan drop and kernel blankness that directly affects yield. Various reasons for premature pecan drop assigned are varietal

*Corresponding author. E-mail: naira.ashraf@gmail.com.

character, poor pollination, water stress and nutritional problems. Nitrogen is the element that provides better tree growth, a higher percent kernel, and a healthier tree. When properly maintained, nitrogen can help to provide optimal year to year production. Nitrogen deficiencies result in poor growth and poor tree health. Pecan trees do not absorb zinc from the soil and trees do not make vigorous growth. So, the application of zinc is essential as it has also been reported to improve the nut quality. Also, boron is passively absorbed and transported through the transpiration stream, so deficiencies of boron may be transitory. Premature flower and fruit drop of tree crops has been attributed to boron deficiency. Foliar application of boron has shown to stimulate the normal flow of hormones and enhance pollen grain and pollen tube formation which thereby, increase fruit retention and fruit quality in a number of perennial tree fruit crops, including pecan (Wells et al., 2008). Supramino which is a combination of amino acids, hydrolysed proteins, organic carbohydrates, bioenzymes and organic micronutrients needs to be supplied to the plants as amino acids have a direct role in the regulation of growth and development. These amino acids also help in flower retention which directly affects yield. Also, a small insect pecan case borer makes small hole in the base of pecan which can easily be identified and controlled. Analysis of leaves for their mineral concentration is widely used to predict the nutrient needs of pecan trees. Therefore, our objectives were to study the effect of nutrients and biostimulant on nut quality and leaf macro and micro nutrient status of pecan nut.

MATERIALS AND METHODS

The experiment was conducted in the experimental orchard of the Department of Fruit Science, Dr. Y .S. Parmar University of Horticulture and Forestry, Nauni-Solan during the year 2008 to 2009. The experimental orchard is situated between 31° N latitude and 77° E longitude at an altitude of 1276 m above mean sea level. The experiment was designed in randomized block with three replications. Twelve treatments were used including urea (0.5%), boric acid (0.1%), zinc sulphate (0.5%) and supramino (5 ml/L), their combinations and control. The trees were given foliar application of these nutrients and biostimulant in two equal splits at bud swell stage and one month after fruit set.

The required amount of each nutrient was weighed with an electronic digital balance. The urea was dissolved in warm water, while $ZnSo_4$ solution was prepared by dissolving $ZnSo_4$ salt with half the amount of lime in water to avoid phytotoxicity. Boric acid solution was prepared by dissolving boric acid (17%) in warm water. Similarly, supramino was taken and was dissolved in water. Nutrient solutions were sprayed in the morning hours with a foot sprayer pump. Ten nuts were randomly selected from every treatment for data collection and the observations in respect of nut and kernel characters, leaf nutrient status (macro and micro) were taken. Ten selected nuts were weighed on digital balance and average was expressed in grams (g). The length and breadth of kernels extracted from 10 selected nuts was measured with the help of Vernier Calliper and averages were expressed in centimeters (cm). The Kjeldahl's method as described by Kanwar and Chopra (1967) for estimation of crude protein in plant samples was followed. Oil

content of the kernel was determined on the weight basis and expressed in percentage. The nuts were dried in the oven at 60°C until they were moisture free. Petroleum ether (40 to 60° B.P.) was used as a solvent for oil extraction in the Soxhlet apparatus (Ranganna, 1997).

Opposite leaf let pairs from middle of the terminal shoots were collected from pecan nut trees in the last week of July and analysed as per standard analytical method. The digestion of the samples for the estimation of nitrogen was carried out in concentrated sulphuric acid (AR grade) by adding digestion mixture. For the estimation of leaf P, K, Ca, Mg, Zn, Fe, Cu and Mn, digestion was done in diacid mixture prepared by mixing nitric acid and perchloric acid (AR grade) in the ratio of 4:1.

Total nitrogen content was determined by Micro kjeldahl's method (AOAC, 1980) and the results were expressed in percentage on dry weight basis. Total phosphorus content was determined by Vanadomolybdophosphoric yellow colour method (Jackson, 1975) and the results were expressed in percentage on dry weight basis. Total potassium content was determined on Flame Photometer (Toshniwal, TMF 45) and the results were expressed in percentage on dry weight basis. Total calcium and magnesium contents were determined with the help of atomic absorption spectrophotometer and the results were expressed in percentage on dry weight basis. The micro nutrients zinc, iron, copper and manganese were also determined with the help of atomic absorption spectrophotometer and the results were expressed in parts per million (ppm) on dry weight basis.

RESULTS

The results revealed that the nut weight, kernel weight, kernel length and kernel breadth were significantly improved with combined foliar spray of 0.5% urea + 0.1% boric acid + 0.5% zinc sulphate + 5 ml/L supramino treatment (Table 1). Further, fruit size and weight was also significantly affected by foliar application of zinc. Maximum kernel length was recorded under urea + boric acid + zinc sulphate + supramino treatment. The maximum kernel percentage and minimum blankness percentage was recorded with combined foliar spray of 0.5% urea + 0.1% boric acid + 0.5% zinc sulphate + 5 ml/L supramino treatment. In the present study, zinc improved the nut quality. The data presented in Table 1 revealed that the maximum kernel protein was recorded with combined foliar spray of 0.5% urea + 0.1% boric acid + 0.5% zinc sulphate + 5 ml/L supramino treatment. Also, combined application of urea + boric acid + zinc sulphate + supramino treatment resulted in highest kernel oil (%).

Data presented in Table 2 indicated that the maximum leaf N, P, K, Ca and Mg contents were obtained with treatment 0.5% urea + 0.1% boric acid + 0.5% zinc sulphate + 5 ml/L supramino.

The data presented in Table 3 revealed that the maximum leaf Zn , leaf Mn and leaf Cu contents were recorded with combined foliar spray of 0.5% urea + 0.1% boric acid + 0.5% zinc sulphate + 5 ml/L supramino treatment and maximum leaf Fe content (287.5 ppm) was obtained with treatment urea + supramino. Significantly higher leaf zinc content of pecan nut leaves was observed. Increased leaf copper content was also observed.

Table 1. Effect of nitrogen, boron, zinc and supramino on nut weight, kernel weight, kernel length, kernel breadth, kernel percentage, shell kernel ratio, blankness percentage, kernel protein and kernel oil of pecan nut cv. Western Schley.

Treatment	Nut weight (g)	Kernel weight (g)	Kernel length (mm)	Kernel breadth (mm)	Kernel (%)	Shell kernel ratio	Blankness (%)	Kernel protein (%)	Kernel oil (%)
Urea (0.5%)	6.21	3.65	32.94	13.13	57.38	0.83	6.82 (2.61)[*]	9.36 (3.06)[*]	67.23 (55.08)[**]
Boric acid (0.1%)	6.28	3.54	33.56	13.56	57.49	0.86	7.95 (2.82)	8.33 (2.89)	68.52 (55.87)
ZnSo₄ (0.5%)	5.54	2.97	31.41	11.99	55.31	0.84	9.06 (3.01)	8.44 (2.91)	62.55 (52.27)
Supramino(5 ml/L)	6.30	3.57	33.66	13.62	57.53	0.85	7.76 (2.79)	8.39 (2.89)	65.08 (53.78)
Urea+boric acid (0.5%+0.1%)	6.47	3.70	33.88	13.66	57.51	0.78	6.35 (2.52)	9.02 (3.00)	71.35 (57.65)
Urea+ ZnSo₄ (0.5%+0.5%)	6.11	3.59	32.74	12.58	57.06	0.75	7.31 (2.70)	7.66 (2.77)	67.32 (55.14)
Urea +supramino (0.5%+5 ml/L)	6.52	3.73	33.97	13.68	57.65	0.86	6.45 (2.54)	10.77 (3.28)	67.40 (55.19)
Boric acid+ ZnSo₄ (0.1%+0.5%)	6.64	3.58	33.91	13.63	57.52	0.81	8.44 (2.90)	8.47 (2.91)	65.11 (53.80)
Boric acid+supramino (0.1%+5 ml/L)	7.07	3.62	34.10	13.68	57.68	0.85	7.99 (2.83)	8.52 (2.92)	68.77 (56.03)
ZnSo₄+supramino (0.5%+5ml/L)	6.52	3.59	34.06	13.69	56.58	0.76	7.55 (2.75)	8.53 (2.92)	71.59 (57.79)
Urea+boric acid+ZnSo₄+supramino (0.5%+0.1%+0.5%+5 ml/L)	7.54	4.53	34.66	13.86	60.48	0.75	5.57 (2.36)	13.68 (3.69)	76.52 (61.02)
Control	5.24	2.54	31.35	11.66	55.24	0.88	10.16 (3.19)	5.77 (2.40)	56.06 (48.48)
CD₀.₀₅	0.56	0.03	0.65	0.04	0.28	0.02	0.03	0.05	0.75
SEM	0.19	0.01	0.22	0.01	0.09	0.01	0.01	0.02	0.25

* Figures in parentheses are square root transformed values;   **, figures in parentheses are arc sine transformed values.

DISCUSSION

This increase was observed due to the fact that N is extremely mobile and developing fruits act as metabolic sink for the nutrient elements. The increase in nut weight could be attributed to the central role of N in various metabolic processes in the plant. Further, N application prolongs the phase of fruit cell division resulting in greater number of cells (Hewitt and Smith, 1975). The possible explanation for increase in fruit size and weight was also due to faster movement of simple sugar into fruit and involvement in cell division and cell expansion (Brahmachari et al., 1997). In macadamia nuts, kernel recovery, kernel weight and percentage of first grade kernels were enhanced by foliar boron sprays (Stephenson and

Gallagher, 1990). The minimum blankness percentage was recorded with combined foliar spray of nutrients because of the cumulative effect of nutrients on kernel weight resulting in higher percentage. Although, blankness is a varietal character but reduction in proportion of blank kernels was due to improved internal nutrient status of trees due to foliar application of nitrogen and zinc which increased growth and vigour associated with higher photosynthesis and trans location of assimilated products in nuts leading to minimum blank kernels. The decrease in shell kernel ratio has been attributed to production of kernels with heavier weight due to cumulative effect of foliar fertilization. Foliar application of urea increased total nitrogen in plants tissues, which led to higher protein content in pecan kernels

due to the fact that nitrogen is a constituent of protein (Salisbury and Ross, 1992). Furthermore, these results are in line with Tous et al. (2005) who reported increased kernel size in 'Negret' cultivar of hazelnut by boron treatments. Same findings have also been reported by Bhatia and Yadav (2005), Bybordi and Malakouti (2006), Farid et al. (2007) in jackfruit and Tomar and Singh (2007). Furthermore, the combined spray of nutrients resulted in increased leaf nutrient status. This increase in leaf N content might have occurred due to the fact that nitrogen is highly mobile. Its efficient translocation and nutrient supply from root to tree leaves could have added to its enhanced accumulation in leaves (Smith, 1962). The increase in leaf calcium content was because of the direct positive relationship

Table 2. Effect of nitrogen, boron, zinc and supramino on macronutrient status of leaves of pecan nut cv. Western Schley.

Treatment	N (%)	P (%)	K (%)	Ca (%)	Mg (%)
Urea (0.5%)	2.65 (1.63)*	0.18 (0.43)*	0.78 (0.88)*	1.55 (1.02)*	0.57 (0.75)*
Boric acid (0.1%)	2.27 (1.51)	0.15 (0.39)	0.75 (0.86)	2.09 (1.45)	0.51 (0.71)
ZnSo₄ (0.5%)	2.24 (1.49)	0.16 (0.41)	0.84 (0.92)	2.20 (1.48)	0.45 (0.67)
Supramino (5 ml/L)	2.49 (1.58)	0.17 (0.42)	0.86 (0.92)	2.16 (1.47)	0.51 (0.71)
Urea+ boric acid (0.5%+0.1%)	2.69 (1.64)	0.19 (0.44)	0.82 (0.90)	2.41 (1.55)	0.56 (0.75)
Urea+ ZnSo₄ (0.5%+0.5%)	2.55 (1.59)	0.18 (0.43)	0.91 (0.95)	2.32 (1.52)	0.52 (0.72)
Urea+ supramino (0.5%+5 ml/L)	2.69 (1.64)	0.17 (0.42)	0.82 (0.90)	2.37 (1.54)	0.58 (0.76)
Boric acid+ ZnSo₄ (0.1%+0.5%)	2.46 (1.56)	0.17 (0.42)	0.86(0.92)	2.25 (1.50)	0.56 (0.75)
Boric acid+supramino (0.1%+5 ml/L)	2.53 (1.59)	0.17 (0.42)	0.78(0.88)	2.27 (1.51)	0.60 (0.77)
ZnSo₄+supramino (0.5%+5 ml/L)	2.59 (1.60)	0.16 (0.41)	1.11 (1.05)	2.31 (1.52)	0.58 (0.76)
Urea+boric acid+znso₄+supramino (0.5%+0.1%+0.5%+5 ml/L)	2.88 (1.69)	0.22 (0.47)	1.14 (1.07)	2.44 (1.56)	0.63 (0.79)
Control	2.06 (1.44)	0.14 (0.37)	0.72 (0.85)	2.06 (1.43)	0.45 (0.67)
CD₀.₀₅	0.02	0.03	0.02	0.43	0.02
SEM	0.01	0.01	0.01	0.14	0.01

*Figures in parentheses are square root transformed values.

Table 3. Effect of nitrogen, boron, zinc and supramino on micronutrient status of leaves of pecan nut cv. Western Schley.

Treatment	Zn (ppm)	Mn (ppm)	Fe (ppm)	Cu (ppm)
Urea (0.5%)	224.70	246.80	277.20	38.38
Boric acid (0.1%)	192.80	174.20	249.10	28.73
ZnSo₄ (0.5%)	290.60	174.10	251.50	30.00
Supramino(5 ml/L)	284.50	212.40	262.50	33.13
Urea+ boric acid (0.5%+0.1%)	195.30	266.90	261.20	47.87
Urea+ ZnSo₄ (0.5%+0.5%)	264.00	212.40	269.70	36.40
Urea+ supramino (0.5%+5 ml/L)	196.50	263.20	287.50	47.44
Boric acid+ ZnSo₄ (0.1%+0.5%)	291.70	182.10	262.30	30.78
Boric acid+supramino (0.1%+5 ml/L)	277.80	222.3	250.30	38.03
ZnSo₄+Supramino (0.5%+5 ml/L)	290.80	223.7	247.80	30.82
Urea+boric acid+znso₄+supramino (0.5%+0.1+0.5%+5 ml/L)	294.10	267.4	287.30	48.47
Control	190.50	150.30	239.40	17.44
CD₀.₀₅	5.16	12.55	3.21	1.98
SEM	1.72	4.18	1.07	0.66

Figures in parentheses are square root transformed values.

(synergysm) between leaf nitrogen and leaf calcium (Childers, 1983). Thus, the foliar sprays of urea might have increased the calcium mobility sufficiently during senescence, thereby increasing its calcium level later on. Increase in the concentration of zinc in plants tissues might be because of the effect of application of zinc. Under nitrogen application, the great availability of manganese might have led to greater uptake of manganese. This might be because of the fact that higher absorption, translocation and utilization of nutrients takes place in the plants which resulted in increased plant growth. Furthermore, Johnson and Amdris (2000) reported increased leaf nitrogen concentration after the foliar application of urea. Increase in leaf P content are in conformity with the findings of findings of Shashi (2003) and Raina et al. (2005) who also reported increased leaf phosphorus content with nitrogen application. These results are in accordance with the findings of Singh (2000) in pecan nut.

Conclusion

From the results, it is evident that the application of nutrients and biostimulant significantly improved nut quality (in terms of nut weight, kernel weight, kernel length, kernel breadth, kernel oil and protein) and leaf nutrient status. Hence, foliar application of urea 0.5% + boric acid 0.1% + zinc sulphate 0.5% + supramino 5 ml/L can be given first at pre-bloom stage and repeated after fruit set stage to enhance nut quality and leaf nutrient status of pecan nut.

REFERENCES

AOAC (1980). Official methods of Analysis of the Analytical Chemists, 13th ed. (W Horwitz, ed.). Association of Official Analytical Chemists,Washington, DC, p. 1018.

Bhatia BS, Yadav RK (2005). Effect of foliar spray of urea and NAA on fruit yield and quality of ber (*Z. mauritiana* Lamk). *Nat. Sem. Commercialization of Hort. In Non traditional areas*, organized by C. I. A.H., Bikaner from Feb. 5-6, 2005. p. 119.

Brahmachari VS, Yadav GS, Kumar N (1997). Effect of foliar feeding of calcium, zinc and boron on yield and quality attributes of litchi *(Litchichinensis* Sonn). Orissa J. Hortic. 25(1):49-52.

Bybordi A, Malakouti MJ (2006). Effects of foliar applications ofnitrogen, boron and zinc on fruit setting and quality of almonds. Acta Hortic. 726:351-357.

Childers NF (1983). Modern Fruit Science. Soil management for apples, pp.59-77. Horticultural Publications 3906 NW31, Place Gainesville, Florida, p. 32606.

Farid ATM, Halder NK, Shahjahan M (2007). Effect of boron for correcting the deformed shape and size in jackfruit. Indian J. Hortic, 64(2):144-149.

Hanna JD (1987). In: Rootstocks for fruit crops (eds. R C Rom and R F Carlson), John Wiley and Sons, New York, pp. 401-410.

Hewitt EJ, Smith TA (1975). Plant mineral nutrition. English University Press Ltd., Warvick Lane, London, p. 295.

Jackson ML (1975). Soil Chemical Analysis. Bombay. Asia Publishing House, pp. 10-205.

Johnson RS, Amdris HL (2000). Combining low biuret urea with foliar zinc sulphate sprays to fertilize peach and nectarine trees in the fall. Acta Hortic. 448:321-327.

Kanwar JS, Chopra SL (1967). Analysis of Fertilizers and Manures. In: Analytical Agricultural Chemistry. S. Chand and Co., Delhi, pp. 119-161

Raina JN, Thakur BC, Shashi S, Spehia RS (2005). Effect offertigation through drip system on nitrogen dynamics, growth, yield and quality of *apricot*. Acta Hort. 696:227-229.

Ranganna S (1997). Handbook of Analysis and Quality Control for Fruits and Vegetable Products. 2nd Ed. Tata McGraw Hill Publishing Co. Ltd. New Delhi, p.1112.

Salisbury FB, Ross CW (1992). Plant Physiology. Wads Worth Publishing Co. Belmont, California, p. 682.

Shashi S (2003). Studies on nitrogen fertilization through drip in apricot cv. New Castle. M. Sc. Thesis, Dr. Y. S. Parmar University of Horticulture and Forestry, Nauni, Solan H. P. India.

Singh RR (2000). Studies on the effect of biofertilizers, bioregulators and urea on plant growth and standardization of propagation techniques inpecan. Ph.D. Thesis, Dr. Y.S. Parmar University of Horticulture andForestry, Nauni, Solan H.P. India.

Smith CB (1962). Mineral analysis of plant tissues. Ann. Rev. Plant Physiol. *13*:81-108.

Stephenson RA, Gallagher EC (1990). Nutritional factors affecting production and quality of macadamia nuts in Australia. Acta Hortic. 275:565-570.

Tomar CS, Singh N (2007). Effect of foliar application of nutrients and bioregulators on growth, fruit set, yield and nut quality of walnut. Indian. J. Hortic. 64(3):271-273.

Tous J, Romero A, Plana J, Sentis X, Ferran J (2005). Effect of nitrogen, boron and iron fertilization on yield and nut quality of 'Negret'hazelnut trees. Acta Hortic, 686:277-280.

Wells ML, Conner PJ, Funderburk JF, Price JG (2008). Effects of foliar applied boron on fruit retention, fruit quality, and tissue boron concentration of pecan. HortScience 27:696-699.

Woodroof JG (1979). Tree nuts. AVI Publishing Co. Inc. Westport, Connecticut.

Permissions

All chapters in this book were first published in AJAR, by Academic Journals; hereby published with permission under the Creative Commons Attribution License or equivalent. Every chapter published in this book has been scrutinized by our experts. Their significance has been extensively debated. The topics covered herein carry significant findings which will fuel the growth of the discipline. They may even be implemented as practical applications or may be referred to as a beginning point for another development.

The contributors of this book come from diverse backgrounds, making this book a truly international effort. This book will bring forth new frontiers with its revolutionizing research information and detailed analysis of the nascent developments around the world.

We would like to thank all the contributing authors for lending their expertise to make the book truly unique. They have played a crucial role in the development of this book. Without their invaluable contributions this book wouldn't have been possible. They have made vital efforts to compile up to date information on the varied aspects of this subject to make this book a valuable addition to the collection of many professionals and students.

This book was conceptualized with the vision of imparting up-to-date information and advanced data in this field. To ensure the same, a matchless editorial board was set up. Every individual on the board went through rigorous rounds of assessment to prove their worth. After which they invested a large part of their time researching and compiling the most relevant data for our readers.

The editorial board has been involved in producing this book since its inception. They have spent rigorous hours researching and exploring the diverse topics which have resulted in the successful publishing of this book. They have passed on their knowledge of decades through this book. To expedite this challenging task, the publisher supported the team at every step. A small team of assistant editors was also appointed to further simplify the editing procedure and attain best results for the readers.

Apart from the editorial board, the designing team has also invested a significant amount of their time in understanding the subject and creating the most relevant covers. They scrutinized every image to scout for the most suitable representation of the subject and create an appropriate cover for the book.

The publishing team has been an ardent support to the editorial, designing and production team. Their endless efforts to recruit the best for this project, has resulted in the accomplishment of this book. They are a veteran in the field of academics and their pool of knowledge is as vast as their experience in printing. Their expertise and guidance has proved useful at every step. Their uncompromising quality standards have made this book an exceptional effort. Their encouragement from time to time has been an inspiration for everyone.

The publisher and the editorial board hope that this book will prove to be a valuable piece of knowledge for researchers, students, practitioners and scholars across the globe.

List of Contributors

T. Stoilova
Institute of Plant Genetic Resources, Sadovo, Bulgaria

G. Pereira
Instituto Nacional Recursos Biológicos (INRB/INIA), P. O. Box 6, 7350-951 Elvas, Portugal

P. Dhananchezhiyan
Agricultural Engineering College and Research Institute, Tamil Nadu Agricultural University, Coimbatore - 641 003, Tamil Nadu, India

C. Divaker Durairaj
Agricultural Engineering College and Research Institute, Tamil Nadu Agricultural University, Coimbatore - 641 003, Tamil Nadu, India

S. Parveen
Agricultural Engineering College and Research Institute, Tamil Nadu Agricultural University, Coimbatore - 641003, Tamil Nadu, India

Jiang Che
College of Agriculture, Northwest A&F University, Yang ling, 712100, China

Yun Cheng Liao
College of Agriculture, Northwest A&F University, Yang ling, 712100, China

Xiao Hui Ding
College of Resources and Environment, Northwest A&F University, Yang ling, 712100, China

Yan Liu
College of Agriculture, Northwest A&F University, Yang ling, 712100, China

Douglas Seijum Kohatsu
Professores Doutores, UEM/CCA/ DCA Agronomia Umuarama – PR, Brazil

Valdir Zucareli
Professores Doutores, UEM/CCA/ DCA Agronomia Umuarama – PR, Brazil

Wilian Polaco Brambilla
Pós graduando em Agronomia, Unesp - Botucatu-SP, Brazil

Elizabeth Orika Ono
Professores Doutores, Departamento de Botânica, Instituto de Biociências, Unesp - Botucatu-SP, Brazil

Tiago Roque Benetoli da Silva
Professores Doutores, UEM/CCA/ DCA Agronomia Umuarama – PR, Brazil

João Domingos Rodrigues
Professores Doutores, Departamento de Botânica, Instituto de Biociências, Unesp - Botucatu-SP, Brazil

S. Najeeb
SHER-I-KASHMIR University of Agricultural Sciences and Technology, Kashmir Mountain Research Centre for Field Crops Khudwani, Anantnag, 192102, Ashmir, India

M. Ashraf Ahangar
SHER-I-KASHMIR University of Agricultural Sciences and Technology, Kashmir Mountain Research Centre for Field Crops Khudwani, Anantnag, 192102, Ashmir, India

S. H. Dar
SHER-I-KASHMIR University of Agricultural Sciences and Technology, Kashmir Mountain Research Centre for Field Crops Khudwani, Anantnag, 192102, Ashmir, India

Qurban Ali
Department of Plant Breeding and Genetics, University of Agriculture, Faisalabad, Pakistan

Muhammad Ahsan
Department of Plant Breeding and Genetics, University of Agriculture, Faisalabad, Pakistan

Muhammad Hammad Nadeem Tahir
Department of Plant Breeding and Genetics, University of Agriculture, Faisalabad, Pakistan

Shahzad Maqsood Ahmed Basra
Department of Crop Physiology, University of Agriculture, Faisalabad, Pakistan

D. Rajalakshmi
Agronomy Agro Climate Research Centre, Tamil Nadu Agricultural University, Coimbatore – 641003, India

R. Jagannathan
Agronomy Agro Climate Research Centre, Tamil Nadu Agricultural University, Coimbatore – 641003, India

V. Geethalakshmi
Agronomy Agro Climate Research Centre, Tamil Nadu Agricultural University, Coimbatore – 641003, India

Ahmad Reza Ommani
Department of Agricultural Management, Shoushtar Branch, Islamic Azad University, Shoushtar, Iran

Azadeh N. Noorivandi
Department of Agricultural Management, Shoushtar Branch, Islamic Azad University, Shoushtar, Iran

Eduardo Romeiro Filho
Industrial Engineering Departament, Universidade Federal de Minas Gerais, Integrated Laboratory of Design and Product Engineering, Escola de Engenharia, bloco 1, sala 3104, Av. Presidente Antônio Carlos, 6627 - Campus Pampulha, 31270-901 - Belo Horizonte – MG, Brazil

Aline Capanema Barros
Industrial Engineering Departament, Universidade Federal de Minas Gerais, Integrated Laboratory of Design and Product Engineering, Escola de Engenharia, bloco 1, sala 3104, Av. Presidente Antônio Carlos, 6627 - Campus Pampulha, 31270-901 - Belo Horizonte – MG, Brazil

Hui ZHOU
College of Horticulture and Landscape, Yunnan Agricultural University, Kunming 650201, China

Ming-hua DENG
College of Horticulture and Landscape, Yunnan Agricultural University, Kunming 650201, China
Institute of Vegetable Crops, Hunan Academy of Agricultural Science, Changsha 410125, China

Jin-fen WEN
Faculty of Modern Agricultural Engineering, Kunming University of Science and Technology, Kunming 650224, China

Jin-long HUO
Faculty of Animal Science and Technology, Yunnan Agricultural University, Kunming 650201, China

Xue-xiao ZOU
Institute of Vegetable Crops, Hunan Academy of Agricultural Science, Changsha 410125, China

Hai-shan ZHU
College of Horticulture and Landscape, Yunnan Agricultural University, Kunming 650201, China

Vishal Gupta
Division of Plant Pathology, Faculty of Agriculture, Sher-e-Kashmir University of Agricultural Sciences and Technology of Jammu, Chatha-180 009, Jammu, India

R. A. Ahanger
Division of Plant Pathology, Faculty of Agriculture, Sher-e-Kashmir University of Agricultural Sciences and Technology of Jammu, Chatha-180 009, Jammu, India

V. K. Razdan
Division of Plant Pathology, Faculty of Agriculture, Sher-e-Kashmir University of Agricultural Sciences and Technology of Jammu, Chatha-180 009, Jammu, India

B. C. Sharma
Division of Agronomy, Faculty of Agriculture, Sher-e-Kashmir University of Agricultural Sciences and Technology of Jammu, Chatha-180 009, Jammu, India

Ichpal Singh
Division of Plant Pathology, Faculty of Agriculture, Sher-e-Kashmir University of Agricultural Sciences and Technology of Jammu, Chatha-180 009, Jammu, India

Kavaljeet Kaur
Division of Plant Pathology, Faculty of Agriculture, Sher-e-Kashmir University of Agricultural Sciences and Technology of Jammu, Chatha-180 009, Jammu, India

M. K. Pandey
Division of Genetics and Plant Breeding, Faculty of Agriculture, Sher-e-Kashmir University of Agricultural Sciences and Technology of Jammu, Chatha-180 009, Jammu, India

Sruti Karmakar
Department of Earth and Environmental Studies, National Institute of Technology, Durgapur – 713209, Burdwan, West Bengal, India

Koushik Brahmachari
Department of Agronomy, Bidhan Chandra Krishi Viswavidyalaya, Mohanpur – 741252, Nadia, West Bengal, India

Aniruddha Gangopadhyay
Department of Earth and Environmental Studies, National Institute of Technology, Durgapur – 713209, Burdwan, West Bengal, India

E. P. Venkatasalam
Central Potato Research Institute, Shimla-171 001 (H.P), India

Richa Sood
Central Potato Research Institute, Shimla-171 001 (H.P), India

K. K. Pandey
Central Potato Research Institute, Shimla-171 001 (H.P), India

Vandana Thakur
Central Potato Research Institute, Shimla-171 001 (H.P), India

Ashwani K. Sharma
Central Potato Research Station, Kufri, Shimla-171 012 (H.P), India

B. P. Singh
Central Potato Research Institute, Shimla-171 001 (H.P), India

Liu Yuelian
Agricultural College, Guangdong Ocean University, Zhanjiang, Guangdong 524088, China

Lu Qingfang
Agricultural College, Guangdong Ocean University, Zhanjiang, Guangdong 524088, China

Satya Prakash Kumar
Agricultural and Food Engineering Department, IIT Kharagpur-721 302 WB, India

K. P. Pandey
Agricultural and Food Engineering Department, IIT Kharagpur-721 302 WB, India

Ranjeet Kumar
Agricultural and Food Engineering Department, IIT Kharagpur-721 302 WB, India

Man Singh
Agricultural and Food Engineering Department, IIT Kharagpur-721 302 WB, India

Biswajit Deb Roy
Microbiology Laboratory, Department of Life Science, Assam University, Silchar-788011, Assam, India

Bibhas Deb
Microbiology Laboratory, Department of Botany, Gurucharan College, Silchar-788004, Assam, India

Gauri Dutta Sharma
Microbiology Laboratory, Department of Life Science, Assam University, Silchar-788011, Assam, India

R. Thiyagarajan
Agricultural Engineering College and Research Institute, Tamil Nadu Agricultural University, Kumulur, Tamil Nadu, India

K. Kathirvel
Agricultural Engineering College and Research Institute, Tamil Nadu Agricultural University, Coimbatore, India

G. C. Jayashree
Indian institute of Crop Processing Technology, Tanjavur, Tamil Nadu, India

Hanieh Davodi
Rural Development, Tehran University, Iran

Tahmasb Maghsoudi
Department of Agricultural Management, Shoushtar Branch, Islamic Azad University, Iran

Hossien Shabanali Fami
Economics and Agricultural Development, Tehran University, Tehran, Iran

Khalil Kalantari
Economics and Agricultural Development, Tehran University, Tehran, Iran

P. O. Oviasogie
Date Palm and Shea Tree Research and Development Department, Nigerian Institute for Oil palm Research, P. M. B. 1030 Benin City, Edo State, Nigeria

J. O. Odewale
Date Palm and Shea Tree Research and Development Department, Nigerian Institute for Oil palm Research, P. M. B. 1030 Benin City, Edo State, Nigeria

N. O. Aisueni
Date Palm and Shea Tree Research and Development Department, Nigerian Institute for Oil palm Research, P. M. B. 1030 Benin City, Edo State, Nigeria

E. I. Eguagie
Date Palm and Shea Tree Research and Development Department, Nigerian Institute for Oil palm Research, P. M. B. 1030 Benin City, Edo State, Nigeria

G. Brown
Date Palm and Shea Tree Research and Development Department, Nigerian Institute for Oil palm Research, P. M. B. 1030 Benin City, Edo State, Nigeria

E. Okoh-Oboh
Date Palm and Shea Tree Research and Development Department, Nigerian Institute for Oil palm Research, P. M. B. 1030 Benin City, Edo State, Nigeria

Intikhab Aalum Jehangir
Division of Agronomy, SKUAST-Kashmir, Shalimar-191 121, India

H. U. Khan
Division of Agronomy, SKUAST-Kashmir, Shalimar-191 121, India

M. H. Khan
Central Institute of Temperate Horticulture, ICAR, Srinagar (J&K) - 190 007, India

F. Ur-Rasool
Division of Agronomy, SKUAST-Kashmir, Shalimar-191 121, India

R. A. Bhat
Division of Agronomy, SKUAST-Kashmir, Shalimar-191 121, India

T. Mubarak
Division of Agronomy, SKUAST-Kashmir, Shalimar-191 121, India

M. A. Bhat
Division of Agronomy, SKUAST-Kashmir, Shalimar-191 121, India

S. Rasool
Division of Agronomy, SKUAST-Kashmir, Shalimar-191 121, India

Phurailatpam Arunkumar
Central Agricultural University, Pasighat, Arunachal Pradesh, India

Bishoyi Ashok
Directorate of Medicinal and Aromatic Plants Research, ICAR, Anand, Gujarat, India

Maiti Satyabrata
Directorate of Medicinal and Aromatic Plants Research, ICAR, Anand, Gujarat, India

N. Deepa Devi
Department of Horticulture, Agricultural College and Research Institute, Tamil Nadu Agricultural University, Madurai- 625104, India

S. Mariappan
Department of Horticulture, Agricultural College and Research Institute, Tamil Nadu Agricultural University, Madurai- 625104, India

A. Manivannan
Department of Plant Breeding and Genetics, AC&RI, Madurai, -625104, India
Directorate of Maize Research, Pusa campus, New Delhi, -110012, India
Department of Plant Breeding and Genetics, College of Agriculture, CCS HAU, Hisar, -125004, India

Somveer Nimbal
Department of Plant Breeding and Genetics, College of Agriculture, CCS HAU, Hisar, -125004, India

A. K. Chhabra
Department of Plant Breeding and Genetics, College of Agriculture, CCS HAU, Hisar, -125004, India

S. B. Meitei
College of Post Graduate Studies (CAU), Barapani-793 103, Meghalaya, India

R. K. Patel
Division of Horticulture, ICAR Research Complex for NEH Region, Umiam-793 103, Meghalaya, India

Bidyut C. Deka
Division of Horticulture, ICAR Research Complex for NEH Region, Umiam-793 103, Meghalaya, India

N. A. Deshmukh
Division of Horticulture, ICAR Research Complex for NEH Region, Umiam-793 103, Meghalaya, India

Akath Singh
CAZRI, Jodhpur, India

Hazem S. Kassem
Department of Agricultural Extension and Rural Society, Faculty of Agriculture, Mansoura University, Egypt

Ahmed M. Sarhan
Department of Agricultural Extension, Faculty of Agriculture, South-Valley University, Egypt

Naira Ashraf
Department of Fruit Science, Dr Y S Parmar University of Horticulture and Forestry, Nauni, Solan, Himachal Pradesh 173230, India

Moieza Ashraf
P. G. Department of Environmental Science, Kashmir University, J&K, India

Gh Hassan
Division of Fruit Science, S.K. University of Agricultural Sciences and Technology, Kashmir, India

Munib-U-Rehman
Division of Fruit Science, S.K. University of Agricultural Sciences and Technology, Kashmir, India

N. A. Dar
S.K. Universities of Agricultural Sciences and Technology, Kashmir, India

Inayat. M. Khan
S.K. Universities of Agricultural Sciences and Technology, Kashmir, India

Umar Iqbal
Division of Fruit Science, S.K. University of Agricultural Sciences and Technology, Kashmir, India

S. A. Banday
Division of Fruit Science, S.K. University of Agricultural Sciences and Technology, Kashmir, India